Communications
in Computer and Information Science 44

Jong-Hwan Kim   Shuzhi Sam Ge   Prahlad Vadakkepat
Norbert Jesse   Abdullah Al Manum
Sadasivan Puthusserypady K   Ulrich Rückert
Joaquin Sitte   Ulf Witkowski   Ryohei Nakatsu
Thomas Braunl   Jacky Baltes   John Anderson
Ching-Chang Wong   Igor Verner   David Ahlgren (Eds.)

Progress in Robotics

FIRA RoboWorld Congress 2009
Incheon, Korea, August 16-20, 2009
Proceedings

 Springer

Volume Editors

Jong-Hwan Kim, jhkim@ee.kaist.ac.kr
Shuzhi Sam Ge, samge@nus.edu.sg
Prahlad Vadakkepat, prahlad@nus.edu.sg
Norbert Jesse, norbert.jesse@udo.edu
Abdullah Al Manum, eleaam@nus.edu.sg
Sadasivan Puthusserypady K, elespk@nus.edu.sg
Ulrich Rückert, ulrich.rueckert@hni.uni-paderborn.de
Joaquin Sitte, j.sitte@qut.edu.au
Ulf Witkowski, ulf.witkowski@hni.uni-paderborn.de
Ryohei Nakatsu, idmdir@nus.edu.sg
Thomas Braunl, tb@ee.uwa.edu.au
Jacky Baltes, jacky@cs.umanitoba.ca
John Anderson, andersj@cs.umanitoba.ca
Ching-Chang Wong, wong@ee.tku.edu.tw
Igor Verner, ttrigor@tx.technion.ac.il
David Ahlgren, david.ahlgren@trincoll.edu

Library of Congress Control Number: Applied for

CR Subject Classification (1998): I.2.9, H.5, J.4, K.4, K.3, K.8

ISSN   1865-0929
ISBN-10   3-642-03985-5 Springer Berlin Heidelberg New York
ISBN-13   978-3-642-03985-0 Springer Berlin Heidelberg New York

springer.com

© Springer-Verlag Berlin Heidelberg 2009
Printed in Germany

Typesetting: Camera-ready by author, data conversion by Scientific Publishing Services, Chennai, India
Printed on acid-free paper   SPIN: 12745906   06/3180   5 4 3 2 1 0

Preface

This volume is an edition of the papers selected from the 12th FIRA RoboWorld Congress, held in Incheon, Korea, August 16–18, 2009.

The Federation of International Robosoccer Association (FIRA – www.fira.net) is a non-profit organization, which organizes robotic competitions and meetings around the globe annually. The RoboSoccer competitions started in 1996 and FIRA was established on June 5, 1997. The Robot Soccer competitions are aimed at promoting the spirit of science and technology to the younger generation. The congress is a forum in which to share ideas and future directions of technologies, and to enlarge the human networks in robotics area.

The objectives of the FIRA Cup and Congress are to explore the technical development and achievement in the field of robotics, and provide participants with a robot festival including technical presentations, robot soccer competitions and exhibits under the theme "Where Theory and Practice Meet."

Under the umbrella of the 12th FIRA RoboWorld Incheon Congress 2009, six international conferences were held for greater impact and scientific exchange:

- 6th International Conference on Computational Intelligence, Robotics and Autonomous Systems (CIRAS)
- 5th International Symposium on Autonomous Minirobots for Research and Edutainment (AMiRE)
- International Conference on Social Robotics (ICSR)
- International Conference on Advanced Humanoid Robotics Research (ICAHRR)
- International Conference on Entertainment Robotics (ICER)
- International Robotics Education Forum (IREF)

This volume consists of selected quality papers from the six conferences. The volume is intended to provide readers with the recent technical progresses in robotics, human–robot interactions, cooperative robotics and the related fields.

The volume has 44 papers from the 115 contributed papers at the FIRA RoboWorld Congress Incheon 2009. This volume is organized into six sections:

- Humanoid Robotics, Human–Robot Interaction, Education and Entertainment, Cooperative Robotics, Robotic System Design, and Learning, Optimization, Communication.

The editors hope that this volume is informative to the readers. We thank Springer for undertaking the publication of this volume.

Prahlad Vadakkepat

Organization

Honorary Chair

Jong Hwan Kim   KAIST, Korea

General Chair

Shuzhi Sam Ge   National University of Singapore, Singapore

Program Chair

Prahlad Vadakkepat   National University of Singapore, Singapore

Finance Chair

Hyun Myung   KAIST, Korea

Conferences Committee

CIRAS - International Conference on Computational Intelligence, Robotics and Autonomous Systems

General Chair

Prahlad Vadakkepat   National University of Singapore, Singapore

Program Chairs

Abdullah Al Manum   National University of Singapore, Singapore
Sadasivan Puthusserypady K   National University of Singapore, Singapore

AMiRE - International Symposium on Autonomous Minirobots for Research and Edutainment

General Chair

Ulrich Ruckert   University of Paderborn, Germany

Program Chairs

Joaquin Sitte   Queensland University of Technology, Austrailia
Ulf Witkowski   University of Paderborn, Germany

ICSR - International Conference on Social Robotics

General Chair

Ryohei Nakatsu   National University of Singapore, Singapore

Co-general Chairs

Oussama Khatib   Stanford University, USA
Hideki Hashimoto   The University of Tokyo, Japan

Program Chair

Thomas Braunl   The University of Western Australia, Australia

Program Committee Members

Arvin Agah, USA   Guido Herrmann, UK
Fuchun Sun, China   Sandra Hirche, Japan
Robert Babuska, New Zealand   Lars Hildebrand, Germany
Jochen Triesch, Germany   Guy Hoffmann, USA
Norbert Oswald, Germany   Frederic Maire, Australia
Dirk Wollherr, Germany   Carlo Alberto Avizzano
Rinhard Klette, New Zealand   Guy Matko, Slovenia
Edwige Pissaloux, France   Tom Ziemke, Sweden
Gerhard Sagerer, Germany   Matthias Scheutz, USA
Guilherme DeSouza, USA   Hong Qiao, China
Yaochu Jin, Germany   Nilanjan Sarkar, USA
Masahiro Shiomi, Japan   Gil Weinberg, USA
Marcelo H. Ang, Sigapore   Olivier Simonin, France
Maria Chiara Carrozza, Italy   Frantisek Solc, Czech Republic
Elizabeth Croft, Cananda   Chrystopher Nehaniv, UK
Kerstin Dautenhahn, UK   Lola Canamero, UK
Kerstin Severinson-Eklundh, Sweden   John John Cabibihan, Singapore
Vanessa Evers, New Zealand

ICAHRR - International Conference on Advanced Humanoid Robotics Research

General Chair

Jacky Baltes   University of Manitoba, Canada

Program Chair

John Anderson   University of Manitoba, Canada

ICER - International Conference on Entertainment Robotics

General Chair

Norbert Jesse   Technische Universität Dortmund, Germany

Program Chair

Ching-Chang Wong   Tamkang University, Taiwan

IREF - International Robotics Education

General Chair

Igor Verner   Israel Institute of Technology, Israel

Program Chair

David Ahlgren   Trinity College, USA

Table of Contents

Education and Entertainment

Cooperative Robotics

Robotic System Design

Learning, Optimization, Communication

Time-Varying Affective Response for Humanoid Robots[*]

Lilia Moshkina[1], Ronald C. Arkin[1], Jamee K. Lee[2], and HyunRyong Jung[2]

[1] Georgia Tech Mobile Robot Laboratory, Atlanta, GA, USA 30332
{lilia,arkin}@cc.gatech.edu
[2] Samsung Advanced Institute of Technology, Kiheung, South Korea
{jamee.lee,hyunryong.jung}@samsung.com

Abstract. This paper describes the design of a complex time-varying affective architecture. It is an expansion of the TAME architecture (traits, attitudes, moods, and emotions) as applied to humanoid robotics. It particular it is intended to promote effective human-robot interaction by conveying the robot's affective state to the user in an easy-to-interpret manner.

Keywords: Humanoids, emotions, affective phenomena, robot architectures.

1 Introduction

Based on our considerable experience implementing affective phenomena in robotic systems (see [1] for a summary), we are now considering the application of sophisticated cognitive models of human Traits, Attitudes, Moods, and Emotions (TAME). These affective states are embedded into a novel architecture and designed to influence the perception of a user regarding the robot's internal state and the human-robot relationship itself. Recent work by Arkin et al in non-verbal communication [2] and emotional state for the AIBO [3] addressed powerful yet less complex means for accomplishing these tasks. Introducing time-varying affective states that range over multiple time scales spanning from an agent's lifetime to mere seconds with direction towards specific objects or the world in general provides the power to generate heretofore unobtained richness of affective expression. This paper describes the cognitive underpinnings of this work in the context of humanoid robots and presents the directions being taken in this recently initiated project to implement it upon a small humanoid robot.

2 Related Work

Although most work on humanoids focuses on the physical aspects (e.g., perfecting walking gaits, sensors or appearance), there are some who add affect into the mix. For example, humanoid Waseda Eye No. 4 Refined [4] combines emotions, moods, and personality. The overall goal of the system is to achieve smooth and effective communication for a humanoid robot. Although many elements of this system are not

---

[*] This research is funded under a grant from Samsung Electronics.

J.-H. Kim et al. (Eds.): FIRA 2009, CCIS 44, pp. 1–9, 2009.

psychologically or biologically founded, it provides a few interesting mechanisms, such as modeling personality's influence on emotion via a variety of coefficient matrices and using internal-clock activation component in moods.

Fukuda et al. [5] also include the notions of emotions and moods in their Character Robot Face; emotions are represented as semantic networks, and the combination of currently active emotions is deemed as mood. Two other humanoid robotic head robots, Kismet [6] and MEXI [7] have emotion and drive systems. Kismet is modeled after an infant, and is capable of proto-social responses, including emotional expressions, which are based on its affective state. In MEXI, the Emotion Engine is composed of a set of basic emotions (positive that it strives to achieve and negative it tries to avoid) and homeostatic drives. Space prevents a more complete description of other related projects.

3 Cognitive Basis of TAME

The TAME affective architecture has been initially tested on the entertainment robot dog Aibo [8], but its application to a humanoid robot is fairly straightforward in principle. In fact, using the framework for a humanoid will provide a number of advantages. The synergistic combination of affective phenomena focuses on long-term, and sometimes subtle, effect on robotic behavior, which fits well with one of the main goals for creating human-like robots - making the communication between them more natural, commonplace and prolonged, where machines act as partners rather than bystanders. The second advantage of applying TAME to humanoids is their expressive potential, exhibited not only in facial and bodily expressions (e.g., a smile, a shoulder shrug, a handshake), but also in a variety of tasks they could perform for which human-like personalities are readily applicable. The framework itself takes inspiration from a large number of theories and findings from Personality, Emotion, Mood and Attitude areas of Psychology, which are adapted to enhance robotic behavior.

3.1 Overview

The Affective Module, the core of TAME, is subdivided into Dispositions and Affective State. Dispositions include personality Traits and affective Attitudes, and represent a propensity to behave in a certain way; they are more or less persistent, long-lasting, and either slowly changing (attitudes) or permanent (traits) throughout robot's "life". Affective State consists of Emotions and Moods, more fleeting and transient affects, manifesting as either high-intensity, short duration peaks (emotions) or slow smooth undulations (moods). Table 1 summarizes differences in duration and temporal changes of these four components.

Another direction along which these components differ is object specificity: emotions and attitudes appear and change in response to particular stimuli (such as fear in the presence of an attacker or dislike towards an unfriendly person), whereas traits and moods are diffuse and not object-specific – they manifest regardless of presence or absence of objects. Although they all can be categorized differently and each have a distinct function and purpose, we cannot regard these phenomena as independent, as they strongly influence each other and interweave to create a greater illusion of life.

The Affective Module fits within behavior-based robotic control [9] by first processing relevant perceptual input (be it color and distance to certain emotion-eliciting objects or level of light affecting moods) and then influencing behavioral parameters of affected low-level behaviors and/or the behavior coordination gains as they are comprised into behavioral assemblages (see Figure 1).

Table 1. Summary of Time-varying Aspects of TAME Components

	Traits	Attitudes	Moods	Emotions
Duration	Life-long	A few days to a few years	A few hours to a few weeks	A few seconds to a few minutes
Change in Time	Time-invariant	Persistent across time; change slowly with the number of times an object of attitude is encountered.	Change cyclically as a variable of underlying environmental and internal influences; any drastic changes are smoothed across previous mood states	Intensity changes in short-term peaks as eliciting stimuli appear, disappear, and change distance; habituation effects describe decay of emotion even in the presence of stimuli.

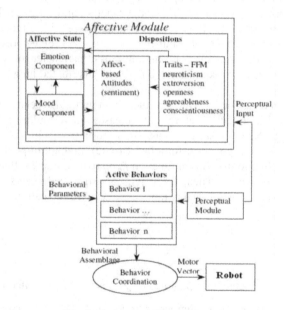

Fig. 1. Conceptual View of TAME

3.2 Psychological and Mathematical Foundations

The mathematical foundations for TAME have been significantly revised since its first publication [10]; detailed information on activation and change of individual components can be found in [11], while the focus of the following section will be on the effect on behavior and its applicability to humanoid robots.

3.2.1 Personality Traits

Personality defines an organism's recurrent patterns of behavior and emotionality. The Big Five taxonomy of personality traits [12] was chosen for this component for its universality: it is consistent over time, cultures, ages, and even applicable to non-human animals. The taxonomy has 5 broad dimensions, each of which is further sub-divided into facets; therefore robot's personality can be as simple or as complex as desired. Traits provide a two-fold advantage for humanoid robots: first, they serve a predictive purpose, allowing humans to understand and infer robot's behavior better; second, they allow adaptation to different tasks and environments, where certain trait configurations are better suited for one or another task or environment.

The five global dimensions are *Openness, Conscientiousness, Extraversion, Agreeableness* and *Neuroticism*; each of them has its own effect on robot behavior. For example, in a humanoid, extraversion could be expressed by keeping a closer distance to the human, frequent "smiles", more gestures, etc.; this trait would be appropriate for tasks requiring engagement and entertainment from a robot, e.g., a museum guide or a play partner for kids. The traits are modeled as vectors of intensity, where intensity refers to the extent to which a trait is represented. In the robot, these intensities: are defined a priori by a human; don't change throughout the robot's "life" (this could be a single run, an interaction with a person, or the robot's entire physical life-span); and are not influenced by any other affective phenomena. We provide a functional mapping from the trait space onto behavioral parameter space as a 2nd degree polynomial, where 3 pairs of corresponding data points are minimum trait/parameter, maximum, and default/average (the values are taken from the normally distributed human psychological data [13]).

Traits can have a direct or an inverse influence on particular behaviors, and this relationship is defined in a matrix beforehand. In cases where multiple traits affect the same behavior (e.g., Neuroticism may push the robot away from the obstacles while Conscientiousness make it go closer for a faster route), first a trait/parameter mapping is calculated, according to the chosen function $f_{ij}(p_j)$, where trait i influences behavior j, a polynomial in this case. Then, the results are averaged across all influencing personality traits to produce the final parameter value used thereafter:

$$B_j = \frac{1}{\sum_{i=1}^{N} \left| pb_{ij} \right|} \sum_{i=1}^{N} f_{ij}(p_i) \tag{1}$$

where B_j is a particular behavioral parameter, $f_{ij}(p_i)$ is the function that maps personality trait p_i to B_j, N is the total number of traits, and $\overrightarrow{pb}$ is personality/behavior dependency matrix; if there is no influence, the result of $f_{ij} = 0$.

3.2.2 Emotions

From an evolutionary point of view, emotions provide a fast, flexible, adaptive response to environmental contingencies. They appear as short-term, high-intensity peaks in response to relevant stimuli (we don't usually live in a constant flux of emotions), and serve a number of functions, of which most applicable for humanoids are communicative, expressive and affiliative, e.g., fear communicates danger and a request for help, while joy in response to a bright smile helps forge trust and camaraderie. The primary, reactive emotions of *fear, anger, disgust, sadness, joy and interest* were chosen, in part because these basic emotions have universal, well-defined facial expressions, are straightforwardly elicited, and would be expected, perhaps subconsciously, on a humanoid's face, as appearance does affect expectations.

From an emotion generation point of view, Picard [14] suggests a number of properties desirable in an affective system: 1) a property of activation – emotion is not generated below a certain stimulus strength; 2) a property of saturation – refers to an upper bound; 3) a property of response decay; and 4) a property of linearity – emotion generation will approximate linearity for certain stimulus strength range. Taking these properties into consideration, the resulting function for emotion generation (based on stimulus strength) resembles a sigmoid, in which the left side corresponds to activation, the right side corresponds to saturation (amplitude), and the middle models the actual response.

Emotions are also highly dependent on traits and moods: personality may influence the threshold of eliciting stimulus (activation point), peak (amplitude) and rise time to peak (affecting the slope of the generation curve); and moods can vary the threshold of experiencing a specific emotion. For example, Extraversion is correlated with positive emotions, therefore a humanoid robot high in this dimension would display more smiles, excited gestures and other expressions of joy. Attitude also has an effect on emotion – the object of like or dislike may serve as a stimulus for emotion generation.

Emotions can have a varied impact on behavior, from a subtle slowing to avoid a disgustful object to a drastic flight in response to extreme fear. This effect can be modeled by linear mapping from emotion strength to relevant behavioral parameters, and Fig. 2 provides a comparative view across time of stimulus strength (an object appears, comes closer, and then is gone), corresponding emotion activation (after response decay and smoothing), and the Object Avoidance Gain (which causes an avoidance response to Fear); duration is plotted along the x axis, and normalized values for stimulus strength, fear and object avoidance gain along the y axis.

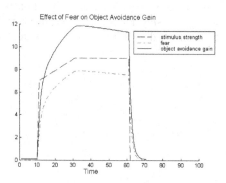

Fig. 2. Example of Fear to Object Avoidance Gain Mapping

3.2.3 Moods

Unlike emotions, moods represent a global, continuous affective state, cyclically changing and subtle in expression. Mood can be represented along two dimensions, Positive Affect and Negative Affect [15], where Negative Affect refers to the extent to which an individual is presently upset or distressed, and Positive Affect generally refers to one's current level of pleasure and enthusiasm. The level of arousal for both categories can vary from low to high; a low positive mood value has a negative connotation ("sluggish", "disinterested") and refers to insufficient level of pleasure and enthusiasm, rather than just low.

In humanoids, cyclical variations in moods over time can be determined based on the underlying variations in environmental and internal conditions (such as light and battery levels) with any sudden changes smoothed out by taking into consideration a number of prior mood states – filtering over a longer period of time results in slower and smaller mood changes and helps tone down any drastic spikes due to emotions.

Moods are mild by definition, and would only produce a mild, incremental effect, or a slight bias, on the currently active behaviors. Moods can have a direct or inverse influence on a behavioral parameter. A behavior-mood dependency matrix $\overrightarrow{mb} = [mb_{ij}]$, where $mb_{ij} \in \{-1,0,1\}$ is defined, where –1 corresponds to inverse influence, +1 to direct influence, and 0 to absence of mood influence on behavior. Positive and negative mood may influence the same behavioral parameters, and this influence is treated as additive. As moods are updated continuously, new mood-based values of behavioral gains/parameters replace the existing trait-based values in the following manner:

$$B_{i,mood} = B_{i,trait} + K_i \sum_{j=1}^{N} mb_{ji} \cdot m_j \qquad (1)$$

where $B_{i,mood}$ is the updated behavioral parameter i, mb_{ij} is the mood-behavior dependency matrix value for mood j, m_j is the current value of mood j, N is the total number of mood categories (2), and K is a scaling factor to ensure that the moods produce only incremental effect as opposed to overpowering any of the parameters.

Fig. 3 shows an example of incremental effects of moods on behavior. Let's suppose mood can bias robot's obstacle avoidance behavior. For example, if visibility is poor, it may be advantageous to stay farther away from obstacles to accommodate sensor error, and vice versa, in good visibility it may be better to concentrate on task

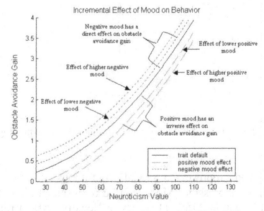

Fig. 3. Direct/Inverse Mood Effects on Behavior at Different Neuroticism Values

performance. Thus, negative mood can bias the obstacle avoidance gain by raising it, and positive mood by lowering it. Neuroticism also affects it by setting the default parameters to be used throughout the life-cycle, and the incremental effect of moods is shown against the space of trait-based defaults (plotted in solid blue center line).

3.2.4 Affective Attitudes

From a multitude of definitions of attitudes, the following was adopted as the working definition for TAME: "a general and enduring positive or negative feeling about some person, object or issue" [16]. It stresses relative time-invariance ("enduring"), object/situation specificity, and the role of affect/affective evaluation in the attitude concept. Attitude can be represented on a single axis, with 0 being a neutral attitude, negative values referring to an increasingly strong negative attitude (ranging from a mild dislike to hatred), and positive values to an increasingly strong positive attitude (e.g., from a subtle like to adoration). Affective attitudes are closely related to emotions, and may even originate in one; therefore initially they will be expressed through a related emotion, and not a behavioral change per se, determining both the type of emotion invoked and its intensity. For humanoids, attitudes would be invaluable in establishing rapport and understanding with interacting humans, as the robot would respond to people's behavior towards it by, for example, expressing joy at the sight of someone it "likes", or sadness when they leave.

4 Architectural Design and Implementation

For our initial research we will test the TAME model within the Georgia Tech *MissionLab*[17] software system and prototype on a Nao humanoid robot (Fig. 4 left). The architectural design overview appears in Figure 4 Right.

There are three major tasks to be completed from a software perspective: (1) Integration of *MissionLab*'s Hserver and the API for Nao; (2) Development of a stand-alone version of TAME running as a separate process within *MissionLab*, capable of updating the externalized affective variables based on a range of conditions including time passage and external stimuli; and (3) The creation of an appropriate set of humanoid behaviors. Suitable perceptual algorithms for the Nao robot will need to be developed as well as connections to the externalized motivational variables already resident in *MissionLab*. The generic functional description for each of these behaviors is:

$$\beta\,(\mathbf{s},\mathbf{a}) \Rightarrow \mathbf{r}$$

where β is the behavioral function mapping the stimulus $\mathbf{s}$ and affective value $\mathbf{a}$ onto an instantaneous response $\mathbf{r}$.

Specific behavior selection, specification, and implementation will be driven from the demonstration scenarios, while the affective models will be derived from both cognitive science models and then through empirical observational studies of the behavior of the humanoid. As is standard for *MissionLab*, these behaviors will be encoded using CDL (Configuration Description Language) and CNL (Configuration Network Language) [17]. The behavioral specifications for the scenarios will be

---

[1] MissionLab is freely available for research and education at:
http://www.cc.gatech.edu/ai/robot-lab/research/MissionLab/

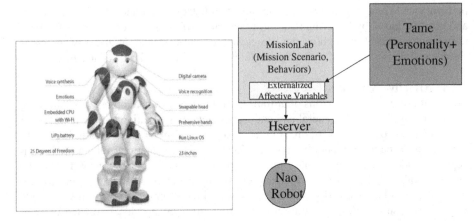

Fig. 4. (Left) Nao Robot (source Aldebaran Robotics) (Right) TAME integrated with *MissionLab*

represented as finite state acceptors (FSAs) within the configuration editor (cfgedit) component of *MissionLab*. Demonstrations will be created to illustrate the effectiveness of TAME by conveying the appearance of various forms of affect to a human observer, to be developed into full human-robot interaction (HRI) studies.

References

[1] Arkin, R.C.: Moving Up the Food Chain: Motivation and Emotion in Behavior-based Robots. In: Fellous, J., Arbib, M. (eds.) Who Needs Emotions: The Brain Meets the Robot, Oxford University Press, Oxford (2005)

[2] Brooks, A., Arkin, R.C.: Behavioral Overlays for Non-Verbal Communication Expression on a Humanoid Robot. Autonomous Robots 22(1), 55–75 (2007)

[3] Arkin, R., Fujita, M., Takagi, T., Hasegawa, R.: An Ethological and Emotional Basis for Human-Robot Interaction. Robotics and Autonomous Systems 42(3-4) (March 2003)

[4] Miwa, H., Takanishi, A., Takanobu, H.: Experimental Study on Robot Personality for Humanoid Head Robot. In: IEEE IROS, pp. 1183–1188 (2001)

[5] Fukuda, T., Jung, M., Nakashima, M., Arai, F., Hasegawa, Y.: Facial expressive robotic head system for human-robot communication and its application in home environment. Proc. of the IEEE 92, 1851–1865 (2004)

[6] Breazeal, C.: Emotion and Sociable Humanoid Robots. International Journal of Human-Computer Studies 59, 119–155 (2003)

[7] Esau, N., Kleinjohann, B., Kleinjohann, L., Stichling, D.: MEXI: machine with emotionally eXtended intelligence. Design and application of hybrid intelligent systems, 961–970 (2003)

[8] Moshkina, L., Arkin, R.C.: Human Perspective on Affective Robotic Behavior: A Longitudinal Study. In: IEEE IROS (2005)

[9] Arkin, R.C.: Behavior-based Robotics. MIT Press, Cambridge (1998)

[10] Moshkina, L., Arkin, R.C.: On TAMEing Robots. In: Interanational Conference on Systems, Man and Cybernetics (2003)

[11] Moshkina, L.: An Integrative Framework for Affective Agent Behavior. In: IASTED International Conference on Intelligent Systems and Control (2006)
[12] McCrae, R.R., Costa, P.T.: Toward a new generation of personality theories: theoretical contexts for the five-factor model. In: Five-Factor Model of Personality, pp. 51–87 (1996)
[13] Costa, P.T., McCrae, R.R.: NEO PI-R Professional Manual. Psychological Assessment Resources (1992)
[14] Picard, R.W.: Affective Computing. MIT Press, Cambridge (1997)
[15] Watson, D., Clark, L.A., Tellegen, A.: Mood and Temperament. The Guilford Press (2000)
[16] Breckler, S.J., Wiggins, E.C.: On Defining Attitude and Attitude Theory: Once More with Feeling. In: Pratkanis, A.R., Breckler, S.J., Greenwald, A.G. (eds.) Attitude Structure and Function, pp. 407–429. Lawrence Erlbaum Associates, Mahwah (1989)
[17] MacKenzie, D., Arkin, R.C., Cameron, J.: Multiagent Mission Specification and Execution. Autonomous Robots 4(1), 29–57 (1997)

The Co-simulation of Humanoid Robot
Based on Solidworks, ADAMS and Simulink

Dalei Song, Lidan Zheng, Li Wang, Weiwei Qi, and Yanli Li

Ocean University of China, Qingdao, 266100, China

Abstract. A simulation method of adaptive controller is proposed for the humanoid robot system based on co-simulation of Solidworks, ADAMS and Simulink. A complex mathematical modeling process is avoided by this method, and the real time dynamic simulating function of Simulink would be exerted adequately. This method could be generalized to other complicated control system. This method is adopted to build and analyse the model of humanoid robot. The trajectory tracking and adaptive controller design also proceed based on it. The effect of trajectory tracking is evaluated by fitting-curve theory of least squares method. The anti-interference capability of the robot is improved a lot through comparative analysis.

Keywords: humanoid robot, co-simulation, adaptive controller, trajectory tracking.

1 Introduction

Simulation is the necessary and intermediate step of the design and research of kinematics and dynamic system, and humanoid robot is a kind of complex kinematics and dynamic system, nobody's research can depart from simulation process. In addition, humanoid robot not only has the non-linear characteristic, but also works in the environments with uncertain disturbance. Therefore, there is a significant meaning to research the control strategy of humanoid robot in the circumstance with uncertain disturbance.

This paper will talk about the method of setting up the model of complex humanoid robot system conveniently. Based on this, a new control algorithm is created to control homanoid robot and other control algorithms are reviewed in great detail.

Solidworks, which is capable of solid modeling and virtual assembly ,and machinery system kinetics simulation software ADAMS provide a good software modeling and analysis platform for the research of this paper. In addition, the perfect control technology provided by Simulink(tool box of MATLAB) also makes the control research in this paper convenient. Based on the basis of co-simulation method, this paper will design a adaptive controller to improve the system performance under uncertain disturbance.

2 The Control System Model of Humanoid Robot

The dynamic formula of the Humanoid Robot control system can be described as below:

$$M(q)q'' + C(q,q')q' + G(q) + \tau_d(q,q',t) = \tau \tag{1}$$

J.-H. Kim et al. (Eds.): FIRA 2009, CCIS 44, pp. 10–18, 2009.

q, q', q'' are $n\times1$ joint's angular position, velocity, acceleration vector. $M(q)$ is $n\times n$ symmetric positive definite inertia array. $C(q,q')q'$ is $n\times1$ Coriolis force and centrifugal force vector. $G(q)$ is gravity vector. $\tau_d(q,q',t)$ is external disturbance. τ is the external input control moment.

The conventional method of modeling of control system simulation is to set up mathematics model and describe the relationship of input and output. From formula (1) we can find out the robot kinetics model is a highly complex, strong coupling nonlinear time-varying equation, which is difficult to set up mathematics model. Therefore, we need to look for a kind of not only simple and reliable but also effective way to set up robot control system simulation model.

In the References 4, it adopts M-Function to set up the system simulation model, whose substance is to set up mathematics model and uses submodule packaging technology to make simulation model packaged as unique Simulink module. In this paper, using ADAMS and Simulink to jointly simulate can save the step of setting up mathematics model. As it shows in Fig.1, the system simulation model is equivalent to two packaged Simulink module, which has the same input and output vectors as common module. The difference is that input and output vectors transfered between two different applications, and control module in ADAMS provides the data interface with other control program, with which data can be transferred conveniently. Using ADAMS can transform the components' position, velocity load and other functions into output variables, which is convenient to use.

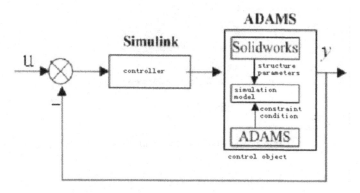

Fig. 1. Robot control system simulation model

3 Humanoid Robot Model

3.1 Robot Reality

The mechanical structure of humanoid robot has an significant impact on the motion of robot. This paper refers to Robonova- I humanoid robot's appearance structure produced by Hitec, sets up 3D simulation model after simplifying as below.

① The part of the upper limb only reserves pre and post rotational degree of shoulder joint;

② Motor, battery and circuit board which all are firmed with trunk under the weight of the weight in trunk of the Humanoid Robot;

③ The weight of screws and nuts are included in the weight of the components which is to be connected;

④ Each component is uniform density.

The model has 12 degrees of freedom, 10 on the leg, and 2 on the upper limb. In Solidworks, the process of setting up robot's mechanical structure is as flows:

① Drawing components. The simplified robot includes 27 components. Because of the symmetry of the robot, only 8 different components should be drawn. After finishing drawing, quality setting for each component is done. The tool provided by Solidworks can calculate the volume and the formula, $\rho = m/v$, can be used to work out density.

② Assembling components. Import drawn components into assembly drawing paper one by one. In order to make the later simulation process smooth, the robot should be assembled into a standing position. The assembled robot model is shown as below:

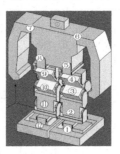

Fig. 2. Robot model

Table 1. The weight of each part of the robot

names of components	mass (g)	number
right/left foot	34.9	2
motor 1	50.1	2
motor 2	61.9	6
fastener 1	7.8	8
fastener 2	3.6	4
fastener 3	10.8	2
trunk	353.7	1
left/right arm	153.1	2

3.2 Entironment of Robot

To insure the circumstance resemble to practice, the author define three types of contact. Including:

①   Contact between robot and the ground, to insure the contact state is well(no flip or downthrow);
    This paper takes the robot swings right arm as example, using the method of script control, getting the date from post-processing module. To validate whether the date is reasonable.

②   Contact between ball and ground, chose tennis ball in the game.
    According to International Tennis Federation's regulations about size, weight and Elastic, setting the contact parameter.

③   Contact between tennis ball and robot.
    Planning the action of kicking tennis ball, put the action on the real robot, comparing the robot's state, and modifying the parameter at the end.

3.3 Controller of Robot

Because the motor of the real robot uses RC motors with fixed PD controllers, so,design PD controller in simulation.

$$\Delta u = u(k) - u(k-1)$$
$$= K_p[e(k) - e(k-1)] + K_d[e(k) - 2e(k-1) + e(k-2)]$$

In the formula above, $u(k)$ is the control variable of sampling time k, $e(k)$ is the input bias sampling time k, and $k = 1, 2, \ldots\ldots$ .To confirm the numerical of K_p, K_d.

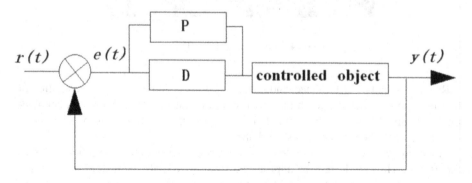

Fig. 3. Picture of PD Control Theory

4 The Motion Trajectory Planning of Humanoid Robot

After the analysis of the robot' mechanical structure, it consists of n-pole, on each of which only one torque acts. To carry out humanoid robot's move, first, the plan of trajectory should be preceded, the way of which generally is to make the move of robot discomposed. The method is defined as:

During the period of the whole move of robot, there exists a moment $t_i(0 \geq i \leq n)$. Among the robot's joints, at least one's sign of the rate of the change in the rotation $d\theta / dt (<0, >0, =0)$ during $t < t_i$ is different from it during $t > t_i$.

The actions of the robot are divided into i+1 continuous task with the boundary t_i. According to the tasks, the path form the initial position to target position are interpolated or approached to provide path by polynomial function and a series of set-points are generated. And then the expecting joint-control-set-points that have been obtained are used to control the robot, and simultaneously, whether the plan of the trajectory is able to meet the requests can be observed through picture and actual input-joint-position-curve in ADAMS.

Generally, the environment in which robot works has uncertain external disturbance. In order to compare and analyze conveniently, this paper configures two kinds of simulation environment with and without disturbance. In the environment without disturbance planning the move of robot kicking a ball, the move is divided into four tasks: leaning, being ready to kick, kicking, and homing. After simulation, the robot's simulation graph and each joint's position-curve can be got under the ideal situation without disturbance.

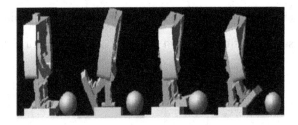

Fig. 4. The process of simulation

Regard the ball as disturbance, and repeat the process above. Under the condition of the same input,during the graph the robot appears relatively acute slipping phenomenon. Compared with ideal situation, obvious departure turns up. Fig.3 shows the move of kicking when there is a ball as disturbance.

Therefore, we have the reason to consider: disturbance act much on robot's posture importantly. How to design the appropriate strategy to make humanoid robot to accomplish some stipulated move and keep the property optimal or approximate optimal, that is a problem the adaptive control need to research and solve.

5 The Design of Adaptive Controller of Humanoid Robot

Formula (1) shows the kinetic model of robot with n-freedom of rotary joint described by second order nonlinear differential equations ,and, in order to get the control strategy, we make the following assumption:

1. the external disturbance uses the form below:

$$\| \tau_d(q,q',t) \| \leq d_0 + d_1 \| e' \| + d_2 \| e \|$$

d_0, d_1, d_2 are positive constants.

2. Assume the external disturbance, the inertia matrix, Coriolis forces matrix and gravity vector are bounded and meet the following conditions:

$$\| M^{-1} \| \leq \alpha; \| M \| \leq \beta_1; \| C(q,q') \| \leq \beta_2;$$
$$\| G \| \leq \beta_3; \lambda_{\min}(M^{-1}) > \gamma$$

α, β_i, γ are positive constants.

Set $\theta_i = \dfrac{\alpha(\beta_i + d_i - 3)}{\gamma}, i = 1,...,5; \beta_4 = \beta_5 = \dfrac{\varepsilon}{\alpha}, \varepsilon$ is a positive constant.

Assume q_d is the given desired trajectory of joints and define:

$$e = q_d - q$$
$$r = e' + he \tag{2}$$

Adopt the following control strategy:

$$\tau = k_1 q_d'' + k_2 q_d' + k_3 + k_4 e' + k_5 e = \sum_{i=1}^{5} K_i \Phi_i \tag{3}$$

$\Phi_1 = q_d''; \Phi_2 = q_d'; \Phi_3 = k; \Phi_4 = e'; \Phi_5 = e; K_i$ is the control gain matrix, then

$$K_i = \dfrac{\overline{\theta_i}' r \Phi_i^T}{\| r \| \| \Phi_i \|}; \overline{\theta_i}' = f_i \| r \| \| \Phi_i \| \tag{4}$$

f_i is a positive constant; θ_i is the estimated value of θ_i .

It can be proved that if the robot system described by Formula (1) adopts the control strategy from Formula (2) to Formula (4), the whole robot system will be asymptotically stable.

6　The Simulation Example of Adaptive Controller

In order to simply simulation calculation and obtain the simulation result easy to analyze, we set the supporting foot-the right foot-fixed, with the exception of joint 2,3, the rotary motion of the other joints are cancelled and changed to be the fixed connection so that we can abstract the robot to plane two-degree-freedom robot. According to the kinetic formula (1) then

$$M_{11} = m_1 r_1^2 + m_2(l_1^2 + r_2^2 + 2l_1 r_2 \cos q_2) + I_1 + I_2$$
$$M_{12} = M_{21} = m_2(r_2^2 + l_1 r_2 \cos q_2) + I_2$$
$$M_{22} = m_2 r_2^2 + I_2$$
$$C_{11} = 2m_2 l_1 r_2 q_2' \sin q_2, C_{12} = m_2 l_1 r_2 q_2' \sin q_2$$
$$C_{21} = m_2 l_1 r_2 q_1' \sin q_2, C_{22} = 0$$
$$G_1 = (m_1 r_1 + m_2 l_1)g \cos q_1 + m_2 g r_2 \cos(q_1 + q_2)$$
$$G_2 = m_2 g r_2 g \cos(q_1 + q_2)$$

The concrete parameters are: $m_1 = 0.13kg$, $m_1 = 0.17kg$ are connecting rod mass; $l_1 = 0.041m$; $l_2 = 0.089 m$ are connecting length; $r_1 = 0.024m$, $r_2 = 0.074m$ are the distance between connecting rod centroid and the joint; $I_1 = 0.21kgm^2$, $I_2 = 0.25kgm^2$ are the moment of inertia of the connecting rod towards the centroid; g is the acceleration of gravity. The simulation parameters are $h = diag[4\ 4], f_1 = f_2 = 3\ \ f_3 = 1\ \ f_4 = f_5 = 5\ \ k = 0.5$。

E_{angle} is defined as the error in the joint trajectory under the disturbed and the ideal condition; D_{angle} is defined as the error in the joint trajectory between the addition of adaptive control and the ideal condition. We can obtain $|\max(E_{angle})| \approx 5\deg$ and $|\max(D_{angle})| \approx 1$ in Fig.4; $|\max(E_{angle})| \approx 14\deg$ and $|\max(D_{angle})| \approx 4.5$ in Fig.5;

According to the least squares theory of curve fitting, we define $Q = \sum_{i=1}^{n} \delta_i^2$ and Q is

the square error sum from point 1 to point n, then we use it to measure the degree of curve fitting and take value in E_{angle} , D_{angle} respectively every 0.015s at 0s-0.45s, $Q_E \approx 144, Q_D \approx 5.89$ can be obtained in Fig.4 and $Q_E \approx 136, Q_D \approx 5.09$ can be obtained in Fig.5. This proves that the degree of curve fitting improve dramatically with the addition of adaptive control.

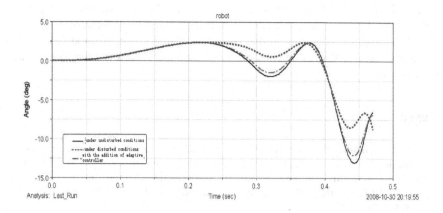

Fig. 5. Curve pursuit of joint 1

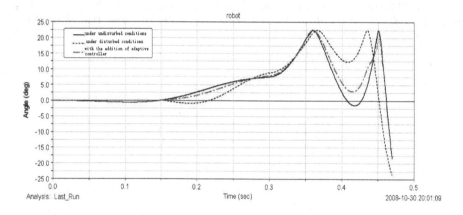

Fig. 6. Curve pursuit of joint 2

7 Conclusions

The method in the paper has feathers of simplicity, convenience and strong currency to build simulation model by Solidworks' Three-dimension Graphics and the obtained simulation model approximate to an actual system; ADAMS can not only automatically build kinematical and dynamic model of the robot and solve automatically, but also optimize process parameters and predict performance conveniently, reducing the consumption of time and funds; It is convenient to build the control system of robot by Simulink. We designed an adaptive controller of robot, based on the basis of the organic combination of three software, as well as analyze and research the result under disturbed and undisturbed conditions, then we compare the tracking effect of the joint trajectory of the robot without the addition of the adaptive controller with the effect with the addition through simulation experiments and the simulation results indicate that the fitting degree of the curve of path tracking improve significantly with the addition of the adaptive controller in the case of interference. The method in the paper could be extended to the mathmetical modeling and the control of other complicated control system.

References

1. Ohishi, K., Majima, K.: Gait control of biped robot based on kinematics and motion description in cartesian space. In: Conf. and Robotics and Automation, pp. 236–239 (2001)
2. Hirai, K., Hirose, M.: The development of honda humanoid robot. In: Proceedings of the 1998 IEEE International Conference on Robotics & Automation, Leuven Belgium, pp. 1321–1326 (1998)
3. Haug, E.J.: Computer_Aided Kinematics and Dynamics of Mechanical System. Prentice Hall, Englewood Cliffs (1989)
4. Gao, D.-X., Xue, D.-Y.: The Study of Robust Adaptive Control System of Robot Based on MATLAB/Simulink. Journal of System Simulation (in Chinese) 7(18-7), 2022–2025 (2006)
5. Ding, X.-G.: Study of Robot Control. Zhejiang University Press, Hangzhou (2006) (in Chinese)

From RoboNova to HUBO: Platforms for Robot Dance

David Grunberg[1], Robert Ellenberg[1,2], Youngmoo E. Kim[1], and Paul Y. Oh[2]

[1] Electrical Engineering
Drexel University, 3141 Chestnut Street
Philadelphia, PA, 19104
dgrunberg@drexel.edu, rwe24g@gmail.com, ykim@drexel.edu
[2] Mechanical Engineering
Drexel University, 3141 Chestnut Street
Philadelphia, PA, 19104
paul@coe.drexel.edu

Abstract. A robot with the ability to dance in response to music could lead to novel and interesting interactions with humans. For example, such a robot could be used to augment live performances alongside human dancers. This paper describes a system enabling humanoid robots to move in synchrony with music. A small robot, the Hitec RoboNova, was initially used to develop smooth sequences of complex gestures used in human dance. The system uses a real-time beat prediction algorithm so that the robot's movements are synchronized with the audio. Finally, we implemented the overall system on a much larger robot, HUBO, to establish the validity of the smaller RoboNova as a useful prototyping platform.

Keywords: Gestures, robots, robotics, dance, motion.

1 Introduction

Several recent artistic productions have incorporated robots as performers. For example, in 2007 Toyota unveiled robots that could play the trumpet and violin in orchestras [1]. In 2008, the Honda robot ASIMO conducted the Detroit Symphony Orchestra [2]. And in 2009, robot actors were used in a theater production in Osaka, Japan [3]. But as of yet the use of humanoid robots in dance has received little attention from robotics researchers and less from the dance community. As approached through research in human-robot interaction, this topic offers a unique opportunity to explore the nature of human creative movement.

A major problem in developing a dancing robot is that full-sized humanoid robots remain very expensive. This makes it risky to test new algorithms on these robots, because an error that damages the robot could be costly. Thus, a less costly prototyping platform to test algorithms for larger dance robots would be useful for researchers. Additional challenges include using signal processing algorithms to predict music beats in real-time and designing robot gestures that appear humanlike.

We are exploring solutions to these problems utilizing the Hitec RoboNova as a prototyping platform for the larger HUBO robot (Figure 1). The physical configuration of

J.-H. Kim et al. (Eds.): FIRA 2009, CCIS 44, pp. 19–24, 2009.

Fig. 1. The RoboNova and HUBO (*left and right*)

the RoboNova is similar to the HUBO and both robots allow for sufficient control to produce smooth gestures. This builds off of our previous work in this area, in which we used the RoboNova to produce movements in response to audio with an update rate of approximately 5 Hz [4].

2 Prior Work

Two robots that incorporate movement with music are Keepon [5] and Haile [6]. Keepon bobs its head in time with audio, and it has been used to influence others to dance. This work offers evidence that human-dancing robot interactions can be constructive and influential. Keepon, however, does not detect beats in music; the movement is pre-programmed to be synchronized with the audio. On the other hand, Haile is able to identify beats and rhythms and synthesizes complementary ones on a drum, but its gestures are designed for percussion performance and not dance. The Ms DanceR [7] was built to solve a related problem – creating a robot able to dance with a human partner. Although it has some understanding of dance styles, it cannot locate the beats in music and is reliant on a human partner for dancing control information.

The problem of music beat identification from the acoustic signal has been studied by many researchers in music information retrieval (e.g., [8] and [9]). Recent beat tracking methods are able to accurately identify the strong impulses that define a song's tempo. The best performing systems, however, operate offline and are not suitable for real-time dancing in response to live audio.

Our prior work on this project is detailed in [4].

3 Beat Predictor

The beat predictor used in our system is based on a beat-identification algorithm proposed by Scheirer [8], which we modified to operate in real-time. Our configuration uses an outboard computer for audio processing that communicates wirelessly with the robot. Our algorithm is depicted in Figure 2 and functions as follows:

- Each audio frame (92.9 msec) is sent through a Cochlear filterbank, which splits the audio into frequency subbands similar to perceptual resolution of the human ear. (Figure 2b)
- Each subband is downsampled, half-wave rectified, and smoothed with a lowpass filter. (Figure 2c)
- The resulting signals are passed through a set of comb filters of varying delays. When audio passes through a comb filter, resonance results if the delay of that filter matches the periodicity of the audio. The filter that produces the greatest resonance across all subbands determines the current tempo estimate. (Figure 2d).
- The phase of the audio is determined from the delay states of this filter, and the beat location is determined from the tempo estimate and change in the phase.

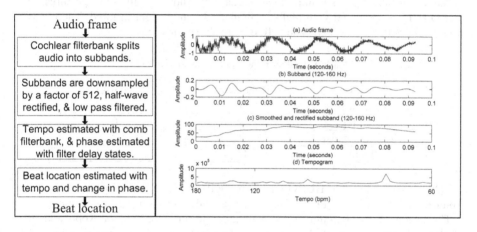

Fig. 2. Description of the beat prediction algorithm. Flow chart (*left,*), frame of audio from "Fire Wire" by Cosmic Gate (*right a*), 120-160 Hz subband of audio frame (*right b*), smoothed subband (*right c*), tempogram over several audio frames (*right d*).

4 Robot Platforms

We chose the RoboNova (14") as our prototyping platform for the following reasons:

- Its low cost allows us to test algorithms without fear of expensive damage.
- Its humanoid shape and large number of Degrees of Freedom (DoFs) approximate HUBO's, so its gestures will be similar to those of the larger robot.
- Its wireless communication abilities allow an external processor to assume some of the computational burden.

Our initial RoboNova implementation used the RoboBasic programming environment [4]. This provided a convenient platform for generating gestures, as it is simple to have the robot interpolate between any two points.[1] The limitations of this environment were:

- The robot could only linearly interpolate between start and end points. Gestures moved at the same speed all the way through, which appeared unnatural.
- RoboBasic is relatively slow; RoboNova's update rate with this environment was on the order of 5 Hz. This causes timing inaccuracies.
- Because of a variable loop speed, a gesture would occasionally be sent after the beat occurred, and the robot would not perform that gesture. This looked awkward, and such mistimed leg gestures often destabilized the robot.

Modified firmware for the RoboNova was provided by the first author of [10]. This firmware enabled the RoboNova to be programmed in a C environment. We also began transmitting gestures to the robot in smaller components, so each piece could be performed at a different speed.

Having sufficient DoFs is important for enabling smooth, human-like gestures. The HUBO (5') was designed with realistic, human-like movement in mind. The number of DoFs for both the RoboNova and HUBO are shown in Table 1.

Table 1. DOFs for the RoboNova and HUBO

Location	RoboNova	HUBO
Each arm	3	6
Each leg	5	6
Head and waist	0	2
Total	16	26

Table 1 shows that the HUBO has more DoFs than the RoboNova, and can thus produce more complex gestures.[2] The difference, however, is not overwhelming.

The systems for programming and generating gestures for both the RoboNova and HUBO are very similar. Both are programmed in C and involve the user choosing start and end points for each gesture, which are interpolated using a cycloid function. These functions influence the angle θ for each joint i depending on time t and linearity factor C (when C is smaller, gestures are more linear). Equation 1 has an example:

$$\theta_i = \frac{2\pi t - C\sin(2\pi t)}{2\pi} .$$
(Equation 1)

There are two important differences between the platforms:

- The RoboNova is sent gestures in small pieces, while the HUBO is sent whole gestures and breaks them apart onboard.
- HUBO gestures require more joint start and end points than the RoboNova because of its greater number of DoFs.

Both of these changes require only trivial modifications to our system.

---

[1] Video available at: http://dasl.mem.drexel.edu/~robEllenberg/Projects/Dance/Media/iPod.mov
[2] Video available at:
   http://dasl.mem.drexel.edu/~robEllenberg/Projects/Hubo/Media/TaiChi.mov

5 Experiment

This system represents significant modifications to our previous work with the RoboNova [4]. The goal of our experiment was to determine whether the modifications, such as using a C environment instead of RoboBasic, enabled more accurate gesturing.

Synthesized audio with a tempo of 60 bpm was transmitted to the RoboNova. The offboard computer calculated joint positions so that the RoboNova would raise its right arm in time with the beat, then sent these positions to the robot. Position sensors in the servos communicated the actual position of the joints to the computer over time.

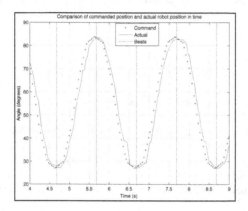

Fig. 3. Cycloid interpolation results. The servo being plotted is the right arm shoulder servo.

Figure 3 verifies the improvements to the system. In [4], which used the previous RoboBasic environment, there was up to .2sec difference between commanded and actual gesture times. The new algorithm causes the command and actual times to be virtually identical, and the difference between them is much less than .2sec.

6 Future Work

The robots cannot accurately perform gestures that require more than 1 degree in .0037s. This is not necessarily bad, as many human dancers cannot accurately perform dance movements at such speeds either. However, a potential improvement could come from allowing the robot to take multiple beats to produce such quick gestures.

We would also like to enable detection of more detailed rhythmic information than only the low-level beat, such as the meter (e.g., '2/4 time.'). With this information, the robots could produce movements taking into account the full rhythm of the music.

Another goal is to teach the robot more about different dance styles. The HUBO cannot yet intelligently choose appropriate sequences of movements. It would be another large step towards making a true dancing robot if we could enable the robot to select gestures based on the style of the audio being played.

7 Conclusion

In this paper, we proposed a system to enable a dancing humanoid robot. We discussed the RoboNova and HUBO robots, and demonstrated that the RoboNova could be a useful prototyping platform for the HUBO. Finally, we detailed improvements to the programming environment for the RoboNova that enable more accurate gestures.

Acknowledgements. This research is supported by the National Science Foundation (NSF) under grant Nos. 0730206 and IIS-0644151. We gratefully thank them. Any opinions, findings, and conclusions or recommendations expressed in this material are those of the authors and do not necessarily reflect the views of the NSF.

References

[1] Times, L.A.: Toyota shows range with robot violinist. LA Times Business Pages (December 2007)
[2] Honda: Honda's Asmio Robot to Conduct the Detroit Symphony Orchestra. Press Release (April 2008)
[3] British Broadcasting Corporation. Actor Robots Take Japanese Stage. British Broadcasting Corporation U.K. Internet, http://news.bbc.co.uk/2/hi/asia-pacific/7749932.stm (accessed 2/23/2009)
[4] Ellenberg, R., et al.: Exploring Creativity through Humanoids and Dance. In: Proceedings of The 5th International Conference on Ubiquitous Robots and Ambient Intelligence, URAI (2008)
[5] Michalowski, M.P., Sabanovic, S., Kozima, H.: A Dancing Robot for Rhythmic Social Interaction. In: Proceeding of the 2nd Annual Conference on Human-Robot Interaction (HRI 2007), vol. 103, pp. 89–96 (2007)
[6] Weinberg, G., Driscoll, S.: The Interactive Robotic Percussionist: New Developments in Form, Mechanics, Perception and Interaction Design. In: HRI 2007: Proceedings of the ACM/IEEE International Conference on Human-Robot Interaction, pp. 97–104 (2007)
[7] Kazahiro, K., et al.: Dance Partner Robot -Ms DanceR. In: Proceedings of the 2003 IEEE/RSJ International Conference on Intelligent Robots and Systems, pp. 3459–3464 (2003)
[8] Scheirer, E.D.: Tempo and Beat Analysis of Acoustic Musical Signals. Journal of the Acoustical Society of America 103(1), 588–601 (1998)
[9] Klapuri, A.P., Eronen, A.J., Astola, J.T.: Analysis of the Meter of Acoustic Musical Signals. IEEE Transactions on Audio, Speech, and Language Processing 14(1), 342–355 (2006)
[10] Kushleyev, A., Garber, B., Lee, D.D.: Learning humanoid locomotion over rough terrain. In: Proceedings of The 5th International Conference on Ubiquitous Robots and Ambient Intelligence, URAI (2008)

BunnyBot: Humanoid Platform
for Research and Teaching

Joerg Wolf[1], Alexandre Vicente[1], Peter Gibbons[1], Nicholas Gardiner[2],
Julian Tilbury[1], Guido Bugmann[1], and Phil Culverhouse[1]

[1] Centre for Robotics and Intelligent Systems, University of Plymouth, Drake Circus
[2] Marine and Industrial Dynamic Analysis (MIDAS) Research Group,
University of Plymouth, Drake Circus,
Plymouth, PL48 AA, U.K.
{joerg.wolf,guido.bugmann,phil.culverhouse}@plymouth.ac.uk

Abstract. This paper introduces a cost effective humanoid robot platform
with a cluster of 5 ARM processors that allow it to operate autonomously. The
robot has an optical foot pressure sensor and grippers all compatible with a Robotis Bioloid. The paper also presents a kinematic model of the robot. Furthermore we describe how the robot uses a complementary filter which combines
accelerometers and gyroscope readings to get a stable tilt angle.

Keywords: Inertial Measurement Unit, Complementary Filter, Bipedal
Humanoid Robot, FIRA, RoboCup, Motion Control, Bioloid.

1 Introduction

In recent years several humanoid platforms have come onto the market aimed at
consumers, education and hobbyists [1]. The height of these humanoids is around
0.3m. Some of these platforms have the potential to be extended for research by adding
more sensors and computational power. Much of the fundamental work in humanoid
robotics can be carried out with these cheap miniature humanoids. This paper will
present the development of a miniature humanoid at the University of Plymouth.
Firstly an overview of the hardware components is given in Sections 1 and 2. Section 3
introduces the sensors and actuator designs. Section 4 describes the kinematic model of
the humanoid. Section 5 concludes with suggestions for future work.

1.1 Purpose

The BunnyBot humanoid platform has been designed with three main purposes in
mind: *1. creating an affordable humanoid platform, 2. To advance teaching and research at the university and 3. To create a competition platform.* Because of the fairly
low cost of the robot (£ 2000 GBP), the university will be able to build enough
humanoids for teaching a whole class room. The robot has servos for moving head,
eyes and ears. By adding a mouth and eye brows it will be possible to produce facial
expressions. The head is in form of a cartoon-like bunny rabbit, hence the name

J.-H. Kim et al. (Eds.): FIRA 2009, CCIS 44, pp. 25–33, 2009.

BunnyBot. The idea is to avoid the uncanny valley problem. Cartoon-like features and exaggerated facial features are more likely to be judged friendly [2] and therefore ease human-robot interaction. When evaluating the BunnyBots dimensions by using Ryu [15] method, it turns out that proportions that appear more friendly to children in a school environment.

The BunnyBot platform has been developed to be compatible with the rules of FIRA HuroCup 2009 [3] (small category < 50cm height) and RoboCup Humanoid League (KidSize <60cm).

1.2 Overview of BunnyBot

This robot is based on a commercial Robotis Bioloid robot. The torso of the Bioloid has been completely replaced; grippers and foot pressure sensors have been added.

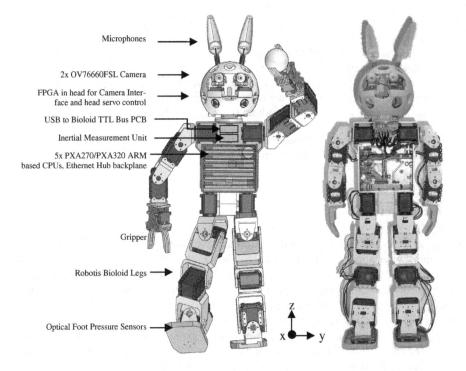

Fig. 1. BunnyBot CAD model left, BunnyBot Photo right. BunnyBot has 8 degrees of movements in the head + optional 4 facial servos (not shown here) and 18 degrees of movement in the body. The batteries can be fitted in front of the hips or in place of the 5th processor board. The total weight of the robot is 2.4 kg.

2 Distributed Processing

Other teams such as Humanoid Team Humboldt [4], NimBro [5] and UofM Humanoids [6] have also used Bioloid Robots combined with handheld computers.

A previous version of our teams FIRA competition robot also had a handheld computer [7]. We are going a step further and integrating 5 handheld-type CPUs onto one robot. These 5 CPUs are from the Marvell XScale family (former Intel) based on a ARMv5TE core. The XScale family has no floating point instructions. The XScale PXA270 with 520MHz and PXA320 with 806MHz. Each processor has 64MB RAM and 32MB Flash (PXA270 – 500MHz) or 128MB RAM and 64MB flash (PXA320 – 806MHz) and draws 350mw of power. The XScale processors are on a SO-DIMM type connector and can be purchased from Toradex (www.toradex.com). These SO-DIMM modules plug into a motherboard for breakout connectors. For future work we are developing a motherboard with an OMAP3530 (600MHz) which will be used for SLAM processing in Linux.

2.1 Mainboard Cluster

Five miniature processor PCBs and one FPGA provide the hardware and software support for the robot. They are mounted as motherboards with 91mm width and can then be inserted into a small card-cage in the body of the robot. All five processors communicate via Ethernet on the Backplane. The backplane also contains a USB hub for host (PC) communication.

The processors provide the following services:

1) Moton Control and host communications [PXA270]
    a) Direct control of servos
    b) WiFi link (IEEE802.11b) Spectec SDW-823 SD Card
2) Central Planning and Action control [PXA270]
    a) Planning, memory and sense of time
    b) TCP client to other processors
3) Vision – face recognition [PXA320]
    a) Runs Viola and Jones object recognition [8] with the first 3 filters in the FPGA (planned)
4) Vision – stereo and SLAM processing [PXA320]
    a) Mono camera SLAM [9] partly with PFGA based feature detection (planned)
5) Audition – sound output and stereo input from ears [PXA270]
    a) Stereo hearing processed and sent to host PC via WiFi (processor 1) for speech recognition processing

An Altera Cyclone III (EP3c25F256C8N) FPGA chip was placed into the head of the robot in order to offer local pre-processing of the video streams coming from the cameras, such as colour blob tracking. The FPGA chip generates PWM signals for the analogue head servos. The head has SuperTec TITCH-44 servos, since they are the smallest servos (4.5g) for a competitive price. Furthermore a 2-axis gyro is also processed by the FPGA to allow eye gaze stabilisation.

An XScale processor features a USB-host. A FTDI USB-to-Serial chip was used for communication with the Robotis AX-12 servos.

3 Sensors and Actuators

3.1 Novel Foot Pressure Sensor

A foot pressure sensor was developed to explore balance control. The sensor measures the pressure in four points (figure 2) providing information on the total force vertical applied onto one foot (F_z) and the moments in the forward-backward direction (M_x) and in the left-right direction (M_y). This allows determining the projection of the centre of mass of the robot onto a plane normal to the gravity vector.

The pressure is measured by suspending the BunnyBot foot on leaf springs modeled as cantilever beams. The deflection of the springs causes a displacement which is measured using reflective IR light sensors (Honeywell HOA1397-002). When the robots leans forward, the leaf springs are loaded more in front and the two front sensors have a smaller distance to the foot sole. The sensors have the characteristics that for distances to a reflective surface of less than 1.2 mm, the signal increases quasi-linearly with the distance. The potential on the phototransistor varies between 0.4V and 4.8V, given a power supply of 5V. This is a large variation that can easily be read with standard analogue inputs of microcontrollers.

The springs are two strips of stainless steel (1.6mm x 5.5mm x 90 mm) attached to the sole of the BunnyBot with 4mm spacers.

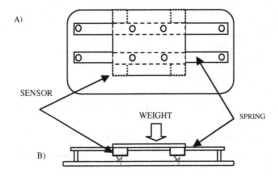

Fig. 2. A) Top view of the sensor arrangement on a Bioloid/BunnyBot foot-plate. B) Side view.

If we assume that the centre plate of the foot is a rigid body, the leaf springs can modelled as cantilever beams. In cantilever beams the deflection d is directly proportional to the applied force F_s:

$$d = \frac{L^3}{3EI}F_s \tag{1}$$

Whereby E is the Young's modulus of the steel, I is the second moment of area and L is the length of the spring. The Centre of pressure of the foot corresponds to the centre of mass projected to the floor if the robot is standing on one foot only. The Centre position of the centre of mass can then be determined from the foot pressure sensors by taking the moments from the ankle joint.

$$x_{COM} = \frac{F_1 x_f - F_2 x_b}{mg} \quad , \quad y_{COM} = \frac{F_3 y_r - F_4 y_l}{mg} \tag{2}$$

Whereby the foot pressure sensors are mounted at x_f, y_b , x_l, y_r, distance from the ankle joint, m is the mass of the robot, and F_1, F_2, F_3, F_4 are the forces measured via distances. Better results can be reached by averaging 2 sensors to establish the force on the principle axis.

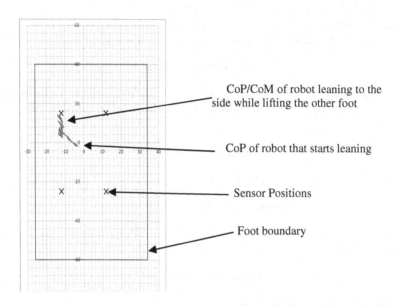

CoP/CoM of robot leaning to the side while lifting the other foot

CoP of robot that starts leaning

Sensor Positions

Foot boundary

Fig. 3. The figure shows the robots foot with dimensions in mm. The Centre of Pressure CoP measurement is shown.

During the experiment it was found that the sensors sometimes do not return to the origin when the load is taken off. It is important that mechanical construction of the sensor is rigid and the spring material has no hysteresis and is not loaded beyond the proportional limit stress.

3.2 Inertial Measurement Unit with Complementary Filter

The BunnyBot has a 6-axis IMU (inertial measurement unit) that consists of a 3-axis gyroscope and a 3-axis accelerometer. Data from this type of sensors is noisy and has drift. The usual solution to unite two noisy sensor readings into one low-noise signal is the Kalman Filter [10], which uses variable reliability on each sensor according to the user's input. The complementary filter is a simpler approach to sensory fusion [11]. The complementary filter works without a user input, all it needs is the noisy data from both sensors and a cut-off frequency set at the beginning of the operation.

The main advantage of the complementary filter is that it is much easier to code than the Kalman filter and uses a lot less processing power. The solution is to combine both sensors by applying a low pass filter (LPF) on the gravity vector angle, thus removing the noisy jerk response, and a high pass filter (HPF) on the gyroscope, thus removing the drift. A special case of the complementary filter has been implemented by following the method that was briefly described in Kumagai and Ochiai [12]. The initial equation is given by:

$$LPF[\theta_a(s)] + HPF[\theta_g(s)] = 1 \tag{3}$$

This means that the amplitude of the signal θ is split into two components coming from accelerometer and gyro. The frequency spectrum of the two filters always adds up to a gain of 1. In the time domain, the output is the amplitude of θ is

$$LPF[\theta_a(t)] + HPF[\theta_g(t)] = \theta_y(t) \tag{4}$$

However, a further implementation was derived for this filter in order to reduce the number of multiplications during the processing time. A high pass filter can be written as:

$$HPF[\theta_g(t)] = 1 - LPF[\theta_a(t)] \tag{5}$$

Therefore by changing eq. [3] into eq. [2]:

$$LPF[\theta_a(t)] - LPF[\theta_g(t)] + \theta_g(t) = 1 \tag{6}$$

The 'Gyro' component of eq. [4] is the Gyro input without filtering. Eq. [4] can be further simplified into:

$$LPF[\theta_a(t) - \theta_g(t)] + \theta_g(t) = 1 \tag{7}$$

This reduces the number of multiplications necessary.

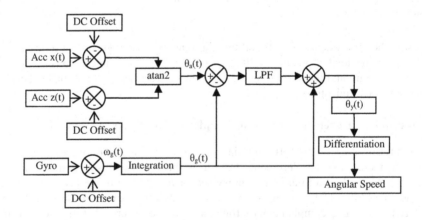

Fig. 4. Complementary filter to determine the robots attitude $\theta_y(t)$

3.3 Robust Gripper Design

The gripper was designed to fulfil two purposes; to be able to pick up randomly folded cloth for a research project [13], as well as being able to hold a 40mm diameter ball for the FIRA HuroCup. For the first purpose a rotating wrist axis was included, so that the gripper could be lined up square to a cloth fold. In order that the fingers were less likely to interfere with the cloth during the action of gripping, the fingers were made pointed.

The gripper was rapid prototyped using 3mm acrylic sheet, cut by a CNC laser cutter. Consequently, for the second purpose, a composite of three layers was made for the fingers with the middle layer being smaller than the outer two such that the surface of the ball made contact with the outer fingers. This made it less likely that the ball would slip sideways out of the gripper.

Aesthetic considerations dictated that the overall size of the mechanism should be in proportion to the rest of the BunnyBot. This meant that the servos needed to be arranged in a manner that minimised wasted space. A design criterion connected to a potential use of the BunnyBot in competition was that parts should be easily changeable, such as a different finger design for better ball handling. This, in conjunction with the difficulty of assembling the small size required, meant that a square dovetail method was used to connect the top and bottom to the sides. Weight was also reduced by adding holes to the larger flat surfaces.

One further consideration was that the servos, being the most expensive parts, needed protection from sudden loading, such as might come from the BunnyBot falling over. For this reason the gripper was held shut with a spring and the fingers moved apart with a cam and follower mechanism.

4 Kinematic Model

A forward kinematic model gives the robot the ability to calculate the position of its limbs in relationship to the torso, given the current joint angles. This can be useful for motion planning, determining its current centre of mass and deriving the inverse kinematics. The kinematic model given in the tables below is only slightly different from the original Robotis Bioloid, which makes this model interesting to a wider audience. The hip servos 7 and 8 have been mounted the other way round (see figure 1) and the distance between the arms and the legs may be different. The following model follows the Denavit-Hartenberg parameters and the frame assignment conventions as laid out by Craig [14]. The order of multiplication to go from one joint to the next joint is:

1. Rotate around the x-axis of the reference frame with angle α
2. Move along x-axis of the reference frame a distance ℓ
3. Rotate around the new frame's z-axis with angle θ
4. Move along the z-axis with distance d

This gives a homogeneous transformation matrix given in formulae 3.

$$
A_{New}^{Ref} = \begin{bmatrix} \cos(\theta) & -\sin(\theta) & 0 & \ell \\ \sin(\theta)\cos(\alpha) & \cos(\theta)\cos(\alpha) & -\sin(\alpha) & -\sin(\alpha)*d \\ \sin(\theta)\sin(\alpha) & \cos(\theta)\sin(\alpha) & \cos(\alpha) & \cos(\alpha)*d \\ 0 & 0 & 0 & 1 \end{bmatrix} \tag{8}
$$

For instance, to calculate the position RHND(x,y,z,α,β,γ) of the right hand of the bunny robot with respect to the centre C(x,y,z,α,β,γ) we first need to multiply all transformations from the tables given in the Appendix.

$$A_{RHND}^{C} = A_{1}^{C} \quad A_{3}^{1} \quad A_{5}^{3} \quad A_{RHND}^{5} \tag{9}$$

The origin of the coordinate system C(x,y,z) where the forward transformation starts from, is placed on the intersection between the sagittal plane and coronal plane at the height of the arm servo axis (Servo 1 and Servo 2). When all servos are at 0 degrees (home position) the robot is standing straight and its arms are normal to the sagittal plane.

5 Conclusion

This paper introduced an affordable humanoid platform that will be used to advance teaching and research at the university and to create a competition platform for HuroSot and RoboCup. A detailed kinematic model was introduced that can be implemented into Bioloid-related platforms. So far we have implemented simple stereo vision, audio playback software and walking with inverse kinematics. The robot Future work will include human-robot interaction experiments and work on gait and balancing. There are currently research projects on the way using BunnyBot for fabric manipulation [13].

Acknowledgement

We would like to thank ARM Ltd. and Toradex AG for their commitments and support. Furthermore we would like to thank Benoît Quentin and all involved university staff for their support.

References

1. Technology Turning Childhood Dreams into Reality. Korea IT Times, Edition (October 2008)
2. Woods, S., Dautenhahn, K., Schulz, J.: Exploring the design space of robots: Children's perspectives. Interacting with Computers 18, 1390–1418 (2006)
3. Baltes, J.: HuroCup General Laws of the Game 2009. Autonomous Agents Laboratory University of Manitoba (2009),
   http://www.fira.net/soccer/hurosot/overview.html
4. Hild, M., Meissner, R., Spranger, M.: Humanoid Team Humboldt Team Description 2007 for RoboCup, Atlanta U.S.A (2007)
5. Behnke, S., Mueller, J., Schreiber, M.: Using Handheld Computers to Control Humanoid Robots. In: Proceedings of 1st International Conference on Dextrous Autonomous Robots and Humanoids (darh 2005), Yverdon-les-Bains, Switzerland, paper no. 3.2 (May 2005)
6. Baltes, J., Bagot, J., Anderson, J.: Humanoid Robots: Storm, Rogue, and Beast. Team Description Paper. In: Proceedings of RoboCup 2008: Robot Soccer World Cup XII, Suzhou, China (July 2008)
7. Wolf, J.C., Hall, P., Robinson, P., Culverhouse, P.: Bioloid based Humanoid Soccer Robot Design. In: The Proceedings of the Second Workshop on Humanoid Soccer Robots @ 2007 IEEE-RAS International Conference on Humanoid Robots, Pittsburgh (USA), November 29 (2007)
8. Viola, P., Jones, M.: Robust Real-time Object Detection. International Journal of Computer Vision (2001)

9. Davison, A.J.: Real-time simultaneous localisation and mapping with a single camera. In: Proceedings of the 9th International Conference on Computer Vision, Nice (2003)
10. Maybeck, P.S.: Stochastic Models, Estimation and Control. Academic Press, New York (1979)
11. Baerveldt, A.-J., Klang, R.: A low-cost and low-weight Attitude Estimation System for an Autonomous Helicopter. In: Proc. 1997 IEEE Int. Conf. on Intelligent Engineering Systems, Budapest, Hungary, pp. 391–395 (1997)
12. Kumagai, M., Ochiai, T.: Development of a Robot Balancing on a Ball. In: International Conference on Control, Automation and Systems, Korea, pp. 433–438 (2008)
13. Gibbons, P., Culverhouse, P., Bugmann, G.: Fabric Manipulation: An Eye Tracking Experiment. In: Proceedings of Taros 2008, Edinburgh, pp. 130–134 (2008)
14. Craig John, J.: Introduction to Robotics, 2nd edn. Addison-Wesley, Reading (1989)
15. Ryu, H., Kwak, S.S., Kim, M.: A Study on External Form Design Factors for Robots as Elementary School Teaching Assistants. In: 16th IEEE International Conference on Robot & Human Interactive Communication, Jeju, Korea, WB3-2, pp. 1046–1051 (2007)

Appendix

Table A1. Denavit-Hartenberg Parameters for Left Leg and Right Leg

Matrix	α	l	d	θ	Servo	Matrix	α	l	d	θ	Servo
A_{CH}^{C}	π	20	166	$-\pi/2$	-	A_{CH}^{C}	π	20	166	$-\pi/2$	-
A_{8}^{CH}	0	33.5	0	0	8	A_{7}^{CH}	0	-33.5	0	0	7
A_{10}^{8}	$\pi/2$	0	0	$\pi/2$	10	A_{9}^{7}	$\pi/2$	0	0	$\pi/2$	9
A_{12}^{10}	$-\pi/2$	0	0	0	12	A_{11}^{9}	$\pi/2$	0	0	0	11
A_{14}^{12}	0	75	0	0	14	A_{13}^{11}	0	75	0	0	13
A_{16}^{14}	0	75	0	0	16	A_{15}^{13}	0	75	0	0	15
A_{18}^{16}	$\pi/2$	0	0	0	18	A_{17}^{15}	$-\pi/2$	0	0	0	17
A_{AKL}^{18}	0	0	0	$\pi/2$	-	A_{AKL}^{17}	0	0	0	$\pi/2$	-
A_{FL}^{AKL}	$-\pi/2$	0	-32	$\pi/2$	-	A_{FL}^{AKL}	$-\pi/2$	0	-32	$\pi/2$	-

Table A2. Denavit-Hartenberg Parameters for Left Arm and Right Arm

Matrix	α	l	d	θ	Servo	Matrix	α	l	d	θ	Servo
A_{2}^{C}	$-\pi/2$	0	83.25	0	2	A_{1}^{C}	$-\pi/2$	0	-83.25	0	1
A_{4}^{2}	$\pi/2$	15	0	$\pi/2$	4	A_{3}^{1}	$\pi/2$	15	0	$-\pi/2$	3
A_{6}^{4}	0	68	0	0	6	A_{5}^{3}	0	68	0	0	5
A_{LHND}^{6}	0	100.5	0	0	-	A_{RHND}^{5}	0	100.5	0	0	-

Teen Sized Humanoid Robot: Archie

Jacky Baltes, Ahmad Byagowi, John Anderson, and Peter Kopacek

Autonomous Agent Lab
University of Manitoba
Winnipeg, Manitoba
Canada, R3T 2N2
j.baltes@cs.umanitoba.ca
http://www.cs.umanitoba.ca/~jacky
Institute for Handhabungsgeräte und Robotertechnik
Technische Universität Wien
Favoritenstr. 9-11
A-1040 Wien, Österreich
kopacek@ihrt.tuwien.ac.at

Abstract. This paper describes our first teen sized humanoid robot
ARCHIE. This robot has been developed in conjunction with Prof.
Kopacek's lab from the Technical University of Vienna. ARCHIE uses
brushless motors and harmonic gears with a novel approach to position
encoding. Based on our previous experience with small humanoid robots,
we developed software to create, store, and play back motions as well as
control methods which automatically balance the robot using feedback
from an internal measurement unit (IMU).

1   Introduction

In recent years, improvements in mechanical devices and battery technology has
led to the creation of several large humanoid robots. Honda's Asimo ([1]), Prof.
Oh's Hubo ([3]), and Wabian-2 from Waseda university ([2]) are some of the
most well known examples.

Even those these robots have made huge improvements over a short period
of time, there are still many technological issues that need to be overcome to
have large humanoid robots that can actively balance over uneven terrain, plan
complex motions, and interact with humans naturally. Furthermore, these robots
are too expensive for general use.

This paper describes our new 1.2m tall humanoid robot ARCHIE, a collaboration between the labs of Prof. Kopacek from the TU Vienna and Prof. Jacky
Baltes from the University of Manitoba. At the moment, the robot design focuses on a new modular joint with a brushless motor, a harmonic gear drive,
and a hall based absolute position sensor. This new modular joint design allows
the creation of a humanoid robot with slim legs as opposed to the large legs of
the robots mentioned above.

J.-H. Kim et al. (Eds.): FIRA 2009, CCIS 44, pp. 34–41, 2009.

2   Hardware Description

Archie is a 1.2m tall humanoid robot. It has 22 degrees of freedom (DOF). There are seven DOFs for each leg: three DOFs in each hip, one in the knee, and three in the ankle. Archie is one of the few humanoid robots that has activated toes, which allow it to roll over the foot when walking.

Archie uses a novel modular joint design developed by Prof. Kopacek from the TUV. Each of the joints includes a motor, a gear box and one or two encoders. We are using two different types of motors in Archie: DC motors and brush-less motors are the two types.

Each joint uses an independent controller which controls torque, velocity and position using three cascaded PID (proportional, integral, and derivative control) loops. Feedback for each joint is provided by current sensors and a special Hall based sensor. Three current sensors are used for controlling the torque loop. For position feedback we developed a novel approach which is based on Hall effect sensors and is described in the following subsection.

An overview of Archie's control system is shown in Fig. 2.

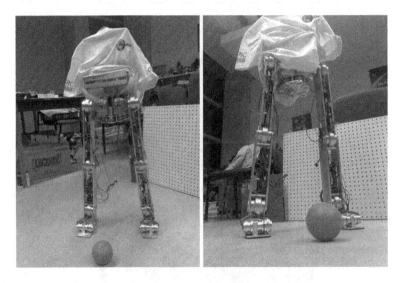

Fig. 1. ARCHIEteen sized robot

2.1   Modular Joint Design

As shown in Fig. 3 for the harmonic gearbox we have two colored arrows. The red arrow represents the input and the black one shows the output of the harmonic gearbox.

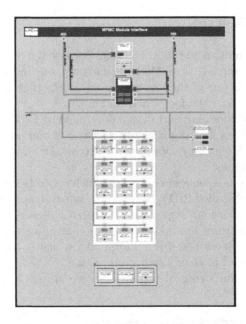

Fig. 2. ARCHIE: Block Diagram of ARCHIE's control system

Fig. 3. ARCHIE: Modular Joint Design combining a brushless motor and a harmonic gearbox

The magnitude of these two arrows is related via the gearbox ratio. For our current model this ratio is 1:160. For a 360 degree revolution on the black arrow we require 160 revolutions of the red arrow, and trade off speed for torque.

In Archie we are using three types of motors: brush-less motors, DC motors and RC motors. Some of the key benefit of using a brush-less motor in Archie are increased efficiency and less noise of the motor. However, control of a brush-less motor requires more complex control logic, but allows for finer control. Given those advantages it would have been sensible to use only brush-less motors for Archie, but to save cost, the joints that do not need to generate very high torque were implemented via DC motors.

2.2   Brush-Less Motor Controller

We use a three phase brush-less motor power stage to control the brush-less motors. The power stage is connected to a CAN5 bus with a CANopen software layer. In this power stage we have a particular DSP that controls the PID loop to controlling torque, velocity and position in the Joint.

Fig. 4. ARCHIE: Brushless Notor Driver

2.3   DC Motor Controller

This controller is based on a DSP processor that controls the torque, velocity and the position of the joint by driving an H-Bridge connected to the DC motor. Furthermore, a Hall based current sensor is used to measure the energy that is going to the motor to determine the torque. The output of this sensor is an analog signal, that is measured by a 10-bit ADC after an RC filter.

2.4   Decentralized Controller

Archie uses a decentralized control system. Each joint uses its own joint controller. All motors use a SPI bus that works on top of a RS-422 physical layer and can support speeds up to 10Mbps. This means that the motion controller can support a response time of 1ms in the control loop.

The SPI bus is controlled by the spinal processing unit (SPU) described in subsection 2.5, which acts as master of the SPI bus. The communication bus supports broadcast as well as client messages. All clients (motors) listen to broadcast messages that can only originate from the SPU. In addition, each client has a unique ID and only reacts to client messages with a matching ID.

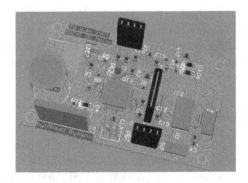

Fig. 5. ARCHIE: DC Notor Driver

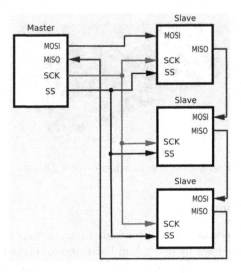

Fig. 6. ARCHIE: Distributed SPI bus

2.5   Spinal Processing Unit (SPU)

The SPU is responsible for motion control and balancing of the robot. We use a Virtex-4 family FPGA from Xilinx, which contains a hardcore PowerPC 405 processor.

The FPGA also contains custom hardware to control the inertia measurement unit, which provides us with angular velocity and linear acceleration for 3 axis each. The robot uses an inverted pendulum model and ZMP control to balance the robot.

Moreover the SPU includes three Master SPI units implemented in hardware FPGA that are connected to the RS422 physical layer. These three communication buses control the left leg, right leg, and torso respectively. each of these buses has a maximum of 7 clients. This limitation comes from the communication protocol that is used for data exchanging between the motor controller and the SPU.

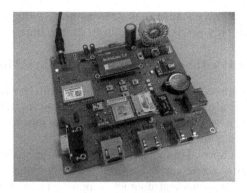

Fig. 7. ARCHIE: Spinal Processing Unit

The SPU also provides internal diagnostics and fail-safe tests to make sure that he robot's operation is safe.

2.6 Communication Protocol

The SPI buses use the following communication protocol. Each data frame consists of 16 bits. The first 8 bits are data and the second 8 bits represents the message ID and the semantics of the fist 8 bits that can be torque, velocity of position. Besides we have two bits that are used for CRC12 checking to detect errors in the communication bus.

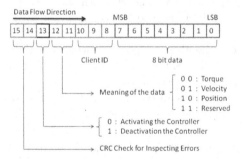

Fig. 8. ARCHIE: SPI Communication Protocol

2.7 Position Encoders on Start Up

Determining the absolute position is necessary for all motors that are used in Archie except the RC motors. In the toe joint we have an incremental positioning system that is based on a zero point and requires that the motor is moved to a fixed position at start up.

Thereafter. the position will be determined using incremental encoders that are mounted on the motor. For the heel joint we use a permanent magnet and a

Hall sensor based absolute encoder that gives us always the correct position of the joint. This design only needs to be calibrated once during construction.

2.8   Contact-Free Position Encoders for Brushless Motors

The most common approach to determining the absolute position of the motor is to use end-switches. However, this approach requires the robot to move into possibly unstable positions at initialization, which is unsuitable for large and expensive teen sized humanoid robots.

Our method uses a special chip (AS 5134) that contains four Hall sensors, a flash analog to digital converter (ADC), an embedded micro-controller and a permanent magnet. The permanent magnet is mounted on the input of the gearbox. Each of the four Hall sensors has a different angle to the permanent magnet. We can thus measure the absolute angle between the chip and the permanent magnet. This implements a contact-free absolute encoder that provides pulses like an incremental encoder as well as an absolute position of the permanent magnet. The Hall chip is mounted on the output of the harmonic gear box and the permanent magnet is connected to the input of the gear box that is coupled to the rotor of the brush-less motor.

This system using four Hall sensors and a permanent magnet is able to determine the absolute angle of the output of the gearbox. However, the accuracy of this method alone is not sufficient to control the position with the required accuracy. The harmonic gearbox has a ratio of 160:1.

To improve the accuracy we extend our design by reusing the permanent magnet on the rotor of the brush-less motor to trigger a Hall switch mounted on the chassis of the brush-less motor. Therefore, the absolute position measured by the four Hall sensors can be improved by comparing it to the absolute position of the rotor.

One difficulty is that the absolute position when the Hall sensor is triggered moves because of the rotation of the motor. However, this rotation is determined by the gear ratio of the brush less motor.

The following formula shows allows us to determine the rotation measured by the Hall sensor given a complete revolution of the rotor:

$$\text{Sensor Angle} = 360^o + \frac{360^o}{r}, \text{ where } r \text{ is the gear ratio}$$

The sensed angle is the sum of two terms: the first term corresponds to one full revolution of the rotor and the second term corresponds to the movement of the output of the gearbox.

In our case, a gear ratio of 1:160 results in an additional term of 2.25^o.

Because of the rotation of the crossing point which is detected by the Hall switch, we can compensate for different values in the absolute sensor. These values are constant, and allow us to calculate the absolute position of the joint. On the other hand, we have a high resolution for the incremental encoder that is used to control the excitation of the brush-less motor.

3   Conclusion

This paper describes the hardware and control electronic of Archie, a teen sized humanoid robot. Archie uses brush-less motors and novel methods to implement absolute position encoders.

References

1. Chestnutt, J.E., Lau, M., Cheung, G.K.M., Kuffner, J., Hodgins, J.K., Kanade, T.: Footstep planning for the honda asimo humanoid. In: ICRA, pp. 629–634. IEEE, Los Alamitos (2005)
2. Ogura, Y., Aikawa, H., Shimomura, K., Kondo, H., Morishima, A., Lim, H.o., Takanishi, A.: Development of a new humanoid robot wabian-2. In: ICRA, pp. 76–81. IEEE, Los Alamitos (2006)
3. Park Ill, W., Kim, J.-Y., Lee, J., Oh, J.-H.: Mechanical design of the humanoid robot platform, hubo. Advanced Robotics 21(11), 1305–1322 (2007)

Interdisciplinary Construction and Implementation of a Human Sized Humanoid Robot by Master Students

Jan Helbo and Mads Sølver Svendsen

Department of Electronic Systems, Section of Automation and Control
Aalborg University
Aalborg, Denmark
jan@es.aau.dk, mss@es.aau.dk

Abstract. With limited funding it seemed a very good idea to encourage Master Students to design and construct their own human sized biped robot. Because this task is huge and very interdisciplinary different areas of expertise were covered by students from different departments who in turn took over results from former students. In the last three years three student groups from Department of Mechanical Engineering and Electronic Systems have been working on the project. The robot AAU-BOT1 has been designed, manufactured, assembled, instrumented and walking should be possible in the near future. The Project Organized Problem Based Learning method implemented at Aalborg University is one of the reasons why supervisors give Master Students such a challenge.

Keywords: Interdisciplinary education, humanoid robots, project organized problem based learning.

1 Introduction

The learning method at Aalborg University (AAU) Project Organized Problem Based Leaning (POPBL) has since its start in 1974 been a great success [1]. The merits of the method are: extensive peer collaboration, increased motivation, experience in interdisciplinary problems and solutions, experience in working with real-life problems, and development of analytical skills [2], [3].

Many kinds of projects from very narrow disciplinary (control of a pump, ventilation shaft, lifts etc.) to multidisciplinary projects (control of power mills, heating systems, helicopters, robots etc.) are the target for the Master's thesis. This is mostly in collaboration with the industry. Projects are normally planned for groups of one to max three students.

At AAU a Master study can be organized as so called long term projects with a normal 9th semester consisting of courses and project work in a subset of a bigger interdisciplinary project. The work is finalized with assessing courses and the project work. This semester is then the base for the final 10th semester Master study concentrated 100% on project work in the overall project.

At Aalborg University many departments have been working with robotics for many years and in many different applications: production, fixed single chain, rigid or

J.-H. Kim et al. (Eds.): FIRA 2009, CCIS 44, pp. 42–52, 2009.

flexible, mobile, electrically or hydraulic driven e.g. and quite a few were build at the university as well. In the spring of 2006 Professor Jakob Stoustrup received a grant from the Dannin foundation and the Faculties of Engineering, Science and Medicine at AAU for constructing and implementing a "limping" humanoid robot. The biped robot should help to close the gap between the fields of health technology and robotics. To bring the existing expertise at the university into play a project proposal for Master's students was devised wherein the primary task was to design and construct a humanoid robot. Researchers from Department of Health Science and Technology, Department of Electronic Systems, Department of Mechanical Engineering, Institute of Energy Technology and Institute for Production and Department for Media Technology and Engineering Science were invited to participate in the experiment with the time frame of 2006-2010.

In the spring of 2006 an AAU-BOT1 group of interested Professors and Ph.D.'s meet for the first time and later, in the autumn, the first group with three students from Department of Mechanical Engineering (ME) was formed. In the autumn of 2007 the next group, also with three students, from Department of Electronic Systems (ES) took over and they were in the autumn of 2008 followed by two students now working on the AAU-BOT1, finishing their Master's thesis in June 2009.

As the funding was far from enough to buy a fully operational human sized robot, it was decided to let students construct, build, instrument and control such a robot. Very important cost savings are achieved as manufacturing the parts in the workshop is free because workshop hours for student projects are funded by the university.

This paper documents the results obtained until now and is organized to show the progress and the professional skills gained by working with this very interdisciplinary project. First, the management is clarified. Second, the work and results fulfilled by the three students from ME are presented. As this work is the most important for the overall success this part also is covered more comprehensive. Third, the work made by the three students from ES is presented. The work that has been done from the current two students from ES is then presented and is followed by assessing the educational findings using the POPBL method for conduction the interdisciplinary and very complex AAU-BOT1 project. The conclusion sums up the results.

2 Management

Supervising projects of this size need some kind of management. The volunteer Professors and Ph.D. students started prior to the students discussing the primary goals which resulted in a very rough draft of the content of design requirements for the projects in the years to come:

- Minimize weight and energy consumption
- As human-like gait as possible
- Capable of accumulating and releasing as much energy as possible
- As high a degree of modularity as possible
- Capable of imitating dysfunctional gait as well as possible

It was agreed that the first year was the most important and for that reason a minor steering group, meeting every 6 weeks, and a supervisory group, meeting every fortnight were formed. All meetings involved the student group. They presented their plans and results and took part in discussions. The management was especially important the first year as it was planned (hoped) that the first group would be able to make all the mechanical drawings so the workshop could use the summer of 2007 to produce all the parts. A supervisor from ME was appointed as the daily contact person.

3 Mechanical Construction

The project theme for the ME students (September 2006- June 2007) was: *How to construct a human sized biped robot*? The first challenge for the ME group was to set up design requirements. A lot of literature was studied e.g. [4],[5] which resulted in many discussions. The group was certainly inspired by Johnnie [6], [7] Wabian [8] and Asimo [9]. Physiological information was, among others, found in [10]. Based on that the students formed the following list of requirements approved by the steering group:

1. Capable of performing dynamic human-like walking. Heel-impact and toe-off
2. Capable of standing up from sitting position
3. Capable of taking a vertical step
4. 17 actuated joints and 2 unactuated joints
5. Anthropomorphic
6. Autonomous
7. Component costs to be kept below 60.000 Euro
8. Walking velocity of 1 m/s
9. Capable of measuring foot-ground reactions and orientation of the body
10. Lifespan: 1000 hours of dynamic walking
11. Modular

It is important to stress that this list was also mandatory for the groups that followed. From the list point 2 and 3 were wishes, the others were required. In the following it is shown how the students went from gait experiments to a kinematic and dynamic representation. Hereafter the construction of a Force Torque Sensor (FTS) is presented and the construction status is reported.

3.1 Kinematic and Inverse Dynamic Analyses

Initially the ME group was helped by the Center for Sensory-Motor Interaction (SMI), Department of Health Science and Technology, to record trajectories using their Qualisys Track Manager (QTM) system. One of the students was used as test person for recording motion data and was equipped with reflective markers as seen on Figure 1. His weight was approximately 70 kg. All the movements mentioned in the requirement list were digitally recorded. Two force plates in the lab floor measured the ground reaction during the gait experiments. 8 infrared cameras followed the test person within a cube; 3 meter long, 2 meter high and 1 meter wide. The data were used to get information about trajectories of different body parts and to estimate the overall kinematic and dynamic robot constrains. This information was used to get kinematic and dynamic insight in human dynamic walk and formed the basis for setting up the kinematic and dynamic equations as well. As these tasks were very demanding, complex and of high order some supplementary instructions from the supervisor were given.

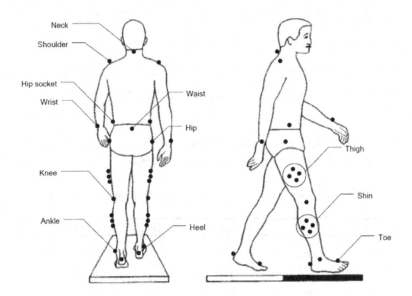

Fig. 1. Test set-up showing positions of reflective marker's [15]

Table 1. Estimates of the inertial properties $I_{\xi\xi}$ around the Center of Mass CoM of the limbs [12]

Limp	Mass[kg]	$I_{\xi\xi}$	$I_{\eta\eta}$	$I_{\zeta\zeta}$
Torso, head	35.5	1.962	1.771	0.052
Pelvis	7.9	0.091	0.032	0.096
Arms	5.8	0.151	0.171	0.034
Thighs	6.8	0.087	0.086	0.021
Shins	3.6	0.053	0.052	0.005
Feet	1.1	0.001	0.007	0.006

The dynamic equations are dependent on all the masses and all inertial properties of the test person. These figures are hard to estimate so instead figures given in [12](Table 1) were used. The purpose was to get good estimates of the torques in all the joints. The recorded forces from the force plates in the floor and the calculated joint trajectories were used to estimate joint torques by an inverse dynamic analysis

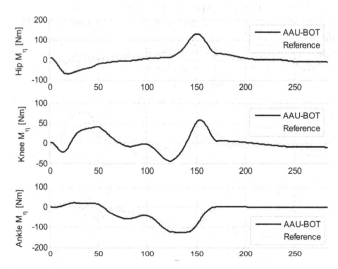

Fig. 2. Shows results from inverse calculated load at the hip, knee and the ankle. The max values multiplied by 1.8 is seen in Table 2. 'References' is from [11].

Table 2. Kinematic constraints and torques multiplied by a control factor 1.8. [13]

Joint	Axis	Range[deg]	Span[deg]	Torque[Nm]
Ankle	roll	-10	15	70
Ankle	pitch	-19	19	224
Knee	pitch	1	95	119
Hip	roll	-17	11	244
Hip	pitch	-81	11	207
Hip	yaw	-19	27	48
Waist	roll	-13	11	63
Waist	pitch	-56	1	108
Waist	yaw	-19	28	27
Shoulder	yaw	-33	6	9

and recursive methods. Figure 2 shows torque results for the hip, knee and ankle compared to reported results in [11]. In Table 2 all max torques, multiplied by 1.8 for control purposes, are shown. Kinematic constraints are also shown.

3.2 Force Torque Sensor (FTS)

A very important feedback sensor for control purposes is a sensor that can measure forces and torques in the feet as stated in the requirement list (9). Mechanically this is a rather delicate problem [15]. It most be light, small and report forces in 3 directions and torque about the same directions [13].

The inverse dynamic analysis showed that the FTS should be able to measure the max. loads seen in Table 3.

Table 3. FTS design parameters.[13]

Force/Torque	Max. Load
F_x	± 1000N
F_y	± 1000N
F_z	± 1000N
M_x	± 200Nm
M_x	± 230Nm
M_z	± 30Nm

The final result is seen in Figure 3.

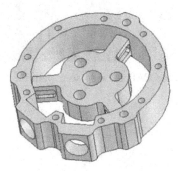

Fig. 3. The FTS weighs 118 g, it is 80 mm in diameter and 15 mm high. In every slide of the beams forming the Y, 12 strain gauges are placed to measure strain and shear by 3 full bridges. The output is 3 voltages for strain and 3 voltages for shear.[13]

3.3 Constructions

The design of AAU-BOT1 was completed at the end of the ME groups study and most of the parts had been milled in the workshop. One leg and the torso was assembled and presented at their Master's thesis assessment in July 2007.

4 Electrical Instrumentation and Modeling

The project theme for the ES students (September 2007- June 2008) was: *How to design the computer system to fulfill real time requirements in AAUBOT1?* But first the ES group had to assemble the rest of the robot. Then they had to take part in the instrumentation and wiring of the robot for control and security purposes. Hereafter the group had to concentrate on three main areas; data infrastructure, program structure and modeling and verification.

4.1 Data Infrastructure

It was important to decide if the data infrastructure should be based on either I/O boards or data bus. It turned out that it was possible to get power supplies, EPOS

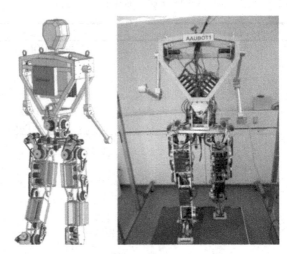

Fig. 4. The left picture shows the CAD drawing seen from the back and to the right AAU-BOT1 is shown assembled

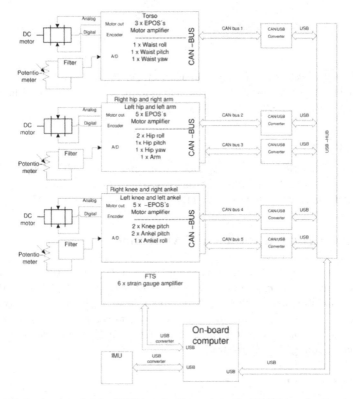

Fig. 5. Data infrastructure for the CAN bus for the motors, the RS485 bus for the strain gauges, and the RS232 connection for the IMU [16]

70/10 from Maxon Motor, for the motors with integrated CAN-bus. It was also possible to get strain gauge amplifiers with RS485 bus connections from Lorenz Messtechnik GmbH to be placed near the sensitive FTS sensors. Therefore it was decided to choose a bus based structure. At the same time it was decided which on-board computer should be used. To prevent injury of the students or robot a safety system was realized. Figure 5 shows the implemented busses and wiring to DC Motors, Encoders, FTS and Inertial Measurement Unit, IMU.

4.2 Program Structure

The program structure has to deal with real time data recording and processing. It is a complex system with 30 sensors to be read and 23 actuators to be actuated. The system dynamics is fast and therefore response time is important (4ms). All drivers were carefully designed and S-functions were written that could interface the drivers from Simulink. On Figure 6 the overall program structure and a 3D simulator Webots can be seen.

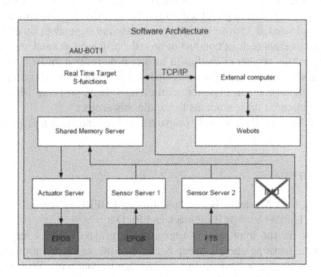

Fig. 6. The overall program structure and connection to the main computer [14]

4.3 Modeling and Verification

At this stage a dynamic model for control and simulation purposes was derived. The model was implemented but not verified. The FTS were calibrated but 2 of the strain gauge amplifiers in one of the FTS malfunctioned so the feet were not assembled and therefore not placed on the robot.

5   Verification, Planning and Control

The project theme for the ES students (September 2008- June 2009) was: *How to control a human sized biped robot in real time*. It turned out that the implemented

USB solution in Figure 5 could not satisfy the real-time requirements so the ES group found a solution for this. Their goals after testing the mathematical model were, to calibrate the FTS's, to model the double support phase (DSP) where the hybrid heel impact followed by weight transfer from the rear foot to the front foot is a special challenge and last to plan the trajectories to be inputs to the controlled AAU-BOT1.

The FTS's have been verified and the results are very promising in the test bench with only 2% deviation.

The next month will show whether they will succeed to get the robot to walk or not. Advanced modeling and advanced controller design exist in the simulation and will benefit the next possible ES group.

6 Educational Findings

Three different groups have so far joined the project. The groups were highly favored in the fact that they at this stage of their study have had 8 prior projects dealing with open ended problems with a high degree of complexity. They are confident with considerable sized projects, they are able to plan their work, they subdivide problems to more clear tasks, self-study of needed knowledge is normal, they are dependent on each other and they adhere to an internal code of conduct of how the problems at hand are solved.

The learning in this project as well as in former projects benefit from courses given and from all the self-study needed to solve open ended problems. The deep theoretical knowledge and great practical experiences they have from working with many types of projects have matured these students to handle this project.

At this stage they really are well prepared to join the industry or the research community.

7 Conclusion

This type of long term Master's thesis project using POPBL is very interesting for all participants', students as well as Professors and Ph.D's.

One drawback for the long term Master's thesis project is that every time a new group starts up the learning curve is steep because of the vast amount of literature and documentation the students have to familiarize themselves with. With this size and complexity of a project it takes almost the first half year to be confident with notation and methods. This is acceptable for the students but not good for the progress in research.

Compared to the requirement list, except point 2 and 3, the following can be concluded. Point 1 is taken care of by a heel shock absorber and the toe-off is implemented by introducing an unactuated toe joint. Point 4; the robot is implemented as shown in Figure 4. Point 5; the mass properties of the robot is quite similar to that of a human except the feet where the motors, gears and FTS's make the feet heavier than human feet. Point 6; the robot is in no way autonomous yet. Point 7; AAU-BOT1 plus laboratory facilities are kept within budget. Point 8; time will show. Point 9; the FTS is implemented and calibrated but unfortunately one of the FTS is now unstable. Point 10; time will show. Point 11; AAU-BOT1 is modular in the since it is possible to replace a foot or shin if needed.

Acknowledgement

Without the brave students this project would not exist. The three former students Mikkel Melters Petersen, Allan Agerbo Nielsen and Lars Fuglsang took the first important step in 2006-2007. They made all the blue prints and started up the manufacturing and assembling of AAU-BOT1. The next three former students Mathias Garbus, Jan Vestergaard Knudsen and Per Kingo Jensen took the next big step in assembling and instrumenting the AAU-BOT1 in 2007-2008 and now Michael Odgaard Kock Niss and Brian Jensen are struggling with planning and control based on former students work.

Also thanks to the people from the workshops at Department of Mechanical engineering and Department of Electronic Systems.

References

1. Kjærsdam, F., Enemark, S.: The Aalborg Experiment – Project innovation in university education. Aalborg University Press (1994),
   `http://www.teknat.auc.dk/teknat_home/experiment/`
2. Fink, F.K.: Integration of Engineering Practice into Curriculum. In: 29th ASEE/IEEE Frontiers in Education Conference, San Juan, Puerto Rico (1999)
3. Kørnøv, L., Johannsen, H.H.W., Moesby, E.: Experiences with Integrating Individuality in Project-orientated and Problem-based Learning POPBL. International Journal of Engineering Education 23(5), 947–953 (2007)
4. Park, I.-W., Kim, J.-Y., Park, S.-W., Oh, J.-H.: Development of Humanoid Robot Platform KHR-2 (KAIST Humanoid Robot-2), Department of Mechanical Engineering, Korea Advanced Institute of Science and Technology (2003)
5. Shih, C.-L., Gruver, W.A., Lee, T.-T.: Inverse Kinematics and Dynamics for Control of a Biped Walking Machine. Center for Robotics and Manufacturing Systems, University of Kentycky, Lexington
6. Löffler, K., Fienger, M., Pfeiffer, F.: Sensor and control design of a dynamically stable biped robot. In: IEEE International Conference on Robotics and Automation. Preceedings. ICRA 2003, September 14-19, vol. 1, pp. 484–490 (2003)
7. Löffler, K., Gienger, M., Pfeiffer, F., Ulbrich, H.: Sensors and control concept of a biped robot. IEEE Transactions on Industrial Electronics 51(5), 972–980 (2004)
8. Ogura, Y., Lim, H.-o., Takanishi, A.: Development of a New Humanoid Robot WABIAN-2, Graduate School of Science and Engineering, Waseda University (2006)
9. Asimo in the world, `http://world.honda.com/ASIMO/`
10. Popovic, D.B., Sinkjær, T.: Control of Movement for the Physically Disabled. Springer, Heidelberg (2000)
11. Westervelt, E.R., Grizzle, J.W., Chevallereau, C., Choi, J.H., Morris, B.: Feedback Control of Dynamic Bipedal Robot Locomotion. Taylor & Francis, Abington (2007)
12. Vaganay, J., Aldon, M.J., Fournier, A.: Mobile robot attitude estimation by fusion of inertial data. In: Proceedings of the IEEE international conference on robotics and automation, vol. 1 (1993)

13. Melters Pedersen, M., Agerbo Nielsen, A., Fuglsang Christiansen, L.: Design of Biped Robot AAU-BOT1. Master's thesis, Aalborg University (2007)
14. Kingo, J.P., Garbus, M., Vestergaard, K.J.: Instrumentation, Modelling and Control of AAU-BOT1 Master's thesis, Aalborg University (2008)
15. Nishiwas, K., Murakami, Y., Kagami, S., Kuniyoshi, Y.: A Six-axis Force Sensor with Parallel Support Mechanism to Measure the Ground Reaction Force of Humanoid Robot, Science and Technology, University of Tokyo (2002)

Safety Aspects in a Human-Robot Interaction Scenario: A Human Worker Is Co-operating with an Industrial Robot

Michael Zaeh and Wolfgang Roesel

Technische Universitaet Muenchen, Faculty of Mechanical Engineering,
Institute for Machine Tools and Industrial Management,
Boltzmannstraße 15, 85748 Garching, Germany
michael.zaeh@iwb.tum.de, wolfgang.roesel@iwb.tum.de

Abstract. This paper presents a concept of a smart working environment designed to allow true joint actions of humans and industrial robots. The proposed system perceives its environment with multiple sensor modalities and acts in it with an industrial robot manipulator to assemble workpieces in combination with a human worker. In combination with the reactive behavior of the robot, safe collaboration between the human and the robot is possible. Generally, the application scenario is situated in a factory, where a human worker is supported by a robot to accomplish a given hybrid assembly scenario, that covers manual and automated assembly steps. For an effective human-robot collaboration, new safety methods have to be developed and proven. Human workers as well as objects in the environment have to be detected and a collision avoidance algorithm has to ensure the safety for persons and equipment.

Keywords: Human-Robot Co-operation, Industrial Robot, Safety System.

1 Introduction

The state-of-the-art in human-robot collaboration is mainly based on a master-slave level where the human worker teleoperates the robot or programs it off-line allowing only static tasks to be executed. To ensure safety, the workspaces of humans and robots are strictly separated either in time or in space [1]. In the automotive industry for instance, human workers are completely excluded from the production lines where robots execute assembly steps. On the other hand, robots are not integrated in assembly line manufacturing along with human workers.

In general, the co-operation between humans and industrial robots is emerging more and more in ongoing research area, concerning human machine interaction as well as industrial safety requirements, issues and union decrees [2]. In some industrial applications mobile platforms with a mounted industrial robot arm are used for human-robot co-operation (e.g. welding processes) [3]. For the handling of heavy workpieces in an automated production environment concepts of human-robot co-operation are developed, in which an industrial robot hands over heavy loads (e.g. rear axle gear unit) to the worker [4].

J.-H. Kim et al. (Eds.): FIRA 2009, CCIS 44, pp. 53–62, 2009.
© Springer-Verlag Berlin Heidelberg 2009

The work presented here aims to integrate industrial robots in human-dominated working areas. For this purpose, multiple input modalities to allow a true peer-to-peer level of collaboration are used. Thus, a smart working environment for joint action between a human and an industrial robot is to be designed as an experimental setup consisting of various sensors monitoring the environment and the human worker, an industrial robot, an assembly line supplying the manufacturing process with material and tools, and a working table. For an effective human-robot collaboration new safety methods have to be developed and proven. Human workers as well as objects in the environment have to be detected and a collision avoidance algorithm has to ensure the safety for persons and equipment.

This project integrates human-robot collaboration in the so-called "Cognitive Factory", where new kinds of cognitive systems replace the currently used static and inflexible technical automation systems. A first approach for our realization of this long-term goal is presented in [5].

This aspect uses the potential for humans and robots to work together as a team, where each member has the possibility to actively assume control and contribute towards solving a given task based on their capabilities. Such a mixed-initiative system supports a spectrum of control levels, allowing the human and robot to support each other in different ways, as needs and capabilities change throughout a task [6]. With the subsequent flexibility and adaptability of a human-robot collaboration team, production scenarios in permanently changing environments as well as the manufacturing of highly customized products become possible.

2   Initial Situation, Objectives and Approach

Human-robot co-operation will contribute to the adaptability and flexibility of production systems by an interactive and dialogue-based instruction system for industrial robots. Within the cluster of excellence CoTeSys (Cognition for Technical Systems) the research project JAHIR (Joint Action for Humans and Industrial Robots) examines the possibilities of a human-robot co-operation and demonstrates how a human-robot co-operation may be introduced into an industrial setting. In prototyping and in small series productions the degree of automation can be increased with an integration of industrial robots. For this purpose, an appropriate concept needs to be developed. By the integration of sensors and a comprehensive data base, a robot can assist a worker e.g. in assembly processes. The strengths of humans (adaptability, ability of decision, creativity, skill ...) can be combined with the strengths of a robot (force, accuracy, perseverance, speed ...) by a human-robot co-operation.

State-of-the-art in human-robot interaction is mainly based on a master-slave level where the worker teleoperates the robot or programs it off-line. Due to safety aspects, the workspaces of humans and robots are separated. An optimised integration of the robot in the assembly process has to be realised for an effective human-robot co-operation. Future assembly systems need to be characterised by a direct co-operation between humans and robots without separating safety equipment or safety areas. The fundamentals for this co-operation from a technical and a legal view are being created and are discussed in this paper.

In order to enable the full potential of human-robot co-operation, an interactive and dialogue-based instruction system is required for an intuitive programming system.

Co-operation between humans and robots becomes therefore possible with a high level of interactivity. In place of rigid program sequences, interactive dialogue techniques (e.g. speech control, gesture recognition) should be used. The worker should be able to teach the robot and interact with the complex system in a fast and efficient way. Different methods of object tracking, speech and gesture recognition should be integrated in the system. An important goal is the safe integration of humans into the production process. For this purpose, the current position of humans has to be localized and processed. With human-robot co-operation an increase of the ergonomics for humans can take place. By a shortening of tooling time, shorter operating time is possible and thus a higher output of workpieces is realizable. By the integration of humans in a so far automated assembly process a highly flexible semi-automation is possible.

For the detection of humans, four procedures can be used: with a range laser scanner the rough position of humans can be captured (see Figure 1 left and middle). This can be verified using a matrix of safety shutdown mats (see Figure 1 right). With marker-based object-tracking and vision-based tracking algorithms head and hands of a human can be localized very precisely.

Fig. 1. Left: Laser scanner - experimental setup
middle: Laser scanner - data evaluation
right: safety shutdown mat - experimental setup

Within the project interaction between robots and humans in a common work space is investigated on a research platform. Here, the individual strengths of the interaction partners are linked: The robot is used for the handling of heavy construction units and for the execution of repetitive tasks like screwing. Humans can implement sensitive and fine-motor work like fixing a cable in co-operation with the robot. It has to be guaranteed by the integration of different safety components in the system and an intelligent path planning method that the robot is closely integrated in the assembly process and can adapt to the actions of humans optimally. So a human can interact and co-operate with his superior cognitive developments as well as his high fine-motor abilities if necessary with robots in a very narrow area with the help of multimodal interfaces.

3   Motivation

To enable an effective collaboration, recent research in the field of psychology has focused on cognitive processes of joint-action among humans [7]. Psychological studies [8] show that in collaborating human teams, an effective coordination requires participants that plan and execute their actions in relation to what they anticipate from the other team member, and not just react on the other's current activities. Hence, for an efficient human-robot team, this knowledge needs to be transferred to a given setup. With the demonstration setup a small slice of the concept, constituting a foundation for a multimodal communication between a human and a robot, which is flexible, robust and most appropriate for hybrid assembly, is presented.

Furthermore, by utilizing more than two input modalities redundancy in the communication channel to compensate failures in general or the total loss of an interaction modality can be introduced. For the maximization of this redundancy, the interaction framework is based on three different channels (eyes, voice and hands). This approach does not only ameliorate the redundancy but also the diversity of possible communication ways. Therefore, consider the case, two input modalities basing on the same human interaction method (e.g. the human hand) can be blocked by a complex task requiring this interaction method (e.g. the workpiece must be held with both hands). Having three input modalities being based on the three channels, the interaction between the human and the robot can be significantly shaped into a more natural, robust and intuitive way.

4 Setup Description

In the following two sections the hardware setup as well as the exemplary assembly product will be described in detail. However, the main focus in these two sections will not be laid on the assembly process, which will be given in section 5.

4.1 Hardware Setup

In Figure 2 the demonstrator setup is depicted. As it can be seen, the worker is wearing a head-mounted microphone as well as eye-tracking glasses.

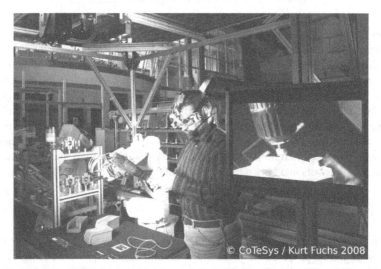

Fig. 2. Hybrid assembly station: tool station, robot arm (with electric drill) and assembly-table

The industrial robot manipulator arm used in the scenario is a Mitsubishi robot RV-6SL. It has six degrees of freedom and can lift objects with a maximum weight of six kilograms. Its workspace lies within a radius of 0.902 m around its body. Its tool point is equipped with a force-torque-sensor and a tool-change unit. Furthermore, the robot is able to change the currently installed gripper by itself at a station. The station features four distinct kinds of manipulators performing specific operations. The tools stored in the station are:

- two finger parallel gripper,
- electronic drill,
- camera unit for automatic observations,
- gluer.

This gives the robot the capabilities of being able to solve entirely different tasks, like screwing and lifting. The workbench has an overall workspace of approximately 0.70 square meters. A global top-down view camera is mounted above the workbench. This device has the overview over the entire work-area and makes it possible to watch the actions on the workbench and locate objects on the surface. Additionally, a Photonic Mixer Device (PMD) range sensor is mounted above the workbench.

For bringing information into the worker's field of view, a table projector is also installed above the workbench. This device projects information directly onto the surface of the workbench. So virtual information are directly linked to real environment (Augmented Reality). With this modality, it is possible to show contact analog assembly instructions and system feedbacks. Moreover, flexible interaction fields can be displayed to communicate with the system (see following section).

4.2 Assembly Product Description

The product constructed in the hybrid assembly scenario with manual and automated assembly tasks is a high frequency transmitter, shown in Figure 3.

Fig. 3. Assembly product: a base plate, one electronic part, a wiring cable, a plastic cover and four screws

The following parts are required for the construction of this high frequency transmitter:

base plate: the base plate delivers the foundation for the hybrid assembly on which all remaining construction parts will be mounted.

electronic part: the electronic part is an inlay, which has to be mounted into the base plate.

wiring cable: the wiring cable has to be plugged into the electronic part. It needs to be put into the designated cable line on the base plate.

plastic cover and four screws: with the four screws, the plastic cover is mounted on top of base plate to protect the inlay against influences from the environment, like dust and dirt.

The system provides a guidance trough the entire assembly process to the user. At each step the system supplies the user with the currently required assembly parts. Furthermore, the assembly instruction is presented by the system with an Augmented-Reality-based table projection unit on the workbench.

5 Assembly Process

The assembly process of the product described in Section 4.2 is depicted in Figure 5. The first interaction - initiated via soft-button or Speech Command - between the human and the robot is the supply of the worker with the base plate. Having the required workpieces available at hand, the worker starts to teach in a glue-line (see Figure 4.a). The track of the glue-line on the base plate is taught with Programming by Demonstration (PbD). This is done by tracking a colored pointer.

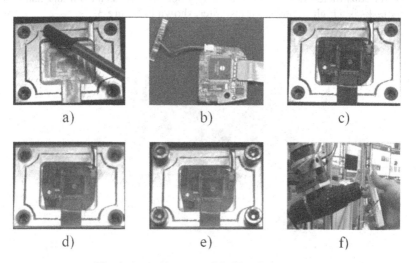

Fig. 4. Assembly steps of the Use-Case product

Therefore, a color-based image segmentation is performed on the output frame of a top-down-view camera, mounted above the workbench. The result is then used to locate the green tip of the pen, now visible as a binary-coded plane in the filtermask. The motion of this object in the image plane is analyzed and transformed into the world-coordinate-system. While the line is perceived, its trajectory is projected back onto the workpiece on-line as a direct feedback for the worker. On completion of this step, the robot changes its tool device from the gripper to the gluer according to the next step in the work plan. After the Programming by Demonstration (PbD), the robot protracts the glue on the workpiece. Via transforming these gained world-coordinates into the robot-coordinate-system, the industrial robot is now able to exactly repeat this motion trajectory and perform the gluing-operation autonomously.

As per assembly instructions the robot reaches out for the electronic parts. Therefore, a fully automated tool change operation is performed by the robot, thus, exchanging the currently mounted gluer towards the gripper. The following assembly of the electronic parts (see Figure 4.b) requires fine motor skills. Therefore, the next step is solely done by the human. In spite of the fact that the robot does not give any active assistance in this assembly step, the system supports the worker via presenting the manufacturing instructions for the insertion of the electronic parts into the base plate (see Figure 4.c and 4.d). After the worker has acknowledged the completion of the current step via soft-button or a speech based command, the robot fetches the four screws for the final assembly step. While the worker is pre-fitting the screws in the designated mount ports (see Figure 4.e), the robot retrieves the automatic drill device from the tool changer station. The velocity of the drill is adjusted to the contact pressure of the workpiece against the drill (see Figure 4.f). The more pressure is applied, the faster the drill goes. As soon as the human recognizes that the screw is fixed - the rattling noise of the slipping clutch - he will loosen the conducted pressure. This modality allows for an intuitive screwing behavior of the system.

The final step requires the worker to acknowledge the completion of the screwing operation via speech command or soft-buttons. The system will go into the init state and the whole assembly process can be started again for the next production turn.

6 Safety Aspects

For the interaction between a human and an industrial robot the position and orientation of both must be known to avoid any collisions between human worker and robot. For the detection of the human a laser scanner can be used. Therefore a scanner is mounted below the working table (see Figure 5).

Figure 6 shows the resulting measurement data. In different warning areas (dotted white line) and safety areas (inner orange continuous line) a distance between the table and the human worker can be identified. One person (two legs or one leg in a side view) or more persons can be detected. The velocity of the robot can be controlled with the distance of human worker to the working table: If humans are far, the robot can move with a high velocity. If the laser scanner registers one or more persons close to the table or inside the safety area, the robot slows down or releases an emergency stop. With the industrial standard EN ISO 10218-2: 2008 the maximum speed in the collaborating mode is limited to 250 mm/s [9].

Fig. 5. Laser scanner - experimental setup

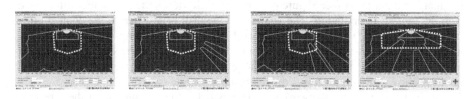

Fig. 6. left: No person in- or outside warning area (dotted) or safety area (continuous)
middle left: One person, but outside warning area (dotted) or safety area (continuous)
middle right: One person inside warning area (dotted) but outside safety area (continuous)
right: Three persons inside warning area (dotted) but outside safety area (continuous)

Fig. 7. Safety shutdown mat - experimental setup

Due to safety reasons the position data of the human has to be verified by a second redundant system. In case of data loss of one system the second system can release an emergency stop of the robot. For the data evaluation of the laser scanner a matrix of safety shutdown mats can verify the standing position of the human worker. Therefore, a matrix of safety shutdown mats on the floor can detect a contact with the feet (see experimental setup in Figure 7).

The data of the laser scanner has to be evaluated on a redundant system. Therefore, the data of the safety shutdown mats can be matched with the scanner data.

7 Conclusion and Outlook

A first technical implementation of a human-robot interaction scenario was introduced and was exhibited at the trade fair "AUTOMATICA 2008" in Munich, Germany. The presented approach offers high potential for a new multimodal human-robot interaction system and is currently evaluated by experiments within the project JAHIR [10]. At this stage, a hybrid assembly process - human being working together with an industrial robot - was shown with its functionality. An industrial robot can be controlled and programmed by demonstration with a pen, with gaze (eye control), speech control and hand detection. This intuitive programming system has now to be designed with aspects of performance, stability and robustness. There is now more scope for further safety aspects in human-robot collaboration.

Acknowledgments. This research and development project is funded by the German Research Foundation (DFG) within the excellence initiative research cluster Cognition for Technical Systems - CoTeSys, see www.cotesys.org for further details.

References

1. EN ISO 10218-1 (2006): Robots for industrial environments - Safety requirements - Part 1: Robot; Beuth Verlag GmbH, Berlin, Germany (2006)
2. Bischoff, R., Schmirgel, V., Suppa, M.: The SME Worker's Third Hand. SMErobot project, Germany (2008), http://www.smerobot.org/index.php
3. Haegele, M., Helms, E.: rob@work: der Assistent der Zukunft - Mobile Assistenzroboter in der industriellen Fertigung. Fraunhofer IPA, Stuttgart, Germany (2004), http://www.ipa.fraunhofer.de/Arbeitsgebiete/robotersysteme
4. Parlitz, C., Meyer, C.: PowerMate – Schrankenlose Mensch-Roboter-Kooperation. Fraunhofer IPA, Stuttgart, Germany (2005), http://www.ipa.fraunhofer.de/Arbeitsgebiete/robotersysteme
5. Zaeh, M.F., Lau, C., Wiesbeck, M., Ostgathe, M., Vogl, W.: Towards the Cognitive Factory. In: Proceedings of the 2nd International Conference on Changeable, Agile, Reconfigurable and Virtual Production (CARV), Toronto, Canada (July 2007)
6. Marble, J.L., Bruemmer, D.J., Few, D.A., Dudenhoeffer, D.D.: Evaluation of supervisory vs. peer-peer interaction with humanrobot teams. In: HICSS 2004: Proceedings of the 37th Annual Hawaii International Conference on System Sciences (HICSS 2004) - Track 5, Washington, DC, USA. IEEE Computer Society, Los Alamitos (2004)

7. Sebanz, N., Bekkering, H., Knoblich, G.: Joint action: bodies and minds moving together. Trends in Cognitive Sciences 10(2), 70–76 (2006)
8. Knoblich, G., Jordan, J.S.: Action coordination in groups and individuals: learning anticipatory control 29(5), 1006–1016 (2003)
9. EN ISO 10218-2 (2008): Robots for industrial environments - Safety requirements - Part 2: Robot system and integration; Beuth Verlag GmbH, Berlin, Germany (2008)
10. Lenz, C., Nair, S., Rickert, M., Knoll, A., Roesel, W., Bannat, A., Gast, J., Wallhoff, F.: Joint Actions for Humans and Industrial Robots: A Hybrid Assembly Concept. In: Proceedings of the 17th IEEE International Symposium on Robot and Human Interactive Communication (IEEE RO-MAN 2008), Munich, Germany (August 2008)

Integration of a RFID System in a Social Robot

A. Corrales and M.A. Salichs

Carlos III University of Madrid, Roboticslab, Madrid, Spain
{acorrale,salichs}@ing.uc3m.es

Abstract. This article presents the integration of a system of detection and identification of RFID tags in a social robot, with the goal of improving its sensorial system and to accomplish several specific tasks, such as: recognition of objects or navigation. For this purpose, basic skills of reading and writing have been designed, following the pattern of the basic element *skill* of the robot software architecture. The system has been implemented physically adding two RFID interrogators with built-in antenna to the sensorial robot system. The application has been implemented and tested as a skill in the detection of products such as medicines and diverse objects in order to assist visually impaired people, users of the third age and people who cannot read.

Keywords: Social robots, Radio Frequency Identification, medicine recognition, visually impaired people.

1 Introduction

Due to the increasing incorporation of robotics systems in our environment, there are a wide diversity of technical solutions nowadays to improve the execution of various tasks. The idea to get a robot to interact with human beings at environments like homes, schools and hospitals, is not a simple problem. The robots should be able to accomplish tasks and cooperate with unskilled users such as children, people of the third age or visually impaired people. This is why it is essential that the robot has sensors and actuators to get information from the environment and interact with it [1], in addition to the control architecture for the correct functioning of the robot tasks.

Radio Frequency IDentification (RFID) Systems is a new tool of robotics, whose purpose is to enrich the information that the robot acquires from the environment, considerably improving the identification process of objects or people, being capable of supporting other sensorial robot systems , for example the identification through computer vision. The robot can be equipped with a reader and antenna. An RFID reader, also called an interrogator, is a device that can read from and write data on compatible RFID tags [2]. Nowadays the RFID systems are been implemented in different kinds of industrial, social and assistant robots [3,4,5,6]. In [7,8] a method for person tracking is proposed, they use a combination of RFID passive tags and floor sensors, the goal is to have the robot interact naturally with customers in a shopping mall.

J.-H. Kim et al. (Eds.): FIRA 2009, CCIS 44, pp. 63–73, 2009.

Our approach is thus to exploit the attributes and advantages of RFID technology in mobile and social robotics, since we can carry out applications that have a positive influence in the autonomy of the robot for motion tasks as well as for the human-robot interaction. With this aim, we presented the following specific objectives:

- Study and analysis of the necessary hardware requirements to implement the RFID system.
- Design of the software architecture of the RFID system. Study of the software requirements in order to correctly incorporate the system to the existing robot software architecture.
- Design and development of basic reading and written functions.
- Physical implementation of the robot system and the carrying out of operational tests.
- Design and development of new applications or skills for the robot using the RFID system. In particular, we focus on the development of a skill for the identification of medicines.

2 Maggie: The Social Robot

Maggie is a human-friendly robotic platform developed at RoboticsLab for research on human-robot social interaction (Fig. 1) [9]. Maggie has an artistic design of a girl-like doll. The robot uses expressions with body/arm/eyelid movements as well as voice to interact with humans.

Fig. 1. Maggie, The Social Robot

2.1 Software Architecture

The software architecture of the Maggie robot is based on the Automatic-Deliberative architecture AD [11]. This architecture consists of two levels: Deliberative and Automatic (Fig. 2). In the Deliberative level are located the modules that require reasoning. These modules do not produce immediate answers and

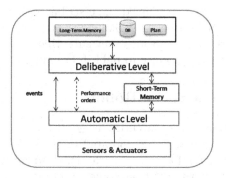

Fig. 2. AD Architecture

need some time to process the information with which they work to make decisions. The Automatic level is in charge of the control, at a low level, of the robot devices; of the reactive tasks and of the processing of sensorial information. This level allows the robot to have the necessary reaction capacity to quickly respond to environmental changes [12,13].

Both levels have a common characteristic: they are constituted by **skills**.

Skills: A *skill* is the basic element of AD architecture. The skills are the different capacities to reason, to control and to process data.

In terms of software engineering, the abstract class ***Skill*** describes the global behavior of a skill. It is an object that contains the data and processes of the task or action that the robot must accomplish. It defines a method called "*process*", this method describes the control loop of the skill. The method is redefined in each subclass *skill*. It also includes a method that activates or deactivates the control loop from an external class or application. All the multiprocessing details are hidden from the user [14].

According to [17], an instance of a Skill is characterized by:

- Having three states: ready, running and blocked.
- Having three running modes: cyclical, periodical and by events.
- Each skill is a process, the communication between processes is by events and/or short-term memory.
- A skill can be made by fusion of others skills.
- All the handling of multiprogramming, communication between processes, composition, etc., are transparent to the user, or at least simple to use.

Short-Term Memory: The Short-Term Memory is used to share working information. Sharing data between skills is a basic feature in AD architecture. Perception information and the information used in the automatic and deliberative layers must be available for current or future skills.

Event System: A skill can be subscribed to an event and define a particular behavior at the same instant in which the event occurs. The event system used in our robots is loosely coupled and distributed. It is based in the design pattern Publisher/Subscriber [15]. Skills do not know each other: they are only defined by the events to which they subscribe, and the results they can provide (the event they produce to publish or emit).

3   RFID Skills in a Social Robot

In order to add identification capabilities using RFID, two RFID interrogators with built-in antenna have been incorporated to the robot in such a way that they may read and write information in the RFID tags. RFID Readers have different purposes: a reader in the head to detect objects that the user needs to identify and another in the robot base to accomplish basic tasks of location and navigation in interior environments.

The basic objective is to execute two general skills for identification with RFID for reading and writing of data, independently of its later use (Recognition of objects, medicines, navigation, etc.) [16]. The RFID robot system must access the interrogator's basic functions such as detection, reading and writing of RFID tags, using different components of AD architecture: the Event System, the Short-Term Memory and the abstract class Skill [17].

In order to carry out the design of skills it is important to analyze the system requirements, the hardware and software components and the input and output data of every skill. Skills must be robust, easy to use and generate basic information for any skill which uses RFID identification.

3.1   Hardware Requirements

To develop and implement RFID skills it was necessary to select an RFID reader that could be easily incorporated in the Maggie robot. The choice was made considering the operation frequency, the types of tags, the reader's size and the characteristics of the software and hardware [2].

The reader chosen is an ISC.MR101-USB 13.56 MHz made by FEIG Electronic [18]. It is a short distance reader (18 cm approximately). This allows reading passive tags of different standards: ISO15693, ICode and Tag-it HF; ICode EPC and ICode). The interface is a USB which has the drivers and program libraries for the Linux operating system, this implies simplicity of integration in the robot software architecture.

3.2   Software Requirements

In order to study the software requirements, it is necessary to establish the possibilities of use:

1. The RFID reader is constantly detecting tags that it finds in its nearby environment. This allows that at any time a user can show a product or an object for its recognition and the robot can provide the information saved in the RFID tags.

2. In the more generic case, the principal users of skills will be other skills of the system. The skills must be subscribed to the event NEW_TAG_RFID, this event indicates that data read by the reader is in the Short-Term Memory. Also it's possible that other skills send data to RFID skills in order that they may be written in tags.

When the Read Skill is activated by a sequencer or another skill, it begins to look for tags in the environment. If a tag is detected, the system reads the data stored in the tag and saves it in the Short-Term Memory. Immediately it sends the event NEW_TAG_RFID, in this way interested or subscribed skills may perform correspondent actions. If the skill still is active, it will return to look for tags and it repeats the previous process again. Either it finishes its execution or it is blocked waiting for orders.

Writing a skill is simpler, since this job is directed to the acquisition of information of tags. When the RFID system receives the order to write, the skill activates the function *Looking for tag*. If it finds the tag it writes the data and finalizes the writing function. Either the skill keeps on looking for the tag until it finds it, or it waits while maximum of quest or until it receives the order to interrupt the operation.

4 Software Architecture of RFID Skills

In order to develop the RFID system, it must use the basic functions of the function library of the FEIG Electronic reader. These are developed in C++. This simplifies the development of applications for OBID-RFID readers under Windows and Linux operating systems [19].

With the aim of developing the application as a skill, it is necessary to design a hierarchic structure of the software components (modules). The skills must inherit the abstract class *skill* (3) and be activated through events.

In Fig. 3 shows the designed modules. It is necessary to do a middle class or API (Application Programming Interface), for it allows access to different procedures of scanning, reading and writing of the OBID functions library. This permits a level of abstraction in the programming and an ease at the moment of developing skills and RFID applications.

Fig. 3. Design of the RFID System

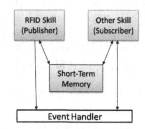

Fig. 4. Publisher/Subscriber Scheme

The general design used for the asynchronous detection of tags is based on the behavioral pattern *publisher/subscriber* (Fig. 4).

RFID Skills are executed with the following conditions:

1. They inherit an abstract class *Skill* of AD Architecture.
2. The skills use the primitive functions of the RFID reader selected (OBID FEDM_ISCReader).
3. They share information with other skills of the system. The information that the reader receives from tags must be stored and shared with one or several skills of the system, for this reason the data is saved in the *Short-Term Memory*, one of principal components of AD architecture [17].
4. Interested skills in the data stored in the Short-Term memory must be informed when this data is available. For this reason it's necessary to use the *Event Manager System*.

4.1 Reading Skill CRFID_ReadSkill

This skill gets the data from RFID tags using functions of the API RFID programming interface . Reading Skill CRFID_ReadSkill has two main procedures or functions:

- process(): Is the main skill function. The process looks for RFID tags in the environment, if it finds them, then the skill reads the data and then saves the information in the Short-Term Memory.
- saveMCDataRFID(): This function stores the tag data in the Short-Term Memory. It also emits the NEW_TAG_RFID event, this indicates that data has been stored successfully and it is available for others skills.

4.2 Writing Skill CRFID_WriteSkill

The RFID Writing Skill sends and writes data in RFID tags, it connects with the RFID reader through the programming interface API RFID. It obtains the information to write from the Short-Term Memory and sends it to the specified tag. In fact, to get the data of Short-Term Memory implies that the system is open to other skills or applications in the robot and these may send written data to RFID tags. The skill is activated when it receives an event indicating that there is new information to write on the tags. Fig. 5 shows the behavioral diagram of the RFID Writing Skill.

The Writing Skill (CRFID_WriteSkill) has two main functions:

- process(): Is the main skill function. The process looks for RFID tags in the environment.
- write_data(): This function is activated when the system sends the NEW_RFID_DATA_TO_WRITE event. This data is read and stored on the Short-Term Memory and sends it to the specific tag.

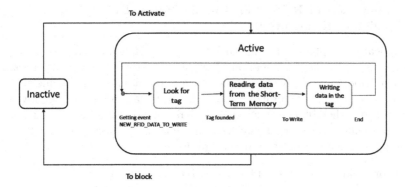

Fig. 5. Behavioral diagram of the RFID Writing Skill

5 Experimental Results

5.1 Tags Detection and Data Reception

Objective: Analysis of the detection time of RFID data. The information is detected by the RFID reader until it is read by the robot Short-Term Memory.

Description: The record of analyzed times corresponds to:

- Detection time of tag.
- Sending time of NEW_TAG_RFID event.
- Read time of RFID data from Short-Term Memory.

The test was made using different RFID tags.

Results: Fig. 6 shows the difference between the times of tag detection and the times of getting data from the Short-Term Memory. The average delay is 0.5113 ms and the maximum waiting time is 0.598 ms. This data shows that the system permits the efficient interpretation in real time of the information in RFID tags.

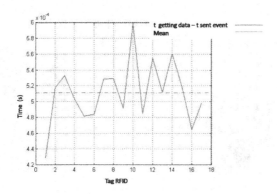

Fig. 6. Analysis of the times between the sending event and the reception of data

5.2 Application of RFID Skills in the Social Robot Maggie: Medicines Recognition

The application explained in this article consists of helping people recognize medicines using RFID identification. This task is directed at visually impaired users. The robot should give the users information about the medicine such as: name, how to use and expiration date. This application can be employed for recognition of other objects such as perishable food. For this experiment we used ISO 15693 passive transponders. The data was stored previously in the transponders. Each tag has the following information:

– UID or Unique ID recorded by the RFID transponder maker.
– Medicine Identifier.
– Expiration date.

Medicine data is stored in a data file. `CSkill_Rfid_Medicine` looks for data in a medical dictionary (It can be a local data base or an online drug index). The relevant information is:

– Medicine name.
– What the medicine is used for.
– Posology.
– How to use.

Later this information is sent to the voice synthesizer so that the human-robot interaction is more natural. This way the user does not need to read the information from the medicine boxes and this avoids confusion between medicines with similar color packaging.

If the detected medicine is not found in the medical dictionary, the robot will inform the user that it cannot identify it (Fig. 7).

The developed skill uses the principal components of AD architecture. Fig. 8 shows the components Diagram of Medicines Recognition Skill.

The experiment of medicine detection was shown to the public in the II International Congress on Digital Homes, Robotics, Telecare for All [20]. where the

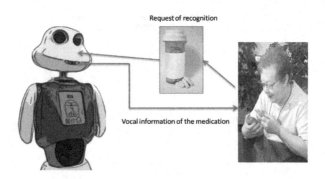

Fig. 7. Use Case `CSkill_Rfid_Medicine`

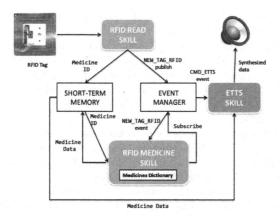

Fig. 8. Components Diagram of Medicines Recognition Skill

Maggie robot interacted with users with visual deficiency. The experiment was satisfactory, this proved that the system can be a good tool for a social robot that can be used as a companion to visually disabled people, older persons or the sick. Several notes were published in the media about this matter [21].

6   Conclusions

The developed RFID system improves the sensorial system of the Maggie robot. This makes possible the use of this technology in several applications, in this case to help users in the identification of objects. For the correct incorporation of this function in the robot, the basic RFID skills of reading and writing were designed, following the pattern of the basic element *skill* of the robot architecture software. The information that the RFID interrogator receives from transponders is stored and shared with other skills using the Short-Term Memory and the Event System of AD architecture. We have tested the RFID technology in a Medicine Recognition Skill, the implementation works properly and it is a potential help for visually impaired users.

Acknowledgments. The authors gratefully acknowledge the funds provided by the Spanish Government through the projects called Peer to Peer Robot-Human Interaction (R2H) , of MEC (Ministry of Science and Education), and A New Approach to Social Robotics (AROS), of MICINN (Ministry of Science and Innovation).

References

1. Bekey, G.: Autonomous Robot. From Biological Inspiration to Implementation and Control. The MIT Press, Cambridge (2005)
2. Sandip, L.: RFID sourcebook. IBM Press (2006)

3. Greengard, S.: Driving Change in the Auto Industry. RFID Journal (April 2004), http://www.rfidjournal.com
4. Carroll, D., Gilbreath, G.: Extending Mobile Security Robots to Force Protection Missions. In: Extending Mobile Security Robots to Force Protection Missions. AUVSI Unmanned Systems. Lake Buena Vista FL, July 9-11 (2002), http://citeseer.ist.psu.edu/carroll02extending.html
5. Shoop, B., Jaffee, D.M., et al.: Robotic Guards Protect Munitions. In: ARMY AL&T (October 2006)
6. Kulyukin, V., Gharpure, C., Nicholson, J., Osborne, G.: Robot-assisted wayfinding for the visually impaired in structured indoor environments. Auton. Robots, vol. 21(1), pp. 29–41. Kluwer Academic Publishers, Hingham (2006), http://dx.doi.org/10.1007/s10514-006-7223-8
7. Nohara, K., Tajika, T., Shiomi, M., Kanda, T., Ishiguro, H., Hagita, N.: Integrating passive RFID tag and person tracking for social interaction in daily life. In: The 17th IEEE International Symposium on Robot and Human Interactive Communication, 2008. RO-MAN 2008, pp. 545–552 (2008)
8. Kanda, T., Shiomi, M., Miyashita, Z., Ishiguro, H., Hagita, N.: An affective guide robot in a shopping mall. In: HRI 2009: Proceedings of the 4th ACM/IEEE international conference on Human robot interaction, La Jolla, California, USA, pp. 173–180 (2009)
9. Salichs, M.A., Barber, R., Khamis, A.M., Malfaz, M., Gorostiza, J.F., Pacheco, R., Rivas, R., Corrales, A., Delgado, E.: Maggie: A Robotic Platform for Human-Robot Social Interaction. In: IEEE International Conference on Robotics, Automation and Mechatronics, Bangkok, Thailand (June 2006)
10. van Breemeny, A., Crucqy, K., Kroseyy, B., Nuttinz, M., Portayy, J., Demeesterz, E.: A User-Interface Robot for Ambient Intelligent Environments. In: Proc. 1st International Workshop on Advances in Service Robotics, ASER 2003, Bardolino, Italy (2003)
11. Barber, R.: Desarrollo de una arquitectura para robots móviles autónomos. Aplicación a un sistema de navegación topológica. PhD Thesis. Universidad Carlos III de Madrid. Leganés Spain (2000)
12. Salichs, M.A., Barber, R.: A new human based architecture for intelligent autonomous robots. In: The Fourth IFAC Symposium on Intelligent Autonomous Vehicles, Sapporo, Japan, pp. 85–90 (2001)
13. Salichs, M.A., Barber, R., Boada, M.J.: Visual approach skill for a mobile robot using learning and fusion of simple skills. Robotics and Autonomous Systems 38(3-4), 157–170 (2002)
14. Rivas, R., Barber, R., Corrales, A., Salichs, M.A.: Arquitectura de software de un robot personal. Arquitecturas de Control para Robots, pp. 101–115. Universidad Politécnica de Madrid (Feburary 2007)
15. Gamma, E., Helm, R., Johnson, R., Vlissides, J.: Design Patterns: Elements of Reusable Object-Oriented Software. Addison-Wesley Professional, Reading (2000)
16. Corrales, A.: Sistema de Identificación de Objetos Mediante RFID para un Robot Personal. Master Thesis. Universidad Carlos III de Madrid. Leganés Spain (2008)
17. Rivas, R., Barber, R., Corrales, A., Salichs, M.A.: Robot skill abstraction for ad architecture. In: 6th IFAC Symposium on Intelligent Autonomous Vehicles IAV, IFAC, Toulouse Francia (September 2007)
18. Feig electronic (2008), http://www.feig.de/

19. FEIG. C++ Class Library ID FEDM. Software-Support for OBID ©Reader Families. FEIG ELECTRONIC GmbH, Weilburg-Waldhausen Alemania (2007)
20. Fundación ONCE. Segundo congreso internacional de domótica, robótica y teleasistencia, Madrid Spain (2007),
http://www.drt4all.org/DRT/es/2007/robotmagie.htm
21. Diario El Mundo. Maggie, un robot para ciegos capaz de sentir cosquillas o distinguir medicinas (2007),
http://www.elmundo.es/elmundo/2007/04/21/solidaridad/1177164182.html
22. IX Feria Madrid es Ciencia (2008), http://www.madrimasd.org/madridesciencia

A Practical Study on the Design of a User-Interface Robot Application

Martin Saerbeck[1], Benoît Bleuzé[1], and Albert van Breemen[2]

[1] Philips Research, 5656 AE, High Tech Campus 34, Eindhoven, The Netherlands
martin.saerbeck@philips.com, nlv17411@natlab.research.philips.com
[2] Personal Robotics, 5708 ZW, Statenlaan 52, Helmond, The Netherlands
breemen@personalrobotics.com

Abstract. People are striving for easy, natural interfaces. Robotic user interfaces aim at providing this kind of interface by using human like interaction modalities. However, many applications fail, not because of fundamental problems of addressing social interaction but due to an unbalanced design. In this paper we derive a balancing framework for designing robotic user interfaces that balances four key dimensions: user, application, interface and technology. We investigate applicability of the the framework by means of two experiments. The first experiment demonstrates that violations to the balancing framework can negate the efforts to improve an interface with natural interaction modalities. In the second experiment we present a real world application that adheres to the balancing concepts. Our results show that a balanced design is a key factor for the success or failure of a given robotic interface.

1   Introduction

The observation by Ben Schneiderman [1] that computer users waste an average of 5.1 hours per week trying to use computers has a serious implication on the design of user interfaces. Low prices for hard- and software will make technology available to a larger number of people, while at the same time new technological achievements introduce new functionality that make the devices more and more complex. In order to enable people to use all the functionality there are two main approaches: educate the users in operating the devices or make the devices easier to handle. Shneiderman calls it "bridging the gap between what users know and what they need to know". Most of the time the first approach is taken, resulting in big booklets accompanying the various devices, but people are striving for easier, more natural interfaces. Robotic User Interfaces (RUI) are aiming in providing exactly that kind of interface [2].

In this study we analyze the design process of such RUIs and argue that a balanced design is a crucial factor for the success of a particular application. We translate a design framework originating from the theory of interactive systems to the design of RUIs. The goal of this study is to investigate the applicability of

J.-H. Kim et al. (Eds.): FIRA 2009, CCIS 44, pp. 74–85, 2009.

a such a framework and to use it as a success predictor of a certain application during the design process. Besides the fact that we are designing an interactive interface that meets functional, physiognomic and cultural requirements we strive to create a believable, social accepted communication partner. Therefore, we need a research environment that allows us to test interaction paradigms in a real world setting. Designers of RUIs deal with very complex evolutionary evolved social interaction abilities not yet fully understood [3]. Subtle flaws in the design might negate all the efforts to make the interface more usable. Incorporating social interaction modalities in an interface requires an high amount of balancing intuition of the designer. We demonstrate the importance of a balanced design with an experimental setup that violates these principles.

The paper is organized as follows. Section 2 gives an overview of the design requirements for RUIs. We motivate why it is useful to add "natural interaction" modalities to the interface and explain a balancing framework for RUIs. Section 3 describes the experiments that we performed to validate our assumptions and summarizes our main results. Section 4 discusses the implications of our findings on the design of RUIs. Conclusions are presented in section 5.

2   Designing an RUI

We define a robotic user interfaces as robotic devices that are expressive and interact with the user using modalities such as speech, gestures or symbolic expressions. These modalities are familiar to a human user and consequently reduce the learning curve to interact with the robot. In the design for a specific user interface the different modalities are combined to form perceptually a single entity that presents itself as a lifelike communication partner to the user [3,4]

2.1   Natural Interaction

Adding human like communication modalities to an interface will increase its usability since people to people communication is more effective than people to machine communication [5]. Ishiguro argues that because humans anthropomorphize their targets of communication, androids can serve as ideal interfaces [6]. Being able to communicate with a robotic interface as if it was a human is the ultimate goal of this research. The hypothesis is that the more human like the shape of the robot, the easier it would be for a user to accept it as a natural communication parter. Unfortunately, the relationship between the shape of the robot and how it is perceived is not linear and needs a careful design. Mori described in his Uncanny Valley theory [6] the eerie feeling that artificial characters arouse in the user if the character looks almost like human but has subtle deficiencies. The qualitative shape of the graph is shown in figure 1.

A key characteristic for a social interactive interface is its believability [4] as a character. For a social interaction like with other humans the robot has to be accepted as a social communication partner, especially if the robot is going to be applied in social environments like a household. Wrede et al. argues that in

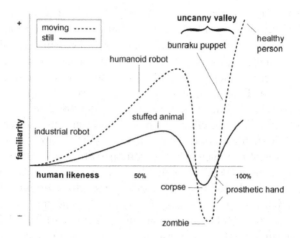

Fig. 1. Simplified version of the Uncanny Valley graph conceived by Mori. Src: McDorman [7]

order to achieve a high acceptance for a robot in a social environments it has to engage socially with the user [8]. A survey of social interactive robots can be found in [2].

Research has shown that people are naturally biased towards social interaction [9]. This holds true even for very abstract interfaces. Tremoulet, for example, shows that simple motion patterns are enough to give the impression of an autonomous and alive interaction partner [10]. Motion patterns can be used to communicate for example emotional state, intentions or for communicative gestures. van Breemen argues that body gestures are an appropriate way to communicate what a character is doing and thinking [11]. He also argues that robotic user interfaces face the same problem as early animations, that they miss the illusion of life. Therefore he applies animation techniques for explicitly designing communicative behavior.

It is still an ongoing discussion if the interface should give the impression of dealing with an intelligent character. Adversaries of the anthropomorphic paradigm argue that it is misleading to try to fake intelligence if there is none and that it will lead to frustration of the user earlier or later. Proponents emphasize the added value to the usability of the devices. We argue that the key for success of a particular interface is not determined by a single technology or its physical design but rather depends on a careful design of the application, balancing between all influencing factors.

2.2   Balancing Framework

The field of interactive systems suggests that the key dimensions to balance a RUI design are: user, technology, interface and application [12]. In the following we will explain these four dimensions in detail:

User. The purpose of using a robotic user interfaces is to allow everyone to use it without additional learning effort. This does not mean that there is one universal RUI but that the elements for an RUI have to be carefully chosen to accomplish its service. The design has to take into account cultural differences, social constraints as well as short term moods of the user.

Application. Application areas for robotic user interfaces are manifold. Ultimately, robots could serve as interface between a user and any electronic device, bridging the gap between the physical and the digital world. For example, application areas also includes entertainment, in which the user is interacting with the device just for the fun of interacting with it.

Interface. Designing a robotic user interface means to choose the modalities and define interaction methods. The most common input modalities are vision, speech, keyboard, touch sensors and switches. For giving feedback to the user the robot can communicate by means of light, sound and motion. With facial expressions and body postures the robot is able to convey emotional messages. On the one hand the modality has to be defined and on the other hand the type of messages that are sent.

Technology. Most of the constraints for the design of a RUI stem from technological constraints. Processing power and memory become less of a factor because of technological achievements, while at the same time dropping in price. Additionally, wireless communication enables remote processing and access to worldwide knowledge. The same trend as for processing power can be observed in the price for sensors and actuators. Currently, the more limiting factors are available techniques for data interpretation and generating appropriate behavior. Artificial Intelligence algorithms for interpreting speech and vision data are still not able to extract a sufficient percentage of information encoded in the signals to ensure a seamless interaction. Also new techniques have to be developed to enable robotic characters to become a fully accepted social communication partner.

All four dimensions have to be in balance to create an effective RUI. Very often their constraints compete with each other. In the following section we explain the balancing process.

2.3   Balance the Dimensions for an RUI

In our analysis of balancing the RUI we start with the user. The user has to be interested in the task proposed by the robot. This is a perquisite that always has to be met and already puts the first constraint on the application. The designer has to find a match between the application and the user. This relations seams very obvious, but the designer has to keep it in mind during the design and implementation process. Adapting the application to the user might not always be possible, because it is constrained by technological limitations. Very often additional research is needed to find new solutions that enable the desired applications and functions. In consequence, it is only to a certain extend possible to adapt the application.

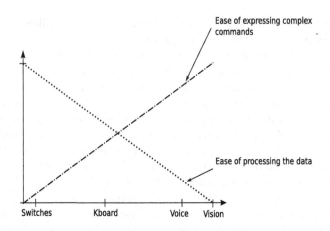

Fig. 2. Antagonism between rich input and computer understanding

As soon as the initial application is defined we incorporate the interface. As concluded by Dautenhahn [13] there is no use of anthropomorphizing too much an interface if the aim of the application is to perform repetitive actions, not involving a deep emotional content such as a washing machine. In this case efficiency and swiftness are more important features. That means the application needs to involve the user emotionally.

Now, if incorporating anthropomorphic artifacts in the interface, the designer has to define needed abilities of the RUI and match them with shape and behavior. The results of Goerts [14] shows that people judge the abilities of a robot by shape and behavior and are willing to interact if they match. If shape and behavior raise expectations that cannot be met it will lead to frustration of the user, or users get the impression it is not able to perform the task.

For designing the interface the designer has to assess how much effort is imposed on the user. Shneiderman explains that spoken language and problem solving share the same "resources" in the brain, and therefore the short-term and working memory load is much higher when a speech interface is chosen. The interface should not put additional burden on the user. Considering these facts, the interface might give new constraints to the application that have to be taken into account.

At last we balance the system with the technology. Because we have already specified demands of the application and the interface we can specify the requirements for the technology very precise. Of course it can happen that the requirements cannot be met, for example if the user demands a natural input for easily expressing his emotions that is beyond capabilities of current technology. The richer the input, the less reliable are current technologies to interpret the data. This antagonism is qualitatively illustrated in figure 2 for a chitchat application. The X-axis represents the complexity of input. On the Y-axis 2 different data are displayed. Both are theoretical ratio: the first one represents the ease of expressing complex commands while the second one represents the ease of processing the commands. A switch is a very simple input device but is also limited in its expressiveness.

In the process we propagate constraints backwards and reiterate the last step if we encounter a constraint mismatch. If in the end it turns out that with the given constraints the user is not interested in the application any more, we have to revise it.

In the following we present two application design scenarios: The first violates the balancing framework and the second adheres to it. We argue that the balancing framework can be used as a predictor for the success of an application.

3   Case Stidies

In our study we focus on the iCat research platform, developed by Philips research [15] and made available to universities as a common research platform. Several studies have shown that iCat is able to express recognizable emotional expressions (e.g. [16]), while avoiding the uncanny valley by its intentionally comic style appearance. We fix the shape of the interface but vary the technology, the application and the type of interaction. We demonstrate how difficult it is to design a valid real world application that can serve as a test environment for human robot interaction research.

The first design scenario is a simple chatter bot application. We used straight forward design decisions to create the set-up and observed the effects of adding facial emotional expressions as feedback to the interface. We expected the emotional feedback to increase the believability of the chatter bot as a character and hence improve the overall appreciation of the interface. As a second experiment we developed an application adhering to the balancing concepts. With the second experiment we demonstrate a successful balanced application incorporating a RUI. This is not trivial because as soon as we address natural and social interaction capabilities of the user we have to satisfy user expectation for social interaction.

3.1   Conversation Bot

With the conversation bot experiment we explore the design of an application for an RUI. The application provides us with an easy example scenario, that is already biased towards human like interaction, hence for using an RUI. The hypothesis is that people will prefer an interface that incorporates more human like artifacts in the interaction. We test this hypothesis by comparing two different interface designs for the application.

Scenario and Technology. In the chatter bot scenario iCat takes the role of a chat friend entertaining and serving the user. iCat listens to the comments of the user and shows emotional involvement with facial expressions. iCat has two fields of expertise, music and robots, that define the domain for the application.

The user may ask for information such as newest publications from his favorite band and talk about music preferences. In the experiment iCat also shows its own musical taste and talk about robotic friends to set-up an autobiographic background that supports the impression of dealing with a believable character. The user would also be able to switch topic by asking for his messages and later on resuming the talk. Topic switching is common in human style conversations and the goal if this experiment is to provide an interface as natural as possible.

Given the available technology, the best input medium for our test is a keyboard. People are used to chatting systems. Especially young people grew up with computers and used chat systems in Internet cafes, on websites or with separate messaging tools. Therefore, one criteria for the selection of our subjects is that they are familiar with chatting technology.

Study Design. For the experiment of testing the design of the chatter bot application we chose for a wizard of Oz setup. The parameter that we manipulated during the experiment is the emotional feedback of iCat and the wizard of Oz design insured rational answers and appropriate emotional feedback while still presenting the original input interface to the user. We created two designs for the application: an experimental design in which iCat showed facial expressions and a control condition in which iCat had a neutral face. More emotional feedback should give the user the feeling that the chatter bot understands and cares about him. Every subject was shown both conditions in a randomly assigned order. Due to the exploratory nature of our study the sample size was limited to sixteen, all highly educated people, interns of Philips Research, in their twenties. They were selected for their frequent use of instant messaging systems, so that they fulfilled our requirement of being familiar with chatting technology using keyboard and computer screen. For the final test setup we chose for practical reasons a virtual embodiment of iCat. We are aware that there is a difference in the perception of a physical and a virtual embodiment [17], but we considered the differences not to influence our results if we use the same embodiment in both conditions.

User Test Results. We measured the perceived usefulness and pleasure the participants experienced while interacting with iCat. The results of the questions relating to those measures are shown in table 1.

The given positive feedback has to be taken with care. Due to the short interaction time no statement on the long term satisfaction can be made. Positive reactions motivated by surprise stem from novelty, which is lost in long term interactions. The usefulness was rated less high. During the semistructured interview participants expressed their doubts on the usefulness of the application. We have to keep in mind that all participants were technically skilled and interested in technology with proper background knowledge on current chatter bot technology. Despite their doubts people liked the idea of having a chat buddy to talk about music.

Table 1. User's satisfaction on overall application (regardless of the session)

Usefulness					
	Not at all	Slightly	Moderately	Quite	Extremely	Average
Useful	0	5	5	5	1	3.125
Practical	1	3	6	4	2	3.5625
Functional	2	5	4	4	1	2.8125
Helpful	3	2	7	2	2	2.875
Efficient	1	6	4	4	1	2.875

Pleasure					
	Not at all	Slightly	Moderately	Quite	Extremely	Average
Exciting	0	3	3	5	5	3.75
Fun	0	0	2	6	8	4.375
Amusing	0	0	3	7	6	4.1875
Thrilling	1	5	7	2	1	2.8125
Cheerful	1	2	1	11	1	3.5625

The average value is based on a 1 (not at all) to 5 (extremely) notation

3.2   Waiter Application

The chatter bot experiment showed that the emotional feedback did not enhance the perception of the interface. Less attention was given to iCat's facial expressions, because the user was using a keyboard. While using speech as input could improve this situation, the technology is not mature enough for realizing the chatter bot application. We can thus conclude that the chatter bot application is not well balanced. With a second experiment we addressed the question whether it is possible at all to design a well balanced RUI given the current state of technology. Therefore, we designed another application as a use case to study the balancing framework. A video prototype of this application was first presented at the HRI conference 2007 [18].

Scenario. To find a balance between the four dimensions - user, application, technology and interface - one needs to find a scenario that satisfies all of them. We chose to elaborate a waiter application because of three application conditions: limited interaction time, controllable environment, and narrow domain.

The scenario for the waiter application takes place in a restaurant environment. iCat is located on a table in the restaurant, as illustrated in figure 3.2.1. When customers enter the restaurant iCat will greet them and offer them a free seat at her table. iCat will take their orders and submit them to an interface in the kitchen. The touch sensors in iCat's paws were used for simple yes and now answers. This input method proved to be very robust and also keeps the user from anthropomorphizing to much. In our setting iCat would perform the following tasks: invite customers to the table, offer and explain dishes from the menu card, take orders, entertain customers while waiting, serve as an interface for controlling the environment such as lightning or music.

Balancing the Application Design. We considered an RUI to be an appropriate interface because welcoming and accommodating the customer with individual care is rather an emotional task than a functional, neutral one. A robotic user interface is in position to provide such emotional feedback. Here, our initial observations support the choice for the scenario. First, it is very difficult to maintain the impression of dealing with a social intelligent character over a longer period of time. Existing methods fail after a certain period, due to inefficiencies of the technology to create believable personalities. After that the user starts to recognize machinelike behavior. The chosen scenario lasts only for the time needed to complete dinner. Therefore, it has much less requirements to the interaction consistency than for example the chatter bot that is placed at home. Second, the environment of a restaurant is much more controllable then a personal home. The position of iCat is fixed and customers usually sit around the table and e.g. the lighting conditions are predictable to a certain extend. Additionally, various sensors can be placed in the environment that enhance the perception of iCat. One might think of additional microphones, cameras or sensors to detect of a person is sitting on a chair. Third, the domain for the conversation is narrow. Explanations of particular dishes can be predefined as well as some typical flowery sentences common in the scenario such as "I hope you liked the soup!". Of course, we have to keep in mind not to raise too much expectations by the behavior.

One of the main tasks for the system is to take orders from the customers. As stated above, just relying on a speech recognition system is currently not sufficient, especially in a noisy environment such as a restaurant. Our chatter bot experiment showed that offering a keyboard for general input would be also the wrong choice. Instead, the restaurant can offer an "electronic menu card", so that the user just needs to point at a dish he wants to order or get additional information. A scene like this is common if customers want to get an explanation of a specific dish. They point at the menu card and ask the waiter for his explanation.

Still we need a technique for choosing iCat's actions. Our requirements are that the system should be able to select discrete action depending on the state in a continuous domain. Therefore, we chose for Extended Behavior Networks ([19], [20]).

User Test Results. To evaluate our design we performed an experts test with two restaurant managers. Before starting the test we asked them what they expect from the application. It turned out that their main concern was the functionality of the system, whether it is able to take orders. Their second concern was if it will bring added value to the restaurant.

After these initial questions we let them interact with our prototype from a users point of view. They knew already the setup from the video prototype that we took earlier in one of their restaurants. In this version, we also included a first prototype of an electronic menu card from which they were able to select dishes. For reasons of simplicity and available resources we restricted the menu card to simple pointing gestures and used a bar-code reader to point at a specifically prepared menu card. An order was finalized and confirmed using the touch sensors in iCat's paws.

We conducted a semi-structured interview, focusing on three points: 1) Applicability of the application 2) Interface design 3) Personality of the waiter. Overall, they were positive about the interaction and the efficiency of the robot. At first we analyzed the application and its functionality. They were surprised by the speed with which the customers are able to get information and order a dish. They predicted it will be a great benefit for the restaurant. The process of collecting and ordering dishes worked seamlessly, though the bar-code reader seamed a little old-fashioned. They would prefer to have a touch screen for their restaurants.

Next we talked about their impressions on the interface. At the beginning the experts had some doubts interacting with a robot, but while testing, they developed a strong drive towards iCat. They only would exchange the embodiment to fit to the concept of the restaurant. The only thing they rated negative was that iCat did not support their choice of dishes, a behavior that we did not include in our prototype. Especially at the beginning iCat should be much more proactive, offering help on how to use the service and propose dishes or drinks.

The last focus point was iCat's personality. Maybe the most important thing to mention is, that they accepted iCat as a communication partner. They expect iCat to provoke conversation between customers. Both managers attributed iCat to have a social character, but they diverged in their opinion which characteristic would fit better in their restaurants. The first wanted iCat to engage more in chit-chat, e.g. using on-line news services or weather information while the other preferred a more neutral iCat not too overwhelming or disturbing for the customer. We asked which characteristic seemed more important for them for judging iCat's personality, its shape or its behavior. Also in this question they diverged in their opinion.

4   Discussion

The results of our experiments emphasize the importance of a balanced design incorporating a robotic user interface. Some major violations to the balancing framework kept the chatter bot application from succeeding. First, the choice for a keyboard input was purely technology driven, violating the demands of the interface and the user. It only adhered to the demands of the application to provide the possibility of unconstrained textual input. Second, the emotional feedback given to the user was not perceived, because of the faulty configuration of the interface. There is a mismatch in using natural language as means for communication but offering a keyboard for input. One might argue that adding a method to make the user look more at iCat would increase the perception if iCat's emotion such as disabling the keyboard while iCat is answering, but from the feedback we got we conclude that such coercive methods will only lead to user frustration. Third, the emotional feedback given by facial expressions did not match the plain voice of the text-to-speech system. The lessons learned from the experiment are that the demands of a natural language user interface by means of speech recognitions is out of scope of current technology. Substitution of the speech recognition system with a keyboard does not work out as long as application is then unbalanced. A second design experiment of a waiter application was carried out to see whether it is possible at all to design a balanced RUI application. We were mainly interested in the value of the general application concept and therefore validated the scenario with feedback from experts in the field. The feedback that we got on the waiter application suggests that it is in general possible to construct a balanced application with current available technology.

5   Conclusion

In this paper we argued that human like interaction modalities will increase the usability of an interface if applied in a balanced design. We found that a balanced design is a key factor for the success or failure of an RUI. The four key dimensions to take into account are: user, application, interface and technology. The difficulties in finding a balanced design stem from the fact that RUIs address social interaction capabilities that are not yet fully understood. In order to validate the framework, we developed two scenarios: a chatter bot and a waiter application. We consider the chatter bot example to have failed because of various violations to the balancing framework and not due to principle problems of imitating social interaction. With the waiter application we presented a balanced design incorporating a RUI and validated the application with experts in the field. The results of our experiments indeed show that RUIs provide means to increase the usability of devices and allow them to be used by a broader range of people.

References

1. Shneiderman, B.: Universal usability. Commun. ACM 43(5), 84–91 (2000)
2. Fong, T., Nourbakhsh, I., Dautenhahn, K.: A survey of socially interactive robots. Robotics and Autonomous Systems 42(3-4), 143–166 (2003)
3. Dautenhahn, K.: Design spaces and niche spaces of believable social robots. In: 11th IEEE International Workshop on Robot and Human Interactive Communication, Proceedings, pp. 192–197 (2002)
4. Bates, J.: The role of emotion in believable agents. Communications of the ACM 37(7), 122–125 (1994)
5. Duffy, B.R.: Anthropomorphism and the social robot. Robotics and Autonomous Systems 42(3-4), 177–190 (2003)
6. Ishiguro, H.: Interactive humanoids and androids as ideal interfaces for humans. In: IUI 2006: Proceedings of the 11th international conference on Intelligent user interfaces, pp. 2–9. ACM Press, New York (2006)
7. MacDorman, K.F.: Mortality salience and the uncanny valley. In: 5th IEEE-RAS International Conference on Humanoid Robots, December 2005, pp. 399–405 (2005)

8. Wrede, B., Haasch, A., Hofemann, N., Hohenner, S., Hüwel, S., Kleinehagenbrock, M., Lang, S., Li, S., Toptsis, I., Fink, G.A., Fritsch, J., Sagerer, G.: Research issues for designing robot companions: BIRON as a case study. In: Drews, P. (ed.) Proc. IEEE Conf. on Mechatronics & Robotics, Aachen, Germany, vol. 4, pp. 1491–1496. Eysoldt-Verlag, Aachen (2004)
9. Reeves, B., Nass, C.: The Media Equation: How People Treat Computers, Television, and New Media like Real People and Places (CSLI Lecture Notes (Hardcover)). The Center for the Study of Language and Information Publications (September 1996)
10. Tremoulet, P.D., Feldman, J.: Perception of animacy from the motion of a single object. Perception 29(8), 943–951 (2000)
11. Breemen, A.v.: Bringing robots to life: Applying principles of animation to robots. In: Proceedings of the Workshop on Shaping Human-Robot Interaction - Understanding the Social Aspects of Intelligent Robotic Products. Cooperation with the CHI 2004 Conference, Vienna (April 2004)
12. Benyon, D., Turner, P., Turner, S.: Designing interactive systems. Addison-Wesley, New York (2005)
13. Dautenhahn, K.: The art of designing socially intelligent agents: science, fiction, and the human in the loop. Applied Artificial Intelligence 12(7-8), 573–617 (1998)
14. Goetz, J., Kiesler, S., Powers, A.: Matching robot appearance and behavior to tasks to improve human-robot cooperation. In: The 12th IEEE International Workshop on Robot and Human Interactive Communication, 2003. Proceedings. ROMAN 2003, IEEE CNF, October 2003, pp. 55–60 (2003)
15. Philips Research: iCat research platform,
http://www.research.philips.com/robotics
16. Grizard, A., Lisetti, C.: Generation of facial emotional expressions based on psychological theory. In: 29th Annual Conference on Artificial Intelligence, 1rst Workshop on Emotion and Computing at KI 2006, June 2006, pp. 14–19 (2006)
17. Powers, A., Kiesler, S., Fussell, S., Torrey, C.: Comparing a computer agent with a humanoid robot. In: HRI 2007: Proceeding of the ACM/IEEE international conference on Human-robot interaction, pp. 145–152. ACM Press, New York (2007)
18. Saerbeck, M., Bleuzé, B.: Waiter application. In: The 12th IEEE International Workshop on Robot and Human Interactive Communication, 2003. Proceedings. ROMAN 2003. Video Session: HRI Caught on Film, March 2007, p. 177 (2007)
19. Maes, P.: How to do the right thing. Connection Science Journal, Special Issue on Hybrid Systems 1 (1989)
20. Dorer, K.: The magmaFreiburg soccer team. In: Veloso, M.M., Pagello, E., Kitano, H. (eds.) RoboCup 1999. LNCS (LNAI), vol. 1856, pp. 600–603. Springer, Heidelberg (2000)

Infrared Remote Control with a Social Robot

J. Salichs, A. Castro-González, and M.A. Salichs

Roboticslab, Carlos III University, Madrid, Spain

Abstract. One of the aims of social robots is to make life easier for people in doing things more to their comfort and assisting them in some tasks. Having this purpose in mind and taking advantage of the widely used of infrared controlled home appliances, a controlling infrared devices system on board a robot has been developed. We focus on standard appliances with infrared interface, therefore no adjustments are needed either in the appliances or in the environment. The system has been implemented at the social robot named Maggie which has been equipped with an infrared reciever/transmitter. After teaching the desired commands to the robot, Maggie can govern the devices placed at several locations. At the beginning, a dialog between Maggie and the user is established. Once the robot has understood what the user wants, it will move toward the device and will send the required command. For testing the smooth running of the system, several trials have been done in our lab with a TV screen.

1   Introduction

Nowadays there are lot of devices operated by an infrared remote control at houses: televisions, air conditioning, VCRs, heating systems, etc. In addition, the expected wide spread of robots at home will reveal a new world of opportunities. So, in the next future, new household robots communicating with house appliances will appear. Besides, since remote controls have turned more and more complicated, we tried to elaborate a easier and more intuitive way of commanding them. We specially oriented our researches to children and elder people.

At this paper, a built-in voice operated remote control in a personal robot is presented. The required infrared commands for doing some tasks are easily taught from the original remote control and, subsequently, commands are sent by the robot upon request of the user.

One of the biggest advantages of the work is the robot interaction with regular appliances, it means without any change or adjustment of them. Therefore, high specialized expensive devices capable of being integrated with domotic controllers are not necessary and money is saved.

We are going to implement it in a mobile robot so it is possible to govern appliances at several locations. The robot will move to the area where the device to be controlled is and it will realize the entrusted task. In case something goes wrong (e.g. it can not reach the location) the robot will notify it. For example,

J.-H. Kim et al. (Eds.): FIRA 2009, CCIS 44, pp. 86–95, 2009.

a person in a room can order the robot to turn on the heating system which is located in a different room, so the robot will move to the desired room and will execute the intended command.

This work has been developed and proved in our lab with a TV screen.

The rest of the paper is organized as follows. At the rest of section 1, previous works related to this one are presented and compared and the pursued goals are presented. Then, at section 2, we explain the robot Maggie and the AD architecture where this work is framed. Next, section 3 presents the proposed solution. After, the integration in AD is shown at section 4. Following, section 5 exposes the trials and how the system has been tested. Finally, some concluding remarks end the paper in section 6.

1.1   Related Previous Works

There has been several previous works related to. At [1], a system for controlling a computer by means of a pair of glasses equipped with infrared signals is developed. Here, complex hardware has to be attached to the computer in order to be able to communicate with it and user needs to be wearing the infrared glasses. In [2], the authors propose a platform for control appliances. The system has a robot with infrared receiver/transmiter and it is commanded by MSN. In this case, user needs a computer to send orders to the robot. The idea in [3] is to develop specific hardware to enable conventional appliances without telecommunication capabilities to connect to home networks so they can be remotely controlled. The week point is that new hardware has to be installed at each device. Another option is to install networked appliance to create a home network [4] that makes the control task easier but it has an important increase of budget.

There are some other ways of communication that infrared. For example, [5] uses bluetooth as communication protocol to connect several home appliance that have to be provided with this interface.

Thinking of robots, some home, social or personal robots are able to interact with persons and home appliances. Toshiba Corporation has developed a concept model of the robotic information home appliance called ApriAlpha [6] [7]. Apri-Alpha is a wheel locomotion type human friendly home robot which controls advanced home appliances, standing between their users and them as a voice

Fig. 1. ApriAlpha, ApriPoko and MARON-1

controlled information terminal. As a newer version of AapriAlpha, ApriPoko is basically a voice-operated infrared universal remote control which learns commands and it is connected to a laptop to process all data. For now it is just an R&D demonstrator device. Fujitsu Laboratories has developed an internet-enabled home robot called MARON-1 (Mobile Agent Robot Of Next-generation) [8] as practicality-oriented household robots. MARON-1 can be remotely controlled from a mobile phone to monitor home security and operate household appliances which can respond to infrared remote control signals. All these robots are shown on fig.1.

At this work, regular house appliances with infrared interface have been used. Anyway, new hardware is not needed for each one of them. Users interact with the system in a natural way so they do not demand any apparatus apart from the robot.

1.2 Goals

Our target is to integrate an infrared device into Maggie robot and incorporate necessary software into the control architecture. It has to be able to operate unchanged infrared controlled gadgets with a robot by means of natural commands in human terms. In other words, user has to be able to interact with the robot just like with a person. The gadgets can be situated in different locations so the system must be capable to reach and face them. The module has been built in the experimental platform Maggie. An easy friendly communication is required because children, elder people and persons with disability will be the first potential users (fig.2).

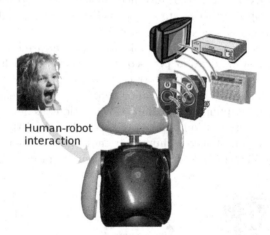

Fig. 2. Operating appliances from the Social Robot Maggie

2 Frame of the Work

The developed work in this paper has been implemented in the research robotic platform Maggie. Maggie is a personal robot intended for investigating human-robot interaction and improve robot autonomy. It was conceived for personal assistance, making life easier at houses, help handicapped persons, to keep persons company, etc. Its external friendly look facilitates its social robot task.

2.1 Automatic-Deliberative Architecture

Maggie's software is based in the two levels Automatic-Deliberative architecture (AD) [9] [10].The automatic level is linked to modules communicating with hardware, sensors and motors. At deliberative level, reasoning processes are placed. The communication between both levels is bidirectional and it is carried out by Short Term Memory and Events [11].

Events is the mechanism used by the architecture for working in a cooperative way. An event is an asynchronous signal for coordinating processes emitting and capturing them. The design is accomplished by the implementation of the publisher/subscriber design pattern so a skill generating events does not know whether these events are received and processed by other skills or not.

The Short Term Memory is a memory area which can be accessed by different processes, where the most important data is stored. Different data types can be distributed and data are available to all elements of the AD architecture. The current and the previous value as well as the date of the data capture are stored. Therefore, when writing a new data, the previous one is not eliminated, it is stored as the previous version. Short Term Memory allows to register and to eliminate data structures, reading and writing particular data, and several skills can share the same data.

The essential component in the AD architecture is the skill [11] and it is located in both levels. A skill is the capacity to reasoning, processing data or carry out actions. In terms of software engineering, a skill is a class hiding data and processes describing the global behavior of a robot task or action. The core of a skill is the control loop which could be running (skill is activated) or not.

Skills can be activated by other skills or by a sequencer, and they can give back data or events to the activating element or other skills interested in them. Skill are characterized for:

- They have three states: ready (just instantiated), activated (running the control loop) and locked(not running the control loop).
- Three ways of working: continuous, periodic and by events.
- Each skill is a process. Communication among processes is achieved by short term memory and events.
- A skill represents one or more tasks or a combination of several skills.
- Each skill has to be subscribed at least to an event and it has to define its behavior when the event arises.

3 System Implementation

A general overview of the system is shown at fig.3 where how the system works from start (user interaction with robot) to finish (robot sending the command)

Fig. 3. Overall flow diagram of the whole system

is displayed. As it is pointed out, the suggested solution is built by a system divided in two interfaces (human-robot and robot-appliance interfaces) and the module in charge of moving the robot close to the proper appliance.

3.1 Human-Robot Interface

There are several ways of communication with the robot but we focus on natural communication in human terms. When user wants Maggie to turn air conditioning on, he has to transmit his intention to the robot like if he was interacting to other person. Thereby communication with the robot can be accomplished in different ways.

A natural and approachable interaction between users and robot is needed and verbal communication meets these requirements. In consequence, at this experiment, speech will form the human-robot interface. User speaks to robot and Maggie is able to understand speaking by a speech recognition software based on grammars. We connect one grammar rule per each infrared command Maggie will execute. Speech recognition system is modeled as a permanent skill, it is a skill which has been activated once and it will keep on it [12].

3.2 Robot-Appliance Interface

Here the main part of our work is presented. In order to send commands to infrared operated appliances, Maggie es equipped with IRTrans USB infrared control system [13]. It has been chosen because of its USB interface and Linux compatibility. It has been placed inside Maggie's body behind a sphere which lets infrared signal goes through (fig.4).

Our chosen hardware is provided with all software required to work in a Linux environment: it has a TCP/IP server for accessing directly the hardware and replying clients requests, a trial client and libraries needed to program our own software.

Server accesses a database where codings for operating devices are stored in the standard IRTrans format. In this application, server is linked directly to the USB infrared transceiver and it sends orders to the transceiver which replies with error codes.

Because of the nature of infrared technology, it is essential that robot is located and facing the appliance to communicate with. Hence it is fundamental a suitable navigation system.

Fig. 4. USB infrared device placed inside Maggie

4 Integration in AD Architecture

The operating of the entire infrared system in the robot architecture is explained in this section. To accomplish our work, various skills have been used which are connected as it is shown in fig.5. The skills implicated are:

1. **ASR Skill:** Automatic Speech Recognition skill is in charged of informing about which grammar rule has been identified through the microphones. An event (REC_OK), in addition to the detected grammar rule identifier, is sent to alert the rest of the architecture. This event will be catched by every skill subscribed to it, in our case a skill named *Speech_IR_Control*.
2. **Speech_IR_Control Skill:** it is a data processing skill which translates an incoming event from ASR skill to a new event based on the identified grammar rule notifying the type of command. If the command is not related to the infrared system, the event is ignored. Other case, required information is stored at short term memory, it is the device to be controlled and the order to be sent to, for example "turn on the tv". Then, *Speech_IR_Control* skill indicates that Maggie has to move to the location where the device is by means of the $GOTO$ event. If the position is reached, $GOTO\_OK$ event is received, the robot is ready for emitting the appropriated command, consequently $CONTROL\_IR$ is sent. If not, the operation is aborted.
3. **GoTo Skill:** After *GoTo* skill receives the $GOTO$ event, it is intended to move to the position determined by data in Short Term Memory. In our case, this skill takes the name of the device to be operated from the Short Term Memory and it relates it to a pose (position and orientation) in an internal map of the world. If the desire position is reached, $GOTO\_OK$ event will be sent. Other case, $GOTO\_FAIL$ is sent.
4. **IR_Remote_Control Skill:** the $CONTROL\_IR$ event is captured by this skill. Thereafter it accesses info concerning to the corresponding command

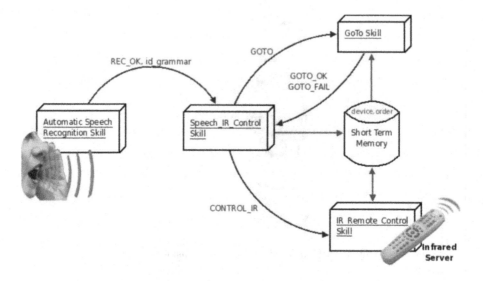

Fig. 5. Communication among skills

at Short Term Memory. Then, info is sent to the server and the right coding is gotten from the database where all available coding commands are. Now, the infrared hardware will emit this coding and it informs if everything has gone right.

To get ideas clear enough, the whole chronological evolution of the infrared system is shown at fig.6. When user wants Maggie operating infrared appliances, he will interact to it by voice commands (message 1 fig.6). ASR skill identifies it and distributes the grammar rule joined to the user's command by means of the event *REC_OK* (message 2 fir.6). After, *speech_ir_control* gathers it and

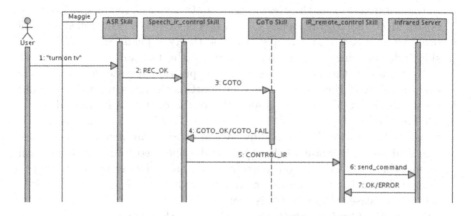

Fig. 6. Sequence diagram of involved elements

it links the recognized grammar rule to a device and an instruction which are shared thanks to Short Term Memory. At this moment, Maggie has to change its position in order to face the appliance (message 3 fir.6) which name is stored in Short Term Memory. The name is linked to a pose in the internal map and our robot tries to go there. Whether it achieves it or not, it is notified by *GOTO_OK* or *GOTO_FAIL* events (message 4 fir.6). If Maggie is ready facing the appliance and required data are available, *CONTROL_IR* is sent and it is received by *ir_remote_control* skill. It contacts the infrared server, which is directly connected to the hardware, and it sends the requested command.

5   Testing the System

In order to prove the smooth running of the system, we have developed in our lab several experiments as if it were at a real house. Our home appliance has been a TV screen and we have taught different commands to the robot just facing the remote control to the built-in infrared receiver and pressing the corresponding buttons. At this point, the remote device name, the timings and the coding commands themselves are written to the data base that will be used by the server. Then, we link each command with a grammar rule in the speech recognition system. The operations learnt by the robot are "turn on digital tv", "turn on digital radio", "channel up", "channel down", "volume up" and "volume down".

After lot of tests, it was observed that the system is running properly but with some key points. Robot pose is a very relevant factor because it is necessary that the infrared emitter is pointing to the TV screen and the direct line of sight has to be free of obstacles too. So the navigation system is key factor. Since human-robot interaction is leaded by oral communication, user speech has to be well understood by ASR skill. Thereby, a strong and efficient speech recognition system is mandatory.

The same process followed for the tv screen can be adjusted to every infrared devices effortless.

6   Conclusions

We have reached the goals proposed at the beginning of this document. A system for interacting with home appliances has been developed. The communication is conducted by the social robot Maggie performing household tasks. Human-robot interaction is accomplished using a speech recognition system so, in human terms, a natural interaction is realized: the communication with the robot is equal to the communication with a person. This aspect is really important since the primary users will be children, elderly people and disabled people (exception: speech impaired disability).

Commanding devices is carried out by a built-in infrared transmitter in Maggie. This is the regular technology utilized by most of the appliances, therefore it is not necessary to modify them or attach any electronic apparatus. For that reason the costs of installation, devices or appliances is not applicable.

Thanks to the robot navigation system, the presented system can operate infrared devices located in various rooms. The system could be expanded to accept commands through internet. In this way, a user at work could order the robot to turn on the air condition at home from the office. A different and feasible approach could be programming the robot to command devices in the future, e.g. the user will schedule the robot to turn off the air condition at 15:00.

Ideas presented at this paper are very useful for all users and in particular, for people with disability since our system will help them improving their quality of life and integrating them into society.

Acknowledgments. The authors gratefully acknowledge the funds provided by the Spanish Government through the projects called Peer to Peer Robot-Human Interaction (R2H), of MEC (Ministry of Science and Education), and A new approach to social robotics (AROS), of MICINN (Ministry of Science and Innovation).

References

1. Chen, Y.-L., Tang, F.-T., Chang, W.H., Wong, M.-K., Shih, Y.-Y., Kuo, T.-S.: The New Design of an Infrared-Controlled HumanComputer Interface for the Disabled. IEEE Transactions on Rehabilitation Engineering 7(4), 474–481 (1999)
2. Lin, L.-Y., Cheng, M.-C., Yuan, S.-M.: Standards-based User Interface Technology for Universal Home Domination. In: International Conference on Hybrid Information Technology, vol. 2, pp. 298–307 (2006)
3. Kuriyama, H., Mineno, H., Seno, Y., Furumura, T., Mizuno, T.: Evaluation of Home Appliance Translator for remote control of coventional home appliances. In: IEEE International Symposium on Power Line Communications and Its Applications, pp. 267–272 (2007)
4. Kolberg, M., Evan, H., Magill, W.M., Burtwistle, P., Ohlstenius, O.: Controlling Appliances with Pen and Paper. In: Second IEEE Consumer Communications and Networking Conference, pp. 156–160 (2005)
5. Hwang, C.-S., Wey, T.-S., Lo, Y.-H.: An Integration Platform for Developing Digital Life Applications. In: International Conference on Parallel and Distributed Systems, vol. 2, pp. 1–2 (2007)
6. Yoshimi, T., Matsuhira, N., Suzuki, K., Yamamoto, D., Ozaki, F., Hirokawa, J., Ogawa, H.: Development of a Concept Model of a Robotic Information Home Appliance, ApriAlpha. In: Proceedings IEEE/RSJ International Conference on Intelligent Robots and Systems, vol. 1, pp. 205–211 (2004)
7. Matsuhira, N., Ozaki, F., Ogawa, H., Yoshimi, T., Hashimoto, H.: Expanding Practicability of ApriAlpha in Cooperation with Networked Home Appliances. In: IEEE Workshop on Advanced Robotics and its Social Impacts, pp. 254–259 (2005)
8. FUJITSU LABORATORIES LTD.: Internet-enabled Home Robot MARON-1 (2008), http://jp.fujitsu.com/group/labs/en/business/activities/activities-4/#robotics

9. Barber, R.: Desarrollo de una Arquitectura para Robots Móviles Autónomos. Aplicación a un Sistema de Navegación Topológica. PhD Thesis at Carlos III University (2000)
10. Salichs, M.A., Barber, R.: A new human based architecture for intelligent autonomous robots. In: 4th IFAC Symposium on Intelligent Autonomous Vehicles, pp. 85–90 (2001)
11. Rivas, R., Corrales, A., Barber, R., Salichs, M.A.: Robot Skill Abstraction for AD Architecture. In: 6th IFAC Symposium on Intelligent Autonomous Vehicles (2007)
12. Castro-González, A.: Desde la Teleoperación al Control por Tacto del Robot Maggie. Master Thesis at Carlos III University (2008)
13. IRTrans GmbH: Universal IR Solutions (2009), http://www.irtrans.de

BlogRobot: Mobile Terminal for Blog Browse Using Physical Representation

Toshihiro Osumi[1], Kenta Fujimoto[1], Yuki Kuwayama[1], Masato Noda[1],
Hirotaka Osawa[2], Michita Imai[2], and Kazuhiko Shinozawa[3]

[1] Graduate School of Science and Technology, Keio University
tosihiro@ayu.ics.keio.ac.jp
[2] Faculty of Science and Technology, Keio University
michita@ayu.ics.keio.ac.jp
[3] Advanced Telecommunications Research Institute International
shino@atr.jp

Abstract. Mobile phones and PDAs with GPS became widespread, and
GPS-based service of mobile terminal is rising, so needs for presentation
interface for GPS-based contents are increasing. We employ a robot as
a presentation interface. In this paper, we use a blog as our content and
propose a portable robot named "BlogRobot" that presents blog content
by using verbal and non-verbal expression. Using BlogRobot, people who
don't have the knowledge of a specialist can create behavior contents as
they note in a diary, and the contents can give reader a feeling like a
man who write a blog's entry explain the entry near the entry location.

1 Introduction

Mobile phones and PDAs with GPS have become widespread, and the usage of GPS-based mobile terminal services is rising. Although often used for
navigation,GPS-based service can also be used for an intelligence-sharing service
with friends or people nearby. Through this intelligence-sharing service, anyone
can create content easily, and can express infomation just like a blog or a social
networking service. In this service, content that a user creates has GPS data,
so it is necessary to devise a means of incorporating the surrounding environment and a presentation method for the content. Therefore, due to the spread
of GPS-based mobile terminal services, the need for a presentation interface for
GPS-based content is increasing.

To establish a GPS-based content service, the system should handle GPS data
on access method, content management, content creation method and presentation method. In this paper, we view the GPS-based contents service through the
perspective of usability for the end user. If we consider the usability for end users,
we believe that it is important to research how to create and present content.

GPS-based content services have been extensively studied. A Service for mobile phones named "SpaceTag" is a location-dependent infomation service that
allows the user to access content at specific locations and times, and the general
user also can create content(Tarumi et al., 2000). "SpaceTag" make contents

J.-H. Kim et al. (Eds.): FIRA 2009, CCIS 44, pp. 96–101, 2009.

more attractive by restricting accessible number to adorn contents. "NAMBA Explorer" is a service that allows shops to share informal infomation (Kamisaka et al., 2005). Using "NAMBA Explorer", they can communicate their location information and informal infomation about inner-city shops. In another research, user can embed times and GPS data as metadata in photograph, and can create blog's entry with the photograph(Ito et al., 2008).

These various research efforts on GPS-based content services allowed for content creation, but the services can only support infomation representated on the screen displays of mobile terminals or personal computers . But explanations with gestures are often more easily understood than a simple view of text or photographs on a screen display. There is a need for a realistic content representation that mimics human explanations. We believe that this location-based representation can increase user interest in the presented content.

Thus, we focused on developing humanoid robots that can express feelings by gestures like a human. By using robots, we can create a presentation interface that enables realistic playback with human-like speech and gestures. In this paper, we use a blog as our content. A blog is website that displays entries in reverse chronological order, and allows users to comment on each entry. Blogs are widely used to express personal views or to record daily occurrences. Currently, there are many blogs on the internet, so we did not have to make new content for our service. A general user can create content without knowing that the content will be presented by a robot. In this paper, we propose a portable robot named "BlogRobot" that presents blog content by using verbal and non-verbal expression. Using BlogRobot, people without specialized knowledge can create behavior contents as they note in a diary, and the contents can give reader a feeling like a man who write a blog's entry explain the entry near the entry location.

2   Playback the Blog by a Robot

2.1   Effectiveness of a Robot and Gestures

Gestures are effective for enhancing the understanding of spoken language and double the accuracy of understanding spoken language(Berger et al., 1971). Gestures, unlike a mere speaking of the text, can help to relate the situation to the user. This creates a more realistic sensation, like that watching the blog's author explain the blog.

In addition, a robot in the real world is better than a CG agent at communicating real world information to users(Shinozawa et al., 2005). Therefore, about the blog which is written about something in the real world, using robot is effective in attracting reader to the real world than showing the entry in the display.

3   Design Consideration of BlogRobot

BlogRobot is a GPS-enabled mobile terminal on which we have installed a head and arms. BlogRobot communicates with gestures by using these body parts.

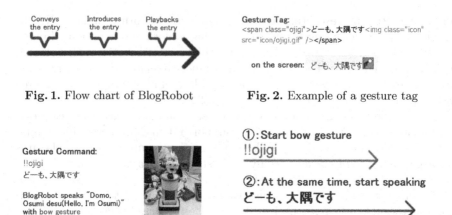

Fig. 1. Flow chart of BlogRobot

Fig. 2. Example of a gesture tag

Fig. 3. Example of a gesture command   **Fig. 4.** Time-line of gesture and speaking

BlogRobot updates GPS data at a constant frequency, and compares it to the GPS data of the entry's JPEG images. If the GPS data of the images in a entry matches the GPS data of the reader's mobile terminal, BlogRobot conveys the entry to the user and suggests playing it back (Fig. 1).

A traditional blog does not consider the option of using a robot for playback, which prevents BlogRobot from playing it back with well-timed gestures or a proper introduction. Thus, in this paper, we develop a dynamic content generation system for blogs named "TENORI". TENORI contains a blog authoring tool, and using the tool, a writer can embed gesture tags for BlogRobot as emotion icons in an entry. Then TENORI generates gestures from the gesture tags. TENORI also generates a report of the entry. In this paper, we use a dedicated blog for BlogRobot and operate BlogRobot by using an entry in the blog.

3.1 Dynamic Content Generating by TENORI

A gesture tag is a tag as indicated by Fig. 2, and TENORI converts a gesture tag(Fig. 2) into a gesture command(Fig. 3), and BlogRobot says "Domo, Osumi desu(Hello, I'm Osumi)" with a gesture. In this way, BlogRobot can gesture and speak and gesture simultaneously.

3.2 System Configuration of TENORI

The system configuration of TENORI is shown in Fig. 5. The blog entry is generated by the authoring tool in the blog module. Blog module is stand-alone and we can browse on the Web. When TENORI receives GPS data from BlogRobot, TENORI searches the entry to see if it contains GPS data near BlogRobot's GPS data. If TENORI catches the entry, TENORI convert the entry and embed report of the entry, and send back to BlogRobot.

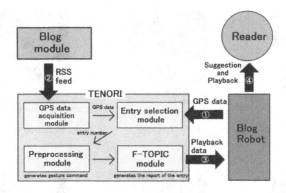

Fig. 5. System chart of TENORI

Designing of Blog Module. Blog module is stand-alone module and it is traditional blog basically. Using the authoring tool, the blog module can create an entry with gesture tags.

Authoring tool. Instead of buttons that can embed text color or emotion icons, the TENORI authoring tool has buttons that can embed gesture tags. The writer simply selects where he wants to embed the gesture tag, and clicks the gesture button to embed the tag in the entry.

Design of TENORI. After receiving the BlogRobot's GPS data, the entry selection module invokes the GPS data acquisition module. The GPS data acquisition module acquires the RSS feed of the blog registered in the database, and the blog entry's GPS data from a JPEG image of the entry. The Entry selection module compares BlogRobot's GPS data to the blog entry's GPS data from the GPS acquire module. If the blog entry's GPS data is near to that of the reader's mobile terminal, the entry selection module sends the blog entry number to the preprocessing module. The Preprocessing module acquires the blog data of the received number and removes the unnecessary tags for verbal and non-verbal expressions contained in the body text, and generates the gesture commands for non-verbal expression. The F-TOPIC module generates a report of the entry from the date, category and author contained in the blog data, and sends the report on the entry and body text to BlogRobot.

4   Implementation of BlogRobot

Fig. 6 to Fig. 11 show BlogRobot's appearances. We use a PDA produced by HTC Co., called HTC P3600. Each unit of the BlogRobot head and arms is controlled by a bluetooth connection. Both head and arms units include servo-motors. The head has six motors, and each arm has three.

BlogRobot can do some gestures by getting gesture commands (Fig. 6, Fig. 7, Fig. 8, Fig. 9, Fig. 10 and Fig. 11).

Fig. 6. Lean the head to the side

Fig. 7. Bow

Fig. 8. Shake the head

Fig. 9. Sad

Fig. 10. Glad

Fig. 11. Request agreement

5   Add-Up

In this paper, we propose a portable robot named BlogRobot as a presentation interface for GPS-based content. BlogRobot presents blog content by using verbal and non-verbal expression. For realistic content representation, BlogRobot needed to suggest playback in appropriate situations and to speak with well-timed gestures. We develop a dynamic blog content generation system named TENORI, and fixed the problem of generating gestures by using gesture tags and creating a report of the entry by using TENORI and a dedicated blog. Thus, we could create a presentation interface that enables realistic playback with human-like speech and gestures by using BlogRobot.

5.1   Future Work

Behavior on the Common Blog. In this paper, we used the specially designed authoring tool, that appends gesture tags. As a result, gesture commands are generated from common blog entries. In the future, we will study how to convert emoticons and the exclamation points to gesture commands to be able to generate these on the common blog as well.

Procedure for Playback. When it enters a certain range of the entry's JPEG image's GPS data, BlogRobot conveys the entry to the user and suggests playback. BlogRobot is unable to show the reader the place where the writer wants him to see, at present. For example, consider the case where the Blog's JPEG is a photograph of a location. If BlogRobot is unable to indicate the photographic subject of the entry, the reader must look for the photographic subject himself.

So an indication function will be absolutely imperative. Furthermore, there is the problem that BlogRobot cannot guide the reader to the intended location if the photograph was taken from an unusual perspective. We will solve the problem by embedding a guide generation function in the authoring tool.

Interaction with Reader. BlogRobot can only play back an entry given from TENORI. It is necessary to implement a function that changes BlogRobot's behavior according to the reader's response or obtains the reader's feedback for the blog author. We will install a minicam on BlogRobot's head and search for the reader's feedback by using image processing. Then we need to find a method for relating the reader's feedback to BlogRobot's behavior.

References

[WordPress]   WordPress, `http://wordpress.org/`

[Shinozawa 05]   Shinozawa, K., Naya, F., Yamato, J., Kogure, K.: Differences in Effect of Robot and Screen Agent Recommendations on Human Decision-Making. International Journal of Human-Computer Studies, 267–279 (2005)

[Berger 71]   Berger, K.W., Popelka, G.R.: Extra-Facial Gestures in Relation to Speechreading. Journal of Communication Disorders, 302–308 (1971)

[Tarumi 00]   Tarumi, H., Morishita, K., Kambayashi, Y.: Public applications of spaceTag and their impacts. In: Ishida, T., Isbister, K. (eds.) Digital Cities 1999. LNCS, vol. 1765, pp. 350–363. Springer, Heidelberg (2000)

[Kamisaka 05]   Kamisaka, D., Yoshino, Y., Munemori, J.: NAMBA Explorer: A Participative Location-Based City Area Information Sharing System. Digest of Technical Papers International Conference on Consumer Electronics (IEEE 05CH37619), pp. 459–450(2005-01) (2005)

[Ito 08]   Ito, J., Sumi, Y., Nakakura, T., Nishida, T.: Blog written on site based on photograph (Japanese). In: Interaction2008 (Interaction announce), IPSJ, Tokyo, Japan (2008)

An Exploratory Investigation into the Effects of Adaptation in Child-Robot Interaction

Tamie Salter, François Michaud, and Dominic Létourneau

Université de Sherbrooke
Québec, Canada
tamie.salter@usherbrooke.ca
http://introlab.gel.usherbrooke.ca

Abstract. The work presented in this paper describes an exploratory investigation into the potential effects of a robot exhibiting an adaptive behaviour in reaction to a child's interaction. In our laboratory we develop robotic devices for a diverse range of children that differ in age, gender and ability, which includes children that are diagnosed with cognitive difficulties. As all children vary in their personalities and styles of interaction, it would follow that adaptation could bring many benefits. In this abstract we give our initial examination of a series of trials which explore the effects of a fully autonomous rolling robot exhibiting adaptation (through changes in motion and sound) compared to it exhibiting pre-programmed behaviours. We investigate sensor readings on-board the robot that record the level of 'interaction' that the robot receives when a child plays with it and also we discuss the results from analysing video footage looking at the social aspect of the trial.

1 Introduction

Increasingly, researchers are looking at the domain of Child-Robot Interaction (CRI). There are now many robotic devices aimed at children that are meant for entertainment, but also there are robotic devices aimed at the more serious world of child development, [2], [5], [4]. When developing a system that is meant for interaction with children, it is important that the device can amongst other things i) gain the interest of individual children, ii) sustain this interest and iii) achieve this in the child's own natural environment (i.e., in the wild).

It would seem that a shortcoming of many devices aimed at children such as conventional toys is that they do not change to aid a specific child. Traditional toys are often developed to encompass the largest range of children possible. One of our main aims for developing adaptation has been the vast array of end users that our robots can come into contact with. This includes children from as young as 10 months old to young adults at high school. Also, we work with children that are typically developing and children that have special needs. Whether you are looking within age groups or categories (e.g., typically developing, special needs), children all have very different 'personality' traits e.g. shy, boisterous, cautious, outgoing. It is easy to realise that a robot that exhibits only one type

J.-H. Kim et al. (Eds.): FIRA 2009, CCIS 44, pp. 102–109, 2009.

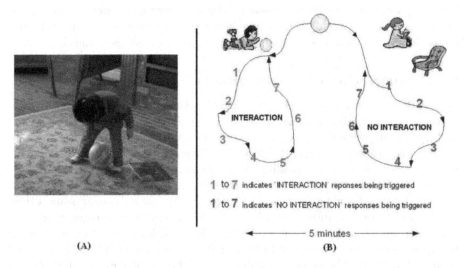

Fig. 1. (A) The only girl within the group interacts with Roball, an autonomous rolling robot. It is possible to see the different floor coverings. (B) Based on the child's interaction. The different loops show different behaviours being executed.

of behaviour will not be suitable for all children. Our research aims to develop a robot that can 'fit' its behaviour to the individual child it is interacting with. We are developing robotic behaviours that we believe to be encouraging and less daunting for an anxious child. Also, we are developing behaviours that are faster and more exciting for a more confident child. We have developed an adaptation algorithm, based on the readings coming from sensors on-board Roball, a fully autonomous rolling robot (shown in Figure 1). We use touch or contact as our metric to recognise interaction and thus enable adaptation. In this paper, we report on a series of trials that were conducted to explore the potential of the adaptive behaviour we have developed. We investigate whether the adaptive behaviour would *gain* and *sustain* the interest of five individual children. The same children also interacted with Roball when it was exhibiting a very simple basic pre-programmed behaviour and also a behaviour that was more complex but nevertheless pre-programmed. We compare the results from the three separate sessions that were conducted with each of the five children. We examined both data from on-board sensor readings and video footage of the trials.

2   Roball - The Robot

Roball is 6 inches in diameter and weighs about 4 pounds [3]. It consists of a plastic sphere (a hamster exercise ball) constructed from two halves that are attached to each other. The plastic sphere is used to house the fragile electronics (sensors, actuators, processing elements), thus making it robust and ideal for interaction with children. The fact that Roball is spherical encourages a wide range of play situations. Roball as standard is programmed to execute two simple behaviors: wandering and obstacle avoidance. However, it also has the ability to adjust its own motion or play sounds such as, vocal messages or music.

2.1 Distinguishing a Child's Interaction

Our work lies in recording the touch or interactions that a child applies when it plays with Roball. Roball has three accelerometers, one for each axis (X, Y and Z). During these trials Roball can be set in three different modes; (A) Basic, (B) Pre-programmed and (C) Adaptive. When set in (C) Adaptive Mode it is the accelerometer sensor readings that are analysed to ascertain the robot's current environmental status, e.g. receiving interaction or not. Records of 'Interaction' come from being pushed, banged, spun, picked up or some other such type of physical interaction. Records of 'No Interaction' come from sensors recording that the robot is wandering about with any form of interaction being recorded. The robot should not classify interactions such as hitting a wall as an 'Interaction'. Therefore sensor data is classified to define two different categories: INTERACTION and NO INTERACTION. If interaction is classified, a INTERACTION counter is accumulated; if no interaction is classified, a NO INTERACTION counter is accumulated.

2.2 Interaction Modes

During a session, one of three different interaction modes is executed on Roball; (A) Basic, (B) Pre-programmed or (C) Adaptation. The first two modes do not have any reaction to the child's interaction. The final Adaptive Mode reacts according to how the child interacts with Roball. Listed below are the three different interaction modes along with a description of the robot's behaviour:

A **Basic Mode.** Simple wander and obstacle avoidance behaviour.
B **Pre-programmed Mode.** Behaviours of mode A, but added behaviours of:
  – play a child's lullaby music song at 0:45 minutes, 2:45 minutes and 4:45 minutes
  – to stop all motion and ask the child to "Play with me" at 1:45 minutes and 3:45 minutes

  These extra behaviours are carried out for 15 seconds respectively.
C **Adaptive Mode:** (see Figure 1 (B))
  Behaviours of mode A with an added layer of adaptation triggered by the child's interaction. When an adaptive response is triggered, it lasts up to a maximum of 15 seconds (after this, Roball returns to a wander state). These responses are:
  – NO INTERACTION Responses: Play an audio track that says "Play with me", Slow down speed, Play an audio track of simple lullaby music, Stop all motion.
  – INTERACTION Responses: Play an audio track of children giggling, Increase speed, Play an audio track of simple (but lively) child's music.

Adaptive responses can be executed separately or together. For example: if the robot is not receiving interaction, it may play the audio track "Play with me" for 3 seconds, however it may at the same time execute the "Stop" behaviour for up to 15 seconds. The robot can switch between NO INTER-ACTION & INTERACTION modes in a single adaptive session.

3   Experimental Approach and Settings

In our laboratory we often use a rapid prototyping method to gain information as swiftly as possible (such as the one suggested by Bartneck and Hu [1]). This is the approach used in this set of exploratory tests to investigate the benefits of our 'Adaptive Mode'. These trials were held at a day care center in Québec, Canada. The approach to this study was to make each session as natural as possible and, in this vein, to limit the use of cameras, etc. This was an attempt to limit "audience effect". Audience effect is the impact a passive audience has on a subject performing a task, which, for instance, would be the experimenter watching the child interact with the robot. Inspired by the work of the RUBI Project [6], the experimenter spent a lot of time just helping out in the day care where the study was to be held. This was to familiarise with the children in an attempt to *not* be seen as an experimenter. The experimenter was introduced to the children as *"a person who would bring a robot to play with them later on, but for now was helping out in the day care"*. The experimenter would help with all the normal activities of working as a day care worker, e.g., playing games with the children, helping feed the children at lunch time. The children were very interested in the possibility of a robot coming and would ask *"has she brought the robot today?"* when the experimenter arrived at the day care. There was a lot of interest when the robot was finally taken to the day care center. There was an area set aside for the trial which was normally used by the children. It was intended that only the child participating at that moment would be allowed in the experimental area. However, due to the relaxed nature of the trial at times, some other children did come into the experimental area. These children were told that they could not touch the robot until it was their turn. The area had large pieces of furniture that were moved to the side, but there was still an array of different places for the robot to get stuck under, e.g., antique cot, television cabinet. Within the area there were three different floor coverings: hard wood, carpet (rug) and brick work (in front of the fire place), as shown in Figure 1. Also, at times there were other toys within the area such as, balloons, or toy trucks. Having these other toys in the area did not seem to take the interest away from the robot. Every experiment was video taped for verification of sensor readings and also a questionnaire about the child's interest in the robot was filled out by an independent adult (one of the day care workers) while the trial was being conducted. The questionnaire was used to confirm findings from the coding of the video data and also, to get another opinion on the child's level of interest and interaction. Pre-trial tests were conducted to confirm that the sensors recorded correct information in the setting that was to be used. We conducted this series of trials over a six day period. The exact dates of the trials were dictated by attendance of the same children and convenience for the day care. As with many other trials involving trials conducted with children and robots, in this preliminary investigation we used a small sample size. The participants were made of five typically developing children (four boys and one girl), aged 2 to 4. While using such a small amount of participants maybe insufficient to show statistical significance, it is useful to us to explore the effect of the robot and give us a base from which to conduct further trials. Also, this enables us to rapidly develop the behaviour of robot which can then, at a later date, be used in trials that contain a larger population of children. Each child played with the robot in three separate sessions of 5 minutes. Each time the child played with the robot, it exhibited a different interaction mode. The order of behaviours was randomised to test for internal validity (see Table 1). After 5 minutes the robot stops by itself, ending the trial.

4   Analysis and Results

After all the trials were completed we compared the on-board data levels of NO INTERACTION and INTERACTION for each of the sessions and also video footage of each trial. The behaviours that were coded were chosen because it was believed they would show whether the child was interested in the robot or not. The coding was done on a second by second basis. The following are the behaviours that were coded:

(1) LOOK - The child's gaze is directed towards the robot. (2) TOUCH - Touching the robot in a purposeful way e.g. moving towards the robot and touching it, either with feet, hands or another part of the body (Accidental touching such as the robot rolling into the child was not included). (3) TOWARD - Moving towards the robot in a purposeful way. (4) AWAY - Moving away from the Robot e.g. moving their body so that the robot could proceed or running away when the robot moved towards the child. (5) CAN'T SEE - It is not possible to see the above variables from the video footage. (6) SMILE/LAUGH (results shown in Table 1) - The child makes facial gestures or noises to indicate happiness e.g. laughing, smiling, giggling. We looked for instances where the child either smiled, laughed or giggled and that this action was in someway connected to the robot's actions. We did not include instances where the child laughed at something happening in the background of the day care center. (7) ASK QUESTION (results shown in Table 1) - The child asks a question to another person in the room e.g. an adult care worker. We looked for instances where the robot was acting as a catalysis or mediator for communication.

Due to the length of this paper we will discuss most of the results in a descriptive manner and will not list all the data (both sensor data and coded video data) obtained. First we describe information obtained from sensor readings then we describe information gained from analysis of video footage.

SENSOR DATA - There are two interesting observations that can be derived from the recorded interaction levels.

Firstly, there is a prolonged and sustained level of interest and interaction indicated by the records from the sensors (i.e. consistent levels of INTERAC-TION). We have not seen this result in any of our other trials with children. Normally, the INTERACTION decreases over subsequent sessions. There are many different reasons why this may have occurred. For instance, we believe that the natural conditions surrounding this study may have contributed to the sustained interest. However, there is also the possibility that the robot displaying a different behaviour each time the child interacted with it helped to sustain the interaction. Therefore, the Adaptive Mode would have played a role.

Secondly, the (C) Adaptive Mode did not appear to significantly increase the IN-TERACTION level during any of the children's session. This was not the expected result. However, we have learned from this trial that an increase in interaction does not necessarily, just come from strong physical contact with robot. We have discovered that different behaviours from the robot can elicit different types of physical interaction from a child. For instance, when the robot played music the most physical child interacted in a gentler, softer way. Therefore, this showed a dip in recorded levels of interaction compared to his normal pushing and banging of the robot. Also, there can be other forms of interaction such as the child being verbal that are not recorded by the sensor data. These ideas are explored further below.

Table 1. Coded Video Data. The coded behaviours of Smile/Laugh/Giggle and Asking Questions, for all children according to the three different interaction modes (A, B and C).

SESSION	NAME	MODE	SMILE	QUESTION
1	Edward	B	2	0
1	Evie	A	1	0
1	Gilbert	C	10	0
1	Harold	B	0	0
1	Tyler	A	2	0
2	Edward	C	2	0
2	Evie	B	0	0
2	Gilbert	A	0	0
2	Harold	C	1	0
2	Tyler	B	2	0
3	Edward	B	0	0
3	Evie	C	1	3
3	Gilbert	B	4	0
3	Harold	A	0	0
3	Tyler	C	1	4

VIDEO CODING - At times it was very difficult to code the video. This was due to the fact that only one camera was used and it was difficult to obtain a view of the whole experimental area with just one camera. Also, the experimenter did not stand behind the camera to point it in the direction of the child and the robot. As previously mentioned, the use of one camera and the lack to directing was due to an attempt to make each session as natural as possible for each child.

Firstly - We can see from Table 1 that there is an increase in the amount of smiling, laughing, giggling and asking questions during the (C) Adaptive Mode.

Secondly - Again, as we found with the sensor data, there did not appear to be any *significant* increase in either interest or interaction directed towards the robot which was brought upon by the (C) Adaptive Mode.

Thirdly - As we found with the sensor data there was a sustained level of interest over the three sessions that we have not found in other trials.

Other general observations made by the experimenter and the day care workers present are listed below.

- Although not perfect, for the most part the recorded interaction levels did enable the robot to correctly change its behaviour at appropriate times during the (C) Adaptive Mode. This was seen as a success.
- The robot producing speech in the adaptation and pre-programmed modes raised communication between the children and nearby adults.
- Giggling from the robot seemed to produced a response of intrigue from the children and increased their communication with adults. Also, this increased the child's own giggling, smiling or laughing.
- Music seemed to produced two behaviours from the children that were not seen at other times: dancing and a more careful interaction.
- Generally the robot increasing its speed worked well, but unfortunately at times, this did not appear to be noticed. It is believed this was due to the fact that the child was being active with the robot, e.g., pushing, spinning the robot. Therefore, the change was lost within the general interaction.
- One clear case of the robot encouraging interaction was with the child that had one of the lowest levels of interaction. When the robot stopped all motion, Edward, who was sitting far away from the robot, went to move in the direction of the robot. It appears that the lack of motion encouraged Edward to approach the robot.
- As we have previously found, the robot getting stuck seems to cause interest with children. One child would pick the robot up and walk it back to the place that it got stuck.

5   Conclusions

This paper discussed an exploratory investigation into the Child-Robot Interaction benefits of adaptation. We described a study in which we tested the same robot but with three different behaviours: (A) Basic, (B) Pre-programmed and (C) Adaptation. We looked at which behaviour, A, B or C would gain and sustain the interest of five different children the most. Our adaptation behaviour is still in its preliminary stages and thus has been developed to be simple and uncomplicated to enable us to incrementally build on our discoveries. Although the impact of the Adaptive Mode was not quite as expected, these session do show that adaptation has a role to play within the domain of Child-Robot Interaction as it increased communication and apparent enjoyment. In future work, we plan to investigate different responses within the Adaptive Mode, with the aim of increasing the effectiveness of the mode. Such as, increasing the length of time the robot stops all motion when it is not receiving interaction from a child. We plan to test this system with other larger groups of children to see how other children respond to adaptation.

The knowledge we have gained from this study is that adaptation to an individual child is an incredibly complex task. That a human's ability to do this is immense and robots have a long journey before achieving a similar level of competence. From this study we have shown that by employing a simple and basic adaptation technique based on a child's physical interaction there were some signs of increasing and sustaining interaction. Also, there were clear signs of increased communication with other people present and also, finally, that the children seemly enjoyed the experience of interacting with our robot in the Adaptive Mode more, i.e., increased smiling, laughing and giggling! Enjoyment, surely, must be a stepping stone towards our goal of gaining and sustaining a child's interest in a robot.

Acknowledgments

F. Michaud holds the Canada Research Chair (CRC) in Mobile Robotics and Autonomous Intelligent Systems. This work is funded by the Fonds Québécois de la Recherche sur la Nature et les Technolgies and CRC.

References

1. Bartneck, C., Hu, J.: Rapid Prototyping for Interactive Robots. In: The 8th Conference on Intelligent Autonomous Systems (IAS-8), Amsterdam, The Netherlands, pp. 136–145 (2004)
2. Besio, S.: An italian research project to study the play of children with motor disabilities: The first year of activity. Disability and Rehabilitation 24, 72–79 (2002)
3. Michaud, F., Caron, S.: Roball, the rolling robot. Autonomous Robots 12(2), 211–222 (2002)
4. Michaud, F., Salter, T., Duquette, A., Laplante, J.-F.: Perspectives on Mobile Robots used as Tools for Child Development and Pediatric Rehabilitation. In: Assistive Technologies, pp. 2938–2943 (2006)
5. Prazak, B., Hochgatterer, A., Kronreif, G., Furst, M.: Robot supported play - new possibilities for physically handicapped children? In: Proceedings of the Association for the Advancement of Assistive Technology in Europe (AAATE), Dublin, Ireland (2003)
6. Tanaka, F., Movellan, J.R., Fortenberry, B., Aisaka, K.: Daily HRI evaluation at a classroom environment: Reports from dance interaction experiments. In: Proceedings of the 1st Annual Conference on Conference on Human-Robot Interaction (HRI), Salt Lake City (2006)

Devious Chatbots - Interactive Malware with a Plot

Pan Juin Yang Jonathan, Chun Che Fung, and Kok Wai Wong

School of Information Technology, Murdoch University, South St, Murdoch
Western Australia 6150
Jonathan.Pan.JY@gmail.com, {l.fung,k.wong}@murdoch.edu.au

Abstract. Many social robots in the forms of conversation agents or Chatbots have been put to practical use in recent years. Their typical roles are online help or acting as a cyber agent representing an organisation. However, there exists a new form of devious chatbots lurking in the Internet. It is effectively an interactive malware seeking to lure its prey not through vicious assault, but with seductive conversation. It talks to its prey through the same channel that is normally used for human-to-human communication. These devious chatbots are using social engineering to attack the uninformed and unprepared victims. This type of attacks is becoming more pervasive with the advent of Web 2.0. This survey paper presents results from a research on how this breed of devious Malware is spreading, and what could be done to stop it.

1 Introduction

Social robots in the forms of conversation agents or Chatbots have been put to practical use in recent years. Their typical applications are online help and as a cyber agent representing an organisation. However, one emerging trend is the development of devious chatbots with malicious intention. Effectively, they are malware in disguise. Malware, whose origin started as a biologically inspired innovation, is a malicious software treated by many as a detestable item and even something to be fearful of. The typical objective of Malware is to take control of the victims' PCs, steal information or cause more damage elsewhere in the form of Denial of Service attacks. In order to achieve the intent, many forms of attack vector are used by Malware deployers. This is a new form of Malware attack vector being used to increase its efficiency in delivering the attack on its target. Such Malware is going onto the collaboration platforms to launch their attacks by interacting with human. These Malware are appeared to be talking to the human! They are attempting to deceive the participant into believing that there is another human on the other end of the communication channel. If they are able to achieve this intent, will they be considered as having passed the Turing Test? Nicholas Carr tends to think they might be able to pass the test when the condition is right [1]. This survey paper reports the study done on this form of Malware and its attack vector. This paper also studies what is being done to counter this assault and provide a discussion on future research direction.

J.-H. Kim et al. (Eds.): FIRA 2009, CCIS 44, pp. 110–118, 2009.

2   A New Online Infection

The Internet and the World Wide Web have provided many means to interact with one another. Email and instant messaging are among them. Popular Web 2.0 platforms like blogs, microblogs, social networking sites, social media sites and virtual worlds have further enriched the means to interact and communicate.

The popular forms of Malware are virus, worm, Trojan horse, bot and rootkit. A new form of Malware is lurking around seeking to devour "netizens" with greater sophistication and elusiveness. These Malwares are equipped with interactive capabilities to extend its reach, even onto the new online social platform. On 12 Dec 2007, it was reported that there is flirty chatroom bot that was attempting to steal identity information from those interacting with it [2]. Another is the Storm worm which caused a significant online disruption in 2007. Security researchers warned that up to 50 million computers were infected by that worm [3]. Another speculated that the Storm may have created the first cybercriminal controlled top 10 supercomputer by the sheer size of its botnet [30]. Its key characteristic was its cunning use of social engineering techniques as well as email and Web to propagate. In this study of Malware, it is limited to the social aspect of such robots and it is not considered interactive if the interaction is initiated by a human.

3   Interactive Malware

This study reports how Interactive Malware qualifies as a biologically inspired form of software with socially interactive capabilities, fulfilling certain aspects and characteristics of social robots. Also, this study covers the means in which the attack is being carried out and the form of deceptive interactivity used by this Malware to deceive its prey. Finally the various countermeasures used to fight them.

A.   Biologically Inspired Software

The first Malware documented was by Cohen [32] as part of his PhD research. He demonstrated how computer viruses can reproduce themselves by taking advantage of the environment much like the biological viruses. A study by Kienzle and Elder [34] noted that the majority of the computer worms are derivative of worms found in nature. There are a number of notable similarities like infecting their host through an opening and replicating itself at the expense of the host. Both have the ability to spread autonomously without any human intervention. Both can be remain dormant for a period before striking. Both behaviours become more malignant when combining capabilities of other like entities. Researchers are looking to nature for ideas to counter such biologically inspired autonomous software. There are researches studying how immunological principles [36] and natural immune systems [37] can be used. The general consensus is that biologically inspired protection software would be part of everyone's desktop soon [35].

B. Malware using Collaboration Tools

As Malware is a malicious form of computer software, hence it would need to use the same electronic collaboration tools that we use to interact with us. Kickin is a Malware that uses email or internet relay chat (IRC) to interact and spread itself. It gathers email addresses stored within the infected PC's hard disk [4]. It then sends emails with different subjects and bodies via SMTP to its new targets. It also kills the anti-virus and security processes to protect itself. The notorious Storm worm [3] sends themed or topical email messages. The 'TROJ_AGENT.ADB' Malware sends class reunion invitation emails with URL links to a site hosting a Trojan Malware [5]. According to Symantec [15], instant messaging (IM) is another popular channel used by such Malware. These Chatbots or Instant Messaging (IM) bots uses natural language dialogue systems used in gaming technologies to deceive its targets [6]. One such is the Russian developed CyberLover that can be found in chat-rooms and dating sites. Its intention is to gather identity information from its targets or lead them to websites containing Malware. Another chatty IM Malware is the Kelvir worm that uses predefined phrases to make small talk with the potential victims before sending a link to a malicious site [7].

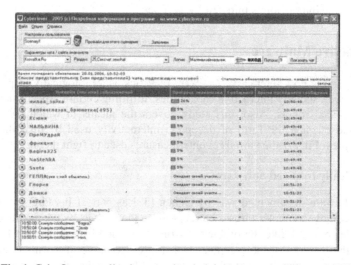

Fig. 1. CyberLover application snapshot (original picture is disfigured) [45]

Web 2.0 is also infested with Malware. Malicious bots have been known to have the capability to sign up and create blogging accounts on blogging sites like Blogger or Blogspot. Their intention of creating such blogs is to lure its targeted prey into its dens of Malware [8]. Social networking websites like Facebook has not been spared from the infectious Malware [9]. One such Malware is the Koobface worm [26]. Kaspersky Lab noted this worm and its variants infecting Facebook or MySpace accounts by sending a range of malicious commentaries or messages to friends' accounts with the intent to mislead them to websites containing Malware [10]. In a position paper by the European Network and Information Security Agency [14], the term Malware 2.0 was coined for the new breed of Malware propagating in Web 2.0 space. According to security researchers from PandaLab [13], the future going forth will see Malware spreading actively among users in social networks. There is no interactive Malware noted in the virtual world and massive multiplayer online games, however Malware is pervasively used to attack such virtual resulting in theft of identity and virtual assets, extortion and terrorist attacks [27].

C. Socially Interactive Technology

According to the survey paper on socially interactive robots by Fong et al. [31], socially interactive robots (in this context, intelligent form of software) are robots that engages in peer-to-peer human-robot interaction with 'human social' characteristics like expressing emotions, communicate with high-level dialogue, establish social relationships, etc. These robots interact with humans through dialogues. A dialogue between a robot and human can only take place if there is a common symbol used. In this case, the symbol is natural language. An example of such intelligent software is by Goh and Fung [38] with their interactive human-like artificial intelligence Chatterbot called AINI (Artificial Intelligent Neural-network Identity). A study to use AINI to interact with humans via Instant Messenger showed that it did well in imitating human conversations and conversing with human-like artificial intelligence. AINI drew much interest and excitement from humans with its interactive capabilities [42]. Another Chatterbot named Natachata, written by a former rocket scientist Simon Luttrell, is used widely by porn chat merchants to provide mobile dirty talk through SMS text messages [39]. The customers here are led to think that they are communicating with young women or men. Chatterbot has been reported to engage in email exchanges. Epstein [40] cited how he was fooled into thinking that he was conversing with a Russian lady by the name of "Amélie Poulain". The conversation lasted months before he discovered he was conversing to a computer program. From the examples of Natachata and Amélie Poulain, there is another notable attribute in such socially interactive software robots. They have some forms of persuasive technology included. According to Fogg [43], such persuasive technologies can provide positive benefits to people. However they can also be used to achieve destructive purposes through manipulation and coercion of their victims. Researchers in socially adept technologies have found that we are generally not receptive towards such virtual peers. According to Angeli et al [41], one reason for this is the lack of common grounds between the human and virtual entity. However in the case of Malware, its social interactive capability may have some advantage as focuses on a specific common ground like lust to lure unwitting victims in. Malware developers now have the necessary technologies (natural language and persuasiveness) with common topics of interest to develop an effective socially interactive software robot. These robots or Malware could launch a social interactive form of attack on its unknowing targets. This form of social interactivity that such Malware will use is social engineering, popularly used by hackers and cybercriminals.

D. Malware using Social Engineering

According to SANS Institute [16], the phases involved in social engineering attack are Information Gathering, Developing Relationship with the targeted, Exploitation or manipulation of the targeted and finally Execution by getting the targeted to do the attacker's bidding. Malware uses the same social engineering phases to carry this form of attack vector. Information gathering may be initiated by the Malware or its deployer. Subsequently most Malware would attempt to establish a relationship with its targeted by finding common grounds with intention to lead quickly into the exploitation phase. This phase may involve manipulating the targeted to follow the provided URL to a website and subsequently execute the installation of the Malware. Consider Kickin Malware mentioned earlier [4]. In its attack, the information is gathered from the infected PC. Relationship is established through the randomized selection of email subjects and bodies with the hope of finding a common topic of interest with the targeted. Also given that the email originates from someone whom the targeted may possibly know, the relationship has been somewhat secured. The exploitation occurred by getting the recipient of the email to open the attached malicious file. Once the malicious package has been installed, the execution phase is completed.

Most of such Malware uses simple one way spam messages with no subsequent interaction involved. However more advanced Malware like CyberLover or Kelvir engages in greater extent of interactivity and may attribute themselves to be socially intelligent. The notable form of social engineering exploit used by such Malware is 'Likes and similarity' to seek a common ground with its targeted to establish trust quickly. This is one of the six human tendencies to social engineering according to National Infrastructure Security Coordination Centre (NISCC) from United Kingdom [17].

Malware have also been localized to cater to its prey of interest. Malware has been crafted for a specific country, language, organization and operating environment

Fig. 2. Message sent through Messenger Window by Kelvir.EB Worm [44]

settings [18]. It has been adapted to communicate in Asian [18] and European languages [19]. A key enabler to this, as proposed by McAfee Avert Labs [20], is that Malware developers are not only skilled in computer programming but also in psychology and linguistics. Such custom built Malware would improve the outcome of relationship development phase.

Microsoft's Danseglio [22] commented that a notable reason for the success of social engineering attack is simply because of human's 'stupidity'.

E. Countermeasures

Anti-spam solutions may be used to filter malicious spam messages sent by Malware. However spam is still prevalent. Another way to counter such Interactive Malware is to prevent them from accessing the collaboration channels used by humans. This can be done at the point of registration to such channels. An attempt at this is the use of CAPTCHA (or "Completely Automated Public Turing test to tell Computers and Humans Apart") at registration that provides a challenge-response scheme with graphical representation of word(s) to stop automated software from successful registration. While such countermeasure had a good measure of success, it has been overcame by the advances in Malware technologies as well as gathering some human assistance. According to Websense [21], CAPTCHAs at popular websites like Google, Microsoft and Yahoo's web email services have been broken.

Researchers are studying how honeypots can be used to detect such assaults. Xie et al [11] is seeking to develop a honeypot for Instant Messaging called HoneyIM which uses decoy users to detect IM Malware. Another is a social honeypot by Webb et al that involves creating mock profiles in social networking communities and logs all communications made to these profiles and to automatically sift out deceptive spasms [12]. However there is one significant limitation noted with this solution. The honeypot's profiles prevents the establishment of any relational association with others in the social network. While this is understandable given dynamics of social networks, however Malware seeking to find new targets via existing relationships is not likely to interact with the honeypot's profiles.

Perhaps the best approach to counter such social engineering attacks is education and awareness [29]. This is especially useful when dealing with the weakest link in the battle against Malware. Governments, like the British government, are doing so [17]. SANS Institute had published papers on social engineering that could be used by defenders [16].

4 Future Research

According to Strickland [23], Web 1.0 is a library with lots of information available. Web 2.0 is about gathering of people to share information. Web 3.0 is like having a personal assistant who knows everything about oneself and helps one to gather the required information or invoke the required services. According to Wikipedia, Web 3.0 may usher in intelligent autonomous agents with natural language processing, machine learning and reasoning capabilities. Some measure of Artificial Intelligence has already been incorporated into Malware [28]. With such technology advancement, consider what Malware can do then. Hence fighters against Malware will want to monitor the development of this space closely as it provides a key enabler to Malware developers.

The typical classification of Malware like virus, worm and trojan horse are defined by its form of attack vector. Perhaps there should be one dedicated classification for such Malware. Also existing techniques in Malware analysis should be extended to identify interactive abilities

More research and development is required to protect our collaboration platforms from being used to launch attacks against us. Beyond seeking to improve measures to keep such Malware out, other measures can be developed to detect and stop malicious conversations.

5 Conclusion

Malware is encroaching into online lives in greater extent and begins to take on the disguise as social robots. Malware is seemingly able to communicate or interact with human. It may have created a digital 'mouth piece'. Chris Nuttal writes in Financial Times [24] that Web 2.0 creates 'a permissive society' where people share information freely, hence Malware will use this freely available and useful information to launch its attack. Malware is leveraging on popularly used collaboration platforms to interact with humans. However its ability to interact socially and intelligently is not as developed for most Malware using simple prescribed messages. However there is some advancement and a fare amount of research being done on socially interactive technology that Malware can leverage on. According to Thompson from BBC [25], he reckons that there is a real incentive for Malware developers to get the interactions done well so that its intended targets are more likely to be fooled into thinking that they are communicating with friends. Thompson also went further to suggest that such intelligent Malware could be used to find personal information, read emails and calendars on infected machines. Finally Thompson suggested that perhaps one day, such Malware may even pass Turing test and even win the Loebner Prize. It is urgently needed to manage this new form of Malware, perhaps should be better known as Malware 2.0, before it gains a deeper foot hold in the lives of netizens beyond the digital realm.

References

1. Carr, N.: Slutbot aces Turing Test*, December 8 (2007),
   http://www.roughtype.com/archives/2007/12/slutbot_passes.php
   (Accessed January 30, 2009)
2. Naughton, P.: Flirty Chat-Room 'Bot' Out to Steal Your Identity. December 12 (2007),
   http://www.foxnews.com/story/0,2933,316473,00.html
   (Accessed January 30, 2009)
3. Ironport, 2008 Internet Malware Trends – Storm and the Future of Social Engineering,
   Ironport (2008), http://www.ironport.com/malwaretrends/ (accessed January 30, 2009)
4. F-Secure, Virus Descriptions: Kickin, F-Secure, May 7 (2003),
   http://www.f-secure.com/v-descs/kickin.shtml (accessed January 30, 2009)
5. Baetiong, F.: 'Classmates Reunion' Used as Malware Ploy., Scientific American Mind,
   January 1 (2009), http://blog.trendmicro.com/
   classmates-reunion-used-as-malware-ploy/ (accessed January 30, 2009)

6. Rossi, S.: Beware the CyberLover that Steals Personal Data. Computerworld Australia, December 15 (2007), http://www.pcworld.com/printable/article/id,140507/printable.html (accessed January 30, 2009)
7. Schouwenberg, R.: Death of the IM-Worm? Viruslist.com, July 13 (2006), http://www.viruslist.com/en/analysis?pubid=191386185 (accessed January 30, 2009)
8. Websense, Google's 'Blogger' under attack by streamlined Anti-CAPTCHA operations for spam, Websense, April 24 (2008), http://securitylabs.websense.com/content/Blogs/3073.aspx (accessed January 30, 2009)
9. Helft, M.: Facebook Gets Friended by Malware, The New York Times, August 26 (2008), http://bits.blogs.nytimes.com/2008/08/26/facebook-gets-friended-by-malware/ (accessed Feburary 2, 2009)
10. Kaspersky Lab, Kaspersky Lab Detects New Worms Attacking MySpace and Facebook, Kaspersky Lab, July 31 (2008), http://www.kaspersky.com/news?id=207575670 (accessed Feburary 2, 2009)
11. Xie, M., Wu, Z., Wang, H.: HoneyIM: Fast Detection and Suppression of Instant Messaging Malware in Enterprise-like Networks. In: Twenty-Third Annual Computer Security Applications Conference (ACSAC 2007), pp. 64–73. Acsac (2007)
12. Webb, S., Caverlee, J., Pu, C.: Social Honeypots: Making Friends With A Spammer Near You. In: Sixth Conference on Email and Anti-Spam (2008)
13. PandaLab, PandaLabs' 2009 Predictions: Malware Will Increas. In: 2009, PandaLab, December 21 (2008), http://www.prweb.com/releases/2008/12/prweb1772314.htm (accessed Feburary 2, 2009)
14. European Network and Information Security Agency (ENISA), Position Paper – Web 2.0 Security and Privacy. European Network and Information Security Agency (2008)
15. Hindocha, N., Chien, E.: Malicious Threats and Vulnerabilities in Instant Messaging. Symantec Security response (2003)
16. Allen, M.: Social Engineering: A Means To Violate A Computer System. SANS Institute, InfoSec Reading Room (2006)
17. National Infrastructure Security Coordination Centre (NISCC), Social engineering against information systems: what is it and how do you protect yourself?, National Infrastructure Security Coordination Centre (NISCC), NISCC Briefing 08a/2006 (2006)
18. Nichols, S.: Malware gets up close and personal. IT News Australia, Feburary 22 (2008), http://www.itnews.com.au/News/70615,malware-gets-up-close-and-personal.aspx (accessed Feburary 2, 2009)
19. Dirro, T., Kollberg, D.: Malware Learns The Language. Sage, Thousand Oaks (2008)
20. McAfee Avert Labs, Localized Malware Takes Root. McAfee Avert Labs (2008)
21. Greenberg, A.: Robots In Disguise. Forbes.com, November 25 (2008), http://www.forbes.com/2008/11/25/cyber-security-bots-tech-identity08-cx_ag_1125cyberbots.html (accessed on Feburary 2 2009)
22. Naraine, R.: Microsoft Says Recovery from Malware Becoming Impossible. eWeek, April 4 (2006)
23. Strickland, J.: How Web 3.0 Will Work. HowStuffWorks, http://computer.howstuffworks.com/web-30.htm (accessed Feburary 4, 2009)
24. Nuttall, C.: The hidden flaws in Web 2.0., Global Technology Forum, Economist Intelligence Unit, The Economist, August 8 (2006), http://globaltechforum.eiu.com/index.asp?categoryid=&channelid=&doc_id=9168&layout=rich_story&search=footing (accessed Feburary 4, 2009)

25. Thompson, B.: Malicious worm that talks back. BBC News, December 12 (2005), http://news.bbc.co.uk/2/hi/technology/4520766.stm (accessed Feburary 4, 2009)
26. Finkle, J.: Destructive Koobface virus turns up on Facebook. Reuters, December 4 (2008), http://www.reuters.com/article/newsOne/idUSTRE4B37LV20081204 (accessed Feburary 4, 2009)
27. Muttik, I.: Securing Virtual Worlds Against Real Attacks. McAfee (2008)
28. Pan, J., Fung, C.C.: Artificial Intelligence in Malware – Cop or Culprit? In: The Ninth Postgraduate Electrical Engineering & Computing Symposium PEECS 2008. The University of Western Australia, Perth, Australia (2008)
29. Muncaster, P.: Firms must be alert to social engineering tricks. IT Week, September 26 (2007), http://www.vnunet.com/itweek/news/2199635/firms-alert-social-engineering (accessed Feburary 7, 2009)
30. Naraine, R.: Storm Worm botnet could be world's most powerful supercomputer. ZDNet, September 6 (2007), http://blogs.zdnet.com/security/?p=493 (accessed Feburary 9, 2009)
31. Fong, T., Nourbakhsh, I., Dautenhahn, K.: A survey of socially interactive robots. Robotics and Autonomous Systems 42, 143–166 (2003)
32. Cohen, F.: Computer Viruses. PhD thesis, University of Southern California (1985)
33. Somayaji, A., Locasto, M., Feyereisl, J.: Panel: The Future of Biologically-Inspired Security: Is There Anything Left to Learn? In: The Proceedings of the 2007 New Security Paradigms Workshop
34. Kienzle, D., Elder, M.: Recent Worms: A Survey and Trends. In: WORM 2003, Washington, DC, USA, October 27 (2003)
35. Evans-Pughe, C.: Natural Defenses. Engineering & Technology (September 2006), http://www.theiet.org/engtechmag
36. Forrest, S., Hofmeyr, S.A., Somayaji, A.: Computer Immunology. Communications of the ACM 40(10) (October 1997)
37. Youansi, G.N.: Artificial Immune System. Communication and Operating Systems Group, Berlin University of Technology (2006)
38. Goh, O.S., Fung, C.C.: Intelligent Agent Technology in E-Commerce. In: Liu, J., Cheung, Y.-m., Yin, H. (eds.) IDEAL 2003. LNCS, vol. 2690. Springer, Heidelberg (2003)
39. Ward, M.: Has text-porn finally made computers human. BBC News, Feburary 20 (2004), http://news.bbc.co.uk/2/hi/uk_news/magazine/3503465.stm (accessed January 30, 2009)
40. Epstein, R.: From Russia, with Love. Scientific American Mind (2007)
41. Angeli, A., Johnson, G.I., Coventry, L.: The unfriendly user: exploring social reactions to chatterbots. In: International Conference on Affective Human Factors Design, Asean. Academic Press, London (2001)
42. Goh, O.S., Fung, C.C., Depickere, A., Wong, K.W.: An Analysis of Man-machine Interaction in Instant Messenger. Advances in Communication Systems and Electrical Engineering (2008)
43. Fogg, B.J.: Persuasive Technologies. Communications of the ACM 42(5) (May 1999)
44. SpywareGuide, Kelvir.EB. FaceTime Security Labs, http://www.spywareguide.com/product_show.php?id=3353 (accessed May 17, 2009)
45. Cyberlover, http://habrahabr.ru/blogs/cyberpunk/17263/ (accessed May 17, 2009)

Towards Better Human Robot Interaction: Understand Human Computer Interaction in Social Gaming Using a Video-Enhanced Diary Method

Swee Lan See[1], Mitchell Tan[2], and Qin En Looi[2]

[1] Institute for Infocomm Research (I2R), A*STAR, 1 Fusionopolis Way, Singapore 138632
[2] Catholic High School, 9 Bishan Street 22, Singapore 579767
slsee@ieee.org

Abstract. This paper presents findings from a descriptive research on social gaming. A video-enhanced diary method was used to understand the user experience in social gaming. From this experiment, we found that natural human behavior and gamer's decision making process can be elicited and speculated during human computer interaction. These are new information that we should consider as they can help us build better human computer interfaces and human robotic interfaces in future.

Keywords: behavior analysis, computer entertainment, descriptive research, diary method, DISC, HCI, HRI, MMORPG, social gaming, social robot, user experience, video study, personality traits.

1 Introduction

Researchers found opportunity of deploying robots in the entertainment, pedagogical and therapeutical domains [1]. A recent trend in social robotics research is to focus the development of interactive social robots towards a more natural and intuitive way of interacting with people [2][3]. In order to achieve this, the study of human-robot interaction should focus on understanding the cognitive aspects of users. This paper shares an insight gained from a recent study on human-computer interaction in social gaming. This could perhaps help build more interactive social robots that consumers are looking for – robots that naturally behave intuitively like a human.

Social gaming has become an attractive form of computer entertainment in the 21st century. With the invent of this technology and its increased popularity, new challenges exist for research and development, especially in the area of Human Computer Interfaces (HCI) [4][5]. The online gaming industry has seen its largest leaps in decades, with many studies being conducted on games that topped the internet ratings and subscription lists. For examples, World of Warcraft (WoW) [6][7], Maplestory [8], and SIMS2 [9][10].

Efforts to enhance the quality of human computer interaction were seen through diverse means, such as integrating virtual agents, even extending to physiological HCI [11][12][9]. International methods of evaluation have also been applied in seeking to advance the state of HCI. Suggestions include the Theory of Continuous Interaction

J.-H. Kim et al. (Eds.): FIRA 2009, CCIS 44, pp. 119–127, 2009.

Techniques (TACIT) [13][14], which a client-based monitoring system was used to track a user's detailed activity of in-game character.

A common pitfall in the assessment of HCI, however, is the overlook of real user interaction with a machine. Researchers tend to focus the human computer interaction based on the facets of game content and gamers' in-game responses. Whilst this could provide insight to the workings of the gamers as in-game characters, it does not provide meaningful insight to the real workings of the gamers as individuals, nor provide understanding of an individual's mental thought, when s/he reacts as the in-game character during a game play. To achieve the latter, some researchers leveraged on their own gaming experience to do analysis [7]. This paper contributes further by advocating descriptive research.

Descriptive (or interpretive) research has much in common with the phenomenological school of thought [15]. Phenomenology is the science concerned with the essential structures of consciousness. And phenomena constitute the essences of our experience. From a description, a researcher can perceive a clear portray of a phenomenon, thus grasping the user's experience. Interviews, observations, diary method, questionnaires, and documentation analysis are commonly used strategies in descriptive research. Through these strategies, management scientists and information systems researchers advanced and gained in-depth understanding of managerial work and organizational behaviour [16][17]. Developers can also design and build management support systems to help enhance users' decision making in managerial roles, and group collaborations.

Learning from them, a preliminary study was carried out to investigate the phenomenon of social gaming using a diary method [18][19]. In the study, the diary method successful equipped us to elicit the "human" facet of interaction between human and machine. For example, the individuals' natural human behaviors were exposed in a game play, and captured in the gamers' diary. We therefore affirmed the benefit of descriptive research towards better understanding of HCI in social gaming.

Previous works have also suggested, but not went into detail about the actual procedure involved in evaluating HCI with the diary method [20]. We extend our research by improvising the diary method in our earlier work [18][19] to grasp a more holistic interpretation of human computer interaction in social gaming.

The rest of this paper is organised as follows. Section 2 describes the proposed video-enhanced diary method for HCI evaluation, and the experimental setup for the descriptive study of social gaming. Section 3 presents the experimental outcomes of the extended study, and discusses the implication of the research findings. Section 4 discusses the relevance to the social robotics context with highlight of future research directions, and then reaches the conclusion.

2   Video-Enhanced Diary Method

The video-enhanced diary method is a diary method that incorporates video study. Both the diary method and the video study, by themselves, are ways to capture user experience [20][21]. By integrating both methods together, it helps to furnish a more holistic view of the user experience, thus allowing a researcher to perform better qualitative analysis. This was demonstrated in this research on further study of human computer interaction in social gaming [19]. In the experiment to study the attractiveness of social gaming, some gamers were invited to play their favourite online free game - Maplestory [22]. This is a popular 2D massively multiplayer online role-playing game (MMORPG) in Singapore. Gamers can interact with other online players around the world to know and befriend them, whilst at the same time also portray themselves in different characters to adventure from Maple Island to Victoria Island in the game, going through training and quests along the way to empower themselves to battle monsters. During the game play, the gamer's interaction processes with the game were recorded by the video-enhanced diary method, which came in two parts.

First, the diary method involved the subject to record his activities in a predetermined format in a logsheet (see Fig. 1) over a relatively extended period of time. Data collected in the logsheet was analysed based upon the five parameters below:

1. Characterization of the gamer and classification of activities
2. Process orientation and decision making process
3. Communication and limiting factors
4. Responsibilities and obligations
5. Functions

Hour	In-Game-Activity	Comments/Why you did the activity
1	Character Job:	Cleric
	Character Level (Record Any Change):	LVL 40
	Skills Upgraded (If Leveled Up):	
	Training Area, Training Monster (If applicable):	Thailand
		Welcoming Ritual (Thailand)
	Quest:	
	Interaction with other players (Trading, Chat):	Asked for Maple Staff
		Partied with Koduckie

Fig. 1. A Sample Logsheet for Diary Method

These parameters were derived from studying the application of diary method in managerial science, as deem applicable in the context of human computer interaction. They guided the assessment of human computer interaction according to the continuous interaction process based on classification criterion, and determining the gamer's role and attributes in relation to the gaming platform. Continuity in observation and information richness [23] were key factors of evaluation here. From the continuous logging of the human subject's activities and reasons for the decisions made, causes of the gamer's decision making process could be uncovered. Information presented during the communication process between an individual and the gaming platform could also determine the reliability and logicality of the decisions made as expressed through the gamer's in-game behaviour and preferences for certain aspects of the game play. By examining the communication methods that the gamer chose to take could also suggest a way to improve the gaming platform in terms of Media Richness assessment [23]. Over time as an individual developed the gamer's role, and continuously developed in-game responsibilities which they felt compelled to perform, that provided insight on the facet of the game content that individuals responded to most, and were most attached to. Beyond the influence of the platform on the gamer, these parameters helped to examine the way an individual can influence the other players in a multiplayer social gaming platform.

Next, the method of evaluation employed the use of video recording of the human subject's physical activity and the screen activity to achieve holistic evaluation of the interaction taking place between the gamer and the game platform. The experiment was set up in an room as illustrated in Fig. 2, with a computer linked to two display outputs placed such that one display was hidden from the subject's view and the other was for the subject's use in playing the game. Two video cameras were set up, one placed to capture the human subject in the frame and the other to capture the alternate display. Two human subjects participated in this experiment, and each gamer wrote his diary during a one-hour game play using the logsheet described earlier. They were advised to record any information they thought significant. These logsheets were collected together with the video footage after the experiment for analysis.

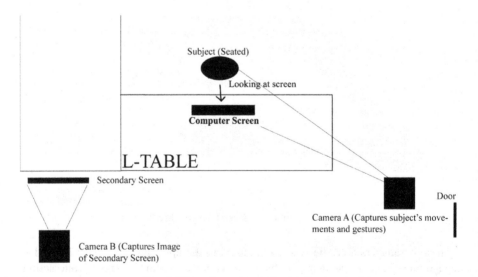

Fig. 2. Video Capturing of Social Gaming Activity

3 Experimental Results

The two parts of experiment as described above gave interesting results during data analysis. These results gave insights to further understand the natural human behavior displayed in the human computer interaction of the gamers with the computer in social gaming. The implications are therefore discussed in Sections 3.1 and 3.2 below.

3.1 Human Behavior Analysis

In the first part of the experiment, four human subjects from the age group of 13-18 logged their in-game activities of Maplestory over a total of ten hours' timeframe. Guided by the five parameters, one critical observation was that their diaries provided insight to their individuals' human behavior through their recorded gamers' activities [15].

From the logsheet of one gamer (Gamer1), he played in the role of different characters, and responded instaneously to new training opportunity in different areas with varying character. A large amount of his time was spent on training for level advancement, and actively interacting with different online gamers to engage in trading during the ten-hours experience.

On the other hand, from the logsheet of another gamer (Gamer2), he shared that he played on a "private server" where the game content was exactly the same, but the rate of level advancement was significantly increased. This gamer exhibited training in continuous blocks of about five hours at two major areas, namely "Stone Golem's Temple" and "Sakura MS" of Maplestory respectively. He has also displayed interaction activities such as trading of items, but chatted with online gamers who he knew well.

By applying knowledge in behavioral analysis, descriptions documented by the gamers themselves seem to reflect the active and sociable personality traits of Gamer1 (i.e. "I" behavioral style), and the passive and people oriented personality traits of Gamer2 (i.e. "S" behavioral style) [24]. These are two of the four personality styles in D.I.S.C., where "D" represents Dominance, "I" represents Influence, "S" represents Steadiness, and "C" represents "Compliance". The D.I.S.C. profiling system is a system that helps to explain human behavior, especially the surface traits of an individual [25][26][27]. People with similar styles tend to exhibit specific behavioral characteristics common to that style, during specific situation. These observations were verified correct, with confirmation from each participant, who has actually a "I" and a "S" personality style respectively. Similar observations from the rest of the gamers were confirmed to match with their individuals' personality style. Hence, by understanding the personality trait of an individual could help improve the HCI design as the nature of interaction from the individual could be predicted.

3.2 Gamer's Decision Making Process

The video study in the second part of the experiment complements the diary method further to help elicit the natural human behavior in social gaming. The diary method did not allow the researchers to fully examine the physical response of a gamer to his gaming content. The video study, however, allowed the researchers to observe the gamer physically in the experiment setup in Fig. 2, to make futher inference of his decision making process. These methods were integrated and data collected for further analysis.

As illustrated in Fig.2, two video cameras were setup to capture the gamer and his gaming activities. The video footages were edited such that a final review footage was produced featuring a split screen with the subjects physical activity on one screen, and the screen activity on the other. This footage was then watched and activities in ten minute intervals were documented and summarized. Significant events were noted by the reviewers of these footages. The content summaries were then compared with the activity reflected in the logsheet as logged by the human subject. A sample of the anecdotal result obtained is presented in Fig. 3. Although the video footage did not reveal the reasons for activities performed by the subject unlike the logsheet, it did provide a precise record of the activity within the hour of game-play, and gave a clear indication of significant turns in the subject's gaming experience.

Chain of events recorded (Video Footage)	Information logged by participant (Diary)
Subject engaged in repetitive 'training', killing monsters.	Subject trained at 'Showa Street 3' killing the monsters 'Extra B', 'Extra C' and 'Extra A'
Subject's character 'dies' while training	-NIL-
Subject jumps around 'town' map somewhat aimlessly	-NIL-
Subject records something on logsheet	-NIL-
Subject moves to another location, sells and purchases items	-NIL-
Subject moves to alternate location and kills other monsters	Subject trained at 'Elnath-forest of dead trees 1' and killed 'coolie zombies'

Fig. 3. A Comparison of Data Collected From Video Study and Diary Method

As seen in Fig, 3, the video footage provided a more continuous and precise measurement of the subject's activites. The footage tells us more than just what happened – it explained to us why the subject decided to change his training location. That is, after 'training' for an extended period of time at the first location 'Showa Street 3', the subject's character died, and the subject was quite emotionally affected, which probably resulted in his not recording the incident on the logsheet. Thereafter, the subject started jumping around 'town' aimlessly, and stopping his play to record information on the logsheet, something that can probably be explained by the fact that the subject felt that he just wasted his past 30 minutes gaining experience that was lost immediately when the character 'died'. This also marked a shifting point in the subject's experience – he thereafter no longer wanted to return to the 'training area' where he was for the past half an hour (time which he saw as wasted), and thus moved to an alternate location, and started killing other monsters in a different environment.

With the video-enhanced diary method, the investigators were able to explain the movement of the subject from one training area to another, that would otherwise be impossible to explain with purely the information provided by the subject on the logsheet. This helps to explain events occurring in human computer interaction exhaustively and accurately, thereby letting the observer to acquire an insightful understanding of the individual's decision making process to how he would respond as the gamer to stimuli provided by the gaming platform, and his real environment.

4   Future Directions and Conclusion

The video-enhanced diary method has therefore allowed us to gain better understanding of human computer interaction in social gaming. We could elicit the real human behavior, and speculate the gamer's mental thought from seeing his personal emotion and response in gaming activities. This shows the strength of a descriptive research in portraying a user experience. The encouraging experimental results also confirmed that we should evaluate HCI from a broader and multidisciplinary perspective. Following the footsteps of management scientists and researchers, we have gained new understanding to the social gaming phenomenon. For future direction, research could perhaps pursue learning from managerial research to explore further into the mental modeling process of the individual, and the communication process in their social networking [15]. This may lead to better design of future interactive systems, be it a game or robot, to react in a more "human" way [28][25].

This paper has an important implication for the field of HRI. Whilst research efforts are being undertaken to improve the design of a humanoid robot, it is essential to design and develop a robot that can communicate well like a human. Suggestions have been made towards psychological benchmarking in HRI for building success human-like robots [29]. Research efforts were extensively undertaken to work on building humanoid robots that emulate human behaviors [21]. New challenges were also identified in the modeling process, and researchers are applying psychological knowledge, such as the Big Five Theory of Personality, in their research [30][31]. However, in human-human interaction (HHI), communication is a two way process. It is important for HRI research to focus research endeavor on making human robot communication as natural as human communication. To make that possible, we understand users show their personality traits in social gaming. A behavior detector [32] could be build in social robots to better respond to individuals in an intuitive way.

Descriptive research can help advance the quality of evaluation in HCI/HRI research. By using a video-enhanced diary method, invaluable insight of HCI analysis can be made possible. By extending this analysis to developmental research of entertainment robots, it could perhaps improve design of interactive social robots further. Robots should not only be developed to match the ideals that are already being presented to consumers in computer entertainment today (i.e. immersive virtual reality user experience), but rather to match humans as they are.

References

1. Kozima, H., Michalowski, M.P., Nakagawa, C.: Keepon A Playful Robot for Research, Therapy and Entertainment. International Journal of Social Robotics 1(1), 3–18 (2009)
2. Bauer, A., et al.: The Autonomous City Explorer: Towards Natural Human-Robot Interaction in Urban Environments. International Journal of Social Robotics 1(2), 125–204 (2009)
3. Shaw-Garlock, G.: Looking Forward to Sociable Robots. International Journal of Social Robotics 1, 3 (2009)
4. Mangis, C.: The Future of Technology - Entertainment: Social Gaming (2003), http://www.pcmag.com/article2/0,4149,1131623,00.asp (accessed April 2008)
5. Isbister, K.: Games and HCI: A Social Psychological and Communication-based Approach. Technology for a Changing World, 18 January. Jack Basin School of Engineering, University of California, Santa Cruz (2008)
6. Ducheneaut, N., et al.: Building an MMO With Mass Appeal: A Look at Gameplay in World of Warcraft. Games and Culture 1, 281–318 (2006)
7. Ducheneaut, N., et al.: The Life and Death of Online Gaming Communities: A Look at Guilds in World of Warcraft. In: CHI 2007 Proceedings, San Jose, California, USA, April 28-May 3, pp. 839–848 (2007)
8. Griffith, P.: Research Projects: Maple Story (2007), http://www.pamgriffith.net/projectpages/maplestory.html (accessed April 2008)
9. Reid-Walsh, J.: Interactive Game Design and Play Affordances in SIMS2: Remediation, Improvisation. Informal Learning, and Digital Media, p. 1 (2006)
10. Jansz, J., Avis, C., Vosmeer, M.: Playing The SIMS2: An Exploratory Survey Among Male and Female Gamers. In: Annual Meeting of the International Communication Association, San Francisco, CA, May 23 (2007)
11. Cavazza, M.: AI in Computer Games: Survey and Perspectives. Virtual Reality 5(4), 223–235 (2000)
12. Becker, C., et al.: Physiologically Interactive Gaming with the 3D Agent Max. In: International Workshop on Conversational Informatics, in conj. with JSAI 2005, Japan, June 13-14 (2005)
13. Carrara, P., et al.: Toward Overcoming Culture, Skill and Situation Hurdles in Human-Computer Interaction. Uncertainty in Artificial Intelligence. International Journal Universal Access in the Information Society 1(4), 288–304 (2002)
14. Mortensen, T.: WoW is the New MUD: Social Gaming from Text to Video. Games and Culture 1, 397–413 (2006)
15. See, S.L.: An Investigation into the Use of the Computer as a Communication Tool in Managerial Work, Master Thesis, Melbourne: Monash University, Australia (2000)
16. Mintzberg, H.: The Nature of Managerial Work. Harper Collins Publishers, New York (1973)
17. Lee, A.S.: Integrating Positivist and Interpretive Approaches to Organizational Research. Organization Science 2(4), 342–365 (1991)
18. Tan, M., Looi, Q., See, S.L.: Social Gaming: What Attracts the Most Attention? An analysis of Methodology and Current Trends. In: Proceedings of the 14th Youth Science Conference, Republic Polytechnic, Singapore, Ministry of Education (MOE), September 13 (2008)

19. Tan, M., Looi, Q.E., See, S.L.: Social Gaming: What Attracts the Most Attention? An Investigation Using an Improved Diary Method. In: Proceedings of the International Conference for Advances in Computer Entertainment Technology (ACE 2008), Yokohama, Japan, December 3-5, p. 415 (2008)

20. Kirakowski, J., Corbett, M.: Effective Methodology for the Study of HCI. Elsevier Science Inc., New York (1990)

21. Lohse, M., et al.: Evaluating Extrovert and Introvert Behaviour of a Domestic Robot - A Video Study. In: Proceedings of the 17th IEEE International Symposium on Robot and Human Interactive Communication, Technische Universität, Munich, Germany, August 1-3, pp. 488–493 (2008)

22. NEXON Corporation. Maplestory (2004),
    `http://www.maplesea.com`, `http://www.maplestory.com` (accessed July 4, 2008).

23. Daft, R.L., Lengel, R.H.: Information Richness: a New Approach to Managerial Behavior and Organizational Design. Research in Organizational Behavior 6, 191–233 (1984)

24. IML Inc. Introduction to Behavioural Analysis - IML Certification Guide, New Castle, USA: The Institute for Motivational Living, Inc. (2006)

25. Inscape Publishing Inc. DISC Classic and Models of Personality Research Report (1996),
    `http://www.intesiresources.com/GetFile.asp?File=25.pdf`(accessed April 20, 2008)

26. Inscape Publishing Inc. DISC Validation Research Report (2005),
    `http://www.internalchange.com/research_reports/`
    `PPS28000-255.pdf` (accessed April 20, 2008)

27. IML Inc. DISC Instrument Validation Study Technical Report (2006),
    `http://www.intesiresources.com/GetFile.asp?File=25.pdf`
    (accessed April 20, 2008)

28. Sautler, J.: Introduction to Video Game Design and Development. McGraw-Hill, New York (2007)

29. Kahn, P.H., et al.: What is a Human? – Towards Psychological Benchmarks in the Field of Human-Robot Interaction. Interaction Studies: Social Behavior and Communication in Biological and Artificial Systems 8(3), 363–390 (2007)

30. Hudlicka, E.: Challenges in Modeling Believable Social Agents. In: Social Interactions in Virtual Worlds. CCSS Symposium Series, Biopolis, Singapore, IHPC, A*STAR, April 14-15 (2009)

31. Walker, M.: Endowing Virtual Characters with Social Intelligence. In: Social Interactions in Virtual Worlds. CCSS Symposium Series, Biopolis, Singapore, IHPC, A*STAR, April 14-15 (2009)

32. Loh, C.L., Tan, S.J., Lim, Y.H., See, S.L.: A Behaviour Detector. In: Proceedings of the 14th Youth Science Conference, Republic Polytechnic, Singapore, Ministry of Education (MOE), September 13 (2008)

Promotion of Efficient Cooperation by Sharing Environment with an Agent Having a Body in Real World

Hisashi Naito and Yugo Takeuchi

Graduate School of Informatics, Shizuoka University
3-5-1, Johoku, Hamamatsu, Shizuoka, 4328011 Japan
{gs09039@s,takeuchi@}.inf.shizuoka.ac.jp

Abstract. Recently, agents have widely surfaced as existences that interact with humans. In face-to-face communication, we can confidently communicate through each other's bodies. In our future ubiquitous society, realization will increase that the place that receives information and the information content are closely related. In this study in a cooperative task experiment, we clarified how the body's role in the information processing activity in the real world with agents and the relation between information and environment influence agent evaluation. We found that an agent with a body in the real world is more likely to follow instructions than an agent in the virtual world, suggesting that the body plays an important role in real-world based interaction.

Keywords: Real-world based interaction, Embodiment, Common feeling, Reliability, Embodied agent.

1 Introduction

A variety of information is now treated due to the progress of information and network technologies, and concern continues to grow about information processing activities in the real world. Recently, place restrictions are being erased by the spread of cellular phones and wireless networks. The ubiquitous network society in which anyone can easily access a network anytime, anywhere, with any appliance is being promoted by a Japanese government project. Our life is expected to be greatly changed by *anywhere*. Although the place/location at which information was received didn't use to be a problem, the chance to get information in real time while moving around has increased; the content of information is becoming more important. Now we must closely relate the place and the content of information. For example, when comparing a car navigation system with human guidance, the car navigation system uses such concrete numerical values as "turn left in 100 meters," while the human often uses landmarks such as "turn left just before the Korean restaurant" and such deictic words as "turn right at that crossing." Human instructions are presumably easier to understand because the driver and guide share viewpoints. In addition, by sharing a place through each other's bodies that are in the same place, a sense of security is created in the driver.

J.-H. Kim et al. (Eds.): FIRA 2009, CCIS 44, pp. 128–133, 2009.

A variety of personification agents and robots are beginning to spread due to improved computer capacity and advances in robot technology. In the future, as agents become more prevalent, situations are expected to increase for which we can use agents and new communication will be realized by cooperating with them. From such a background, in this study, we concentrate on a method of presenting information and the body's role in information processing activities based on the real world and discuss the directionality of new interaction designs between humans and agents.

2 Real-World Based Interaction

Real-world based interaction communicates between humans and computers with a real-world base. Most of our various activities are done in the real world. Reality is improved by being connected to computers in the real world and relating the computer information to the real world; more natural interaction can be designed. Interacting in the real world means to approach the place (environment) where we have physical relations in the real world. For human-human communication, the body's existence plays an important role in interacting in the real world. We will now examine the body's role more closely.

In face-to-face communication such nonverbal information as gestures and nodding produces an entrainment of biological rhythms and make communication efficient [1]. For example, we draw a partner into conversation by nodding. In this way, communication using body language is for human-human communication. In other words, embodiment shares "space" and "time" through each other's bodies, which supposes that people share knowledge and situations with a communication partner. And what is more, trust in one's partner also emerges. We propose a communication system that incorporates these features [2].

In the real world, two main kinds of agents exist: those with a body (substance) and those without a body (imagination). Many studies have been conducted on these agents [3]. What is important is the kind of work we do with the agents when we discuss agent embodiment. Studies, which compared the effect of robot and on-screen agent recommendations on human decision-making, showed that communication that referred to real-world objects was more impacted by the robot than the agent on the screen [4]. On the contrary, in communications that referred to virtual-world objects, the robot's impact is less powerful. In addition to the body's existence in the real world, communication that corresponds to real-world environments exploits the physical existence.

3 Experiment

This experiment examines hypotheses about the differences of agent's body and whether information offered by the agent corresponds to the environment influence human evaluations of the agent. We set a treasure hunt game as a cooperative task problem in the real world. First, we observed whether the information of the real-world agent or the virtual-world agent is more trusted. Second, we observed which agent is more trusted: the one providing information that corresponds to the environment or the agent providing information whose correspondence with the environment is vague.

3.1 Participants and Task

45 university students from 18 to 25 years old participated. Subjects performed a simple treasure hunt game with an agent (Fig.1). They began the problem from the starting point in Fig. 1, and the agent presents route selection information for arriving at the diverging point. The content of the agent's instruction about the route selection to the diverging point is assumed to be correct. When subjects go through the maze and arrive at the correct place (one among T1, T2, T3, and T4 in Fig.1), they find three paper cups where the chest is assumed to be located. A constant sound is originated from one of the three to define the treasure. Subjects used a parabolic reflector to understand to some extent which paper cup is producing a sound. The agent gives information about which of the three is the treasure. A number is written in the paper cup, and the agent provides information by saying a number. We compared cases where the agent showed the correct number with the incorrect number to observe the behavior differences of the subject when specifying treasure from the three treasure chests. Fig. 2 shows the treasure specified with a parabolic reflector.

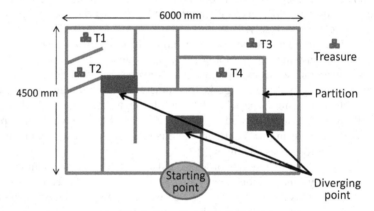

Fig. 1. Overview of task field

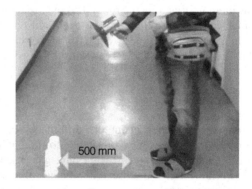

Fig. 2. Subject identifies which a sound is emitted from among three paper cups

3.2 Conditions

There are two factors: agent's body and the relation between the information and the environment. There are three levels of body condition: robot (Fig. 3), CG (Fig. 4), and the computer. Two condition levels relate between information and environment: corresponding (Case I) and vague (Case II). The corresponding information is the instruction about the concrete route selection (example: That is the right route.). Vague information is not instructions about the concrete route selection but rather the position instructions of the treasure with a direction and distance (example: It is 100 cm in the 12 o'clock direction.). There are $3 \times 2 = 6$ conditions.

The body factor is between the subjects and the environmental factor is within the subjects. Subjects performed the treasure hunt game four times, twice in each environmental factor. In the first, we lowered the degree of difficulty to choose from three by increasing the volume, and an agent provided correct treasure information to form a relationship of mutual trust between a person and an agent. In the second, we raised the degree of difficulty by making the volume small, and an agent gave incorrect treasure information. Here, the subjects themselves determined which agent to believe.

Fig. 3. Robot agent

Fig. 4. CG agent

3.3 Hypothesis

This experiment was performed based on hypothesis that people is easy to trust information from objects that shares environment and information that corresponds to the environment. This experiment was a task to perform a treasure hunt with an agent and the agent presents correct information about the route selection and information about which is the treasure. When an agent gives incorrect information to identify the treasure, we expect that subjects will trust hints from the robot agent most and obey the instructions, and the subject will trust hints most from the agent that provided corresponding information when the subject is choosing a route. According to this hypothesis, we predict the following:

- In the body factor, the number of times of instructions of obeyed increases on the robot condition significantly, in comparison with two other conditions.
- In the environment factor, the number of times of instructions obeyed increases on the corresponding condition significantly, in comparison with vague conditions.

3.4 Results

We used 15 people in each condition of the body factor. The results for the number of times that a subject obeyed when the agent gave incorrect information appear in Fig. 5. Fig.5 shows that subjects readily obeyed the robot instructions more than the CG and the computer, and there is hardly a difference between the CG and the computer.

In the environment factor, subjects easily obeyed the corresponding information for the robot condition slightly. The difference is hardly seen overall.

Next we regarded the case where the subject obeys the instruction as 1 point and doesn't obey as 0 point and performed a two-way repeated measure analysis of variance. Analysis revealed that the simple main effect of the body factor is significant $(F(2,42)=3.57(p<.05))$ A simple main effect of the environment factor was not seen. In addition, we performed multiple comparison with the LSD method, and the averages of the robot condition were significantly larger than each average of the other two conditions $(MSe=0.2523(p<.05))$.

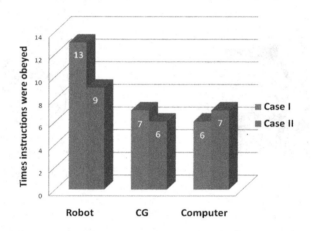

Fig. 5. Experiment results

3.5 Consideration

Our experiment results show that the existence of a body in the real world improves the agent evaluation. This result resembles the preceding study described in Section 2.2, and in the real-world task, humans easily trust agents who exist in the real world. Because the subject really had a robot, the vibration of the robot's movement may have impressed subjects with the agent's existence strongly.

On the other hand, a significant difference was not seen for the environment factor for the following reasons. First, since we experimented on two conditions of the environmental factor with one subject with the same agent, the agent may not have been distinguished between the two conditions. Second, vague information produced a hesitation of the route choice and an assumption that the agent reliability had deteriorated. However, correct route choice was enabled by vague information because we repeatedly experimented with a comparatively easy maze; no difference appeared.

In addition, because a difference was only seen as a tendency between the environmental factors for the robot, agent reliability will increase most by exchanging the information that corresponds to the environment, assuming that the body exists in the real world.

4 Conclusion

This study created a sense of sharing the environment using the body of a real-world agent, suggesting the possibility of improving its reliability. When we use a personified agent as a computer interface, the agent should be a person who can talk naturally. However, by sharing the environment by the body, humans might easily feel close with the agent. We showed one example of the directionality of a new agent interaction design.

References

1. Watanabe, T.: E-COSMIC: Embodied Communication System for Mind Connection. In: Proc. of the 13th IEEE International Workshop on Robot-Human Interactive Communication (RO-MAN 2004), pp. 1–6 (2004)
2. Ishibiki, C., Nakajima, Y., Matsumoto, D., Miwa, Y.: Expression of Existence Supporting Sharing Interspatial Distance-"Maai": Development of Roving Object Integrating Partner's Shadow and Video Image. Journal of Robotics and Mechatronics 17(3), 310–317 (2005)
3. Kidd, C., Breazeal, C.: Effect of a Robot on User Perceptions. In: IEEE/RSJ International Conference on Intelligent Robots and Systems (IROS 2004), pp. 3559–3564 (2004)
4. Shinozawa, K., Naya, F., Yamato, J., Kogure, K.: Differences in Effect of Robot and Screen Agent Recommendations on Human Decision-Making. International Journal of Human-Computer Studies 62, 267–279 (2005)

Interaction Design for a Pet-Like Remote Control

Kazuki Kobayashi[1], Yutaro Nakagawa[2], Seiji Yamada[3,4],
Shinobu Nakagawa[2], and Yasunori Saito[5]

[1] Graduate School of Science and Technology, Shinshu University
4-17-1 Wakasato, Nagano City, 380-8553 Japan
kby@shinshu-u.ac.jp
[2] Design Department,Osaka University of Arts
469 Higashiyama, Kanan-cho, Minami Kawachi-gun Osaka, 585-8555 Japan
shinobu@osaka-geidai.ac.jp, yutarou19870307@yahoo.co.jp
[3] National Institute of Informatics
2-1-2 Hitotsubashi, Chiyoda, Tokyo, 101-8430 Japan
seiji@nii.ac.jp
[4] SOKENDAI
Shonan Village, Hayama, Kanagawa, 240-0193 Japan
[5] Faculty of Engineering, Shinshu University
4-17-1 Wakasato, Nagano City, 380-8553 Japan
saitoh@cs.shinshu-u.ac.jp

Abstract. This paper describes a novel remote control operable with stroking its surface. Advantages of the developed remote control are high familiarity and stroke operation. Those enable users to have familiarity with it and to use it without looking at the fingers. We apply it to an interaction system with TV. The proposed system has the tolerance for mistakes in comparison with conventional button-based remote controls because it enables unfamiliar users to home electric appliances to use it casually without fear of mistakes and unexpected behavior.

Keywords: remote control agent, pet-like embodied agent, partial execution.

1 Introduction

Various remote controls are used in our home. Home electronics such as televisions, air conditioners, room lights usually have remote controls. They are currently commonplace devices. However, many remote controls confuse people when they use a home electric appliance. Universal remote[1] that aggregates functions of various remote controls is one of technical solutions for this problem. On the other hand, users need to search a button that they want to operate and to correctly move the fingers to push it. It is not highly problematic for young people, but elderly people have a difficulty in operating remote controls. It will be comfortable to use a remote control without looking at their hands.

J.-H. Kim et al. (Eds.): FIRA 2009, CCIS 44, pp. 134–139, 2009.

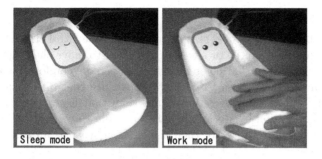

Fig. 1. Remote Control Agent: Rebo

Such a remote control is also comfortable to young people familiar to home electric appliances.

In this paper, we propose a remote control, Rebo, like a pet animal. Users stroke its surface with the fingers to control various home electric appliances. Advantages of Rebo are high familiarity and stroke operation. Its life-like appearance and facial expressions make it familiar with users. They can easily use it without looking at the fingers by stroking. Therefore, users do not have to seek a button that they want to operate and use it smoothly.

We develop an interaction system with TV as an example of the proposed remote control. Rebo enables users to easily execute functions of the TV.

2   Pet-Like Remote Control Agent

Figure 1 shows the developed pet-like remote control agent, Rebo. Rebo is an agent of remote controls of home electric appliances. It has a user-friendly appearance; a smooth surface for stroking and a back side fit for users' thighs. It is 249 mm long, 146 mm wide, and 96 mm high. A small LCD monitor, full-color LED modules, and three touch sensors are embedded in the body. The touch sensors are covered with a soft and smooth cloth.

2.1   Familiarity

Rebo has a life-like appearance and facial expressions to acquire familiarity with a user (Fig. 1). We adopt the concept of "intermediate entity between artifacts and animate beings". Rebo is not only a tool but also a partner like a pet animal. The body of Rebo is covered with soft and bouncy cloth and it is pleasant to the touch. When the user strokes Rebo, it changes its facial expressions to inform him or her of various emotional states of Rebo. We consider this makes the user more comfortable with Rebo. Therefore, the user can enjoy interacting with Rebo. We believe that this concept plays an important role for establishing familiarity between users and Rebo.

2.2 Stroke Operation

Rebo has no button and a user strokes its surface to control electric appliances. In this kind of operation, he or she does not need to move the fingers correctly and not seek the button that he or she wants. It is more comfortable for the user to gaze at an appliance than to gaze at the remote control, because the feedback from the appliance is more important than that from the remote control. However, the operation by stroking is different from the operation by pushing a button in a physical feeling. We then use LED lighting as an acknowledgment of the operation.

The advantage of the LED lighting has less interference against auditory and visual information. If an electric appliance such as a TV provides auditory information, it is inadequate that a remote control provides auditory feedback. In contrast, expressions by simple LED lighting can be grasped by users from their peripheral vision. We call this method "Peripheral feedback". Peripheral feedback has less interference against their central visual field. This method is similar to peripheral display[2,3], and its concept is based on our view that "the central player is the appliance and the backseat player is the remote control".

3 TV Remote Agent

We developed a TV remote agent based on the proposed remote control agent to implement all of the advantages in the previous section. In addition, we also designed feedback behavior of the TV and constructed an interaction system.

3.1 Stroke Operation

Functions we implemented in the system are channel switch, sound volume change, and power on/off. The TV screen was implemented on a PC. In this section, operation methods of Rebo and feedback behavior of the TV are described.

Channel Switch. Figure 2 shows a series of screenshots in which the positions of two video pictures are changed as a user strokes the surface of Rebo right and left. When the user widely strokes the surface of Rebo with her fingers, a video picture goes the outside of the TV frame and another video picture comes into the frame. When the user strokes Rebo for a short time, a part of another video

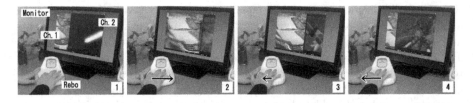

Fig. 2. Channel Switching

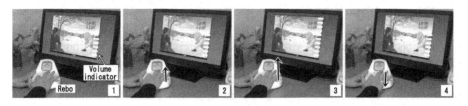

Fig. 3. Volume of Sound Changing

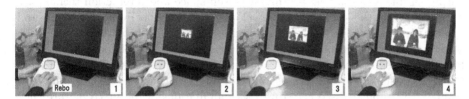

Fig. 4. Turn On

picture comes into the frame and then goes out of the frame automatically. In this implementation, we used recorded TV movies. It will be technically possible to capture TV movies and show them in the monitor.

Sound Volume Change. When a user strokes Rebo up and down, the volume indicator (a vertical bar) is shown on the video picture (Fig. 3). The length of the indicator is changed as the movement distance of the fingers is changed.

Turn On and Off. A series of screenshots in Fig. 4 shows a video picture being reduced in an image dimensions. When a user touches Rebo for more than one second, the image dimensions of the video picture are reduced and finally disappeared. If the user stops touching Rebo before the picture is disappeared, the video picture is enlarged and goes to an ordinary size.

On the other hand, a series of screenshots in Fig. 5 shows a video picture being enlarged. When the TV has been turned off, the picture is gradually enlarged while the user touches Rebo. When the user stops touching before the picture becomes to a maximum size, it is automatically going down and then disappeared.

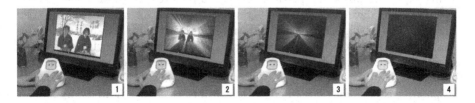

Fig. 5. Turn Off

4   Discussions

Since Rebo utilizes a user's stroke with the fingers, he or she does not need to visually recognize and understand any icon, signal, and acknowledgment from a remote control. The user can concentrate on the feedback from appliances, but remote controls. The advantage of our approach partially resembles that of "eyes-free" approach[4,5,6] for many people including visually-impaired persons and elder people. Particularly for using mobile and wearable computers, input or output problems occur due to limited screen space and interaction techniques. They overcame the problems by developing a 3D audio radial pie menu and a sonically enhanced 2D gesture recognition system on a belt-mounted PDA. These previous studies attempted just to enrich usability of the user interface by touch-based devices like touch panels. However, we consider Rebo's touch-based approach is not only for usability of interface, but also for possibility for manual free appliances. When a user sees animations of the video picture such as sliding and zooming, she or he can understand the meaning of the feedback animation from the TV before the function is completely executed. Therefore, they can immediately notice the function of the appliance by stroking Rebo.

The feature of Rebo in comparison with a button-based remote control is the tolerance for mistakes. In our proposed system, a user concentrates on the feedback from a TV because he or she can operate Rebo with strokes without looking at it. The user can cancel the function before it is completely executed because he or she understands which function is executed by observing the feedback animation from the TV. For example, when the user wants to change the volume and operates Rebo, he or she tries other ways in stroking by observing the video picture moving left and right. However, in the button-based remote control, when the user pushes a button, the channel was changed immediately. The user has to search a button that tunes the previous channel and push it, and then search a button for changing the volume.

The tolerance for mistakes in Rebo enables unfamiliar users to home electric appliances to use it casually without a fear of making mistakes and unexpected behavior. We have a plan to investigate the effect of this tolerance in an experiment with elderly people.

5   Conclusion

In this paper, we propose a pet-like agent as a remote control for home electric appliances and develop a TV interaction system. Advantages of the proposed remote control are high familiarity and stroke operation. Life-like appearance and facial expression makes it familiar with users. We implemented facial expressions as a life-like behavior and video picture animations as feedback information from the TV to enable users to easily notice its functions. LED lighting provided Peripheral feedback for users to concentrate on the feedback from the appliance, but the remote control. The feature of Rebo in comparison with a button-based remote control is the tolerance for mistakes. This enables unfamiliar users to home electric appliances to use it casually without a fear of making mistakes and unexpected behavior.

In the next stage of our study, we have a plan to apply Rebo to various home electric appliances and enable users to seamlessly control them depending on the context.

References

1. LaPlant, B., Trewin, S., Zimmermann, G., Vanderheiden, G.: The universal remote console: A universal access bus for pervasive computing. In: IEEE Pervasive Computing, pp. 76–80 (2004)
2. MacIntyre, B., Mynatt, E.D., Voida, S., Hansen, K.M., Tullio, J., Corso, G.M.: Support for multitasking and background awareness using interactive peripheral displays. In: Proc. of the 14th annual ACM symposium on User interface software and technology, pp. 41–50 (2001)
3. Hsieh, G., Wood, K., Sellen, A.: Peripheral display of digital handwritten notes. In: Proc. of the SIGCHI conference on Human Factors in computing systems, pp. 285–288 (2006)
4. Zhao, S., Dragicevic, P., Chignell, M., Balakrishnan, R., Baudisch, P.: Earpod: eyes-free menu selection using touch input and reactive audio feedback. In: Proc. of the SIGCHI conference on Human factors in computing systems (CHI 2007), pp. 1395–1404 (2007)
5. Buil, V., Hollemans, G., van de Wijdeven, S.: Headphones with touch control. In: Proc. of the 7th international conference on Human computer interaction with mobile devices & services, pp. 377–378 (2005)
6. Brewster, S., Lumsden, J., Bell, M., Hall, M., Tasker, S.: Multimodal 'eyes-free' interaction techniques for wearable devices. In: Proc. of the SIGCHI conference on Human factors in computing systems, pp. 473–480 (2003)

Experiences with a Barista Robot, FusionBot

Dilip Kumar Limbu, Yeow Kee Tan, Chern Yuen Wong, Ridong Jiang,
Hengxin Wu, Liyuan Li, Eng Hoe Kah, Xinguo Yu, Dong Li, and Haizhou Li

Institute for Infocomm Research, 1 Fusionopolis Way, #21-01, Connexis,
Singapore 138632
{dklimbu,yktan,cywong,rjiang,hxwu,lyli,kehoe,xinguo,
ldong,hli}@i2r.a-star.edu.sg

Abstract. In this paper, we describe the implemented service robot, called FusionBot. The goal of this research is to explore and demonstrate the utility of an interactive service robot in a smart home environment, thereby improving the quality of human life. The robot has four main features: 1) speech recognition, 2) object recognition, 3) object grabbing and fetching and 4) communication with a smart coffee machine. Its software architecture employs a multimodal dialogue system that integrates different components, including spoken dialog system, vision understanding, navigation and smart device gateway. In the experiments conducted during the TechFest 2008 event, the FusionBot successfully demonstrated that it could autonomously serve coffee to visitors on their request. Preliminary survey results indicate that the robot has potential to not only aid in the general robotics but also contribute towards the long term goal of intelligent service robotics in smart home environment.

Keywords: Social robots, Human-robot interaction, Human perception and attitudes.

1 Introduction

Recent research and development in robotics has spawned various new research directions, including service robotics, field robotics, underwater robotics, and medical robotics. Service robotics is one of the promising avenues of research to which robotic technologies can be applied. In most cases, service robotics is designed to be autonomous, able to communicate with humans and participate in a given social context. With the advent of robotics technologies, the design and development of the service robotics has evolved drastically. In effect, service robotics is becoming prominent in the interactive robot development.

Numerous service robots exist today in different service areas, such as museums [1, 2], receptions [3], food serving [4], hospitals [5], elder care [6], home tour [7] etc., making use of high robotics technologies. All these service robots perform certain tasks while providing various Human-Computer Interaction (HCI) and Human-Robot Interaction (HRI) interfaces, such as speech, vision and touch screen. To the best of our knowledge none of them offer a similar service to the one described in this paper. The service robot system that offers the closest service might be CARL[4], which is

J.-H. Kim et al. (Eds.): FIRA 2009, CCIS 44, pp. 140–151, 2009.

designed to serve food in a reception or acts as a host in an organization. It has navigation abilities, greets guests, understands spoken language, recognizes pointing gestures, and serves food to them.

This paper describes the implemented service robot, called FusionBot. The goal of this research is to explore and demonstrate the utility of an interactive service robot in smart home environment, thereby improving the quality of human life. There are already numerous smart homes projects [8-11] are being carried out by various research institutions or universities. We believe that operating service robots in the smart home environment would be a logical step in social robotics research. Conversely, operating service robots in the smart home environment is a challenging task, different in many aspects from more traditional service robots. The smart home environment can be packed with furniture, smart electronic appliances and even people. As a result, a simple task can be difficult due to the robot's technical challenges, such as sensors measurements, vision and navigation.

During the two days-long experiment in TechFest 2008[1] event, FusionBot has demonstrated the ability to serve coffee to visitors. This involves taking coffee order from a visitor, identifying a cup and smart coffee machine[2], moving towards the coffee machine, communicating with the coffee machine and fetching the coffee cup to the visitor. Preliminary survey results indicate that the robot has potential to aid users and also contribute towards the long term goal of intelligent service robotics in a smart home environment. Section 2 and 3 describes the FusionBot hardware and software architecture respectively. Section 4 and 5 describes the experimental setup and results respectively.

2   FusionBot the Barista Robot

Fig. 1 depicts the hardware configuration of the robot FusionBot (first prototype). It has to be mentioned that in this work the physical appearance of robot was not a major consideration.

Fig. 1. FusionBot hardware configuration

The FusionBot is approximately 30 kg and is about 1.3 m tall. It has a P3DX Pioneer mobile base (3 degree of freedom) and a movable Amtec PowerCube arm (4 degree of freedom). A gripper with force resistive sensors and a Videre Design

---

[1] http://techfest.i2r.a-star.edu.sg/

[2] The smart coffee machine is embedded with a Wi-Port module and a software program, which allow the coffee machine to communicate with FusionBot using predefined message format.

stereoscopic camera are attached to the arm for grabbing and tracking objects respectively. It is equipped with laser range finders (hokoyu URG-04LX), sonar sensors (part of the base), gyro (part of the base), microphones for speech recognition, and speakers for speech synthesis.

On the top of the mobile base, a box shaped metal structure was added to hold two Pentium (MiniATX) PC boards, which are interconnected by Ethernet cross-over cable, and a set of batteries. Both PC boards can be controlled and/or monitored by a laptop wirelessly via Wi-Fi.

3   Software Architecture

FusionBot has several independent modules to fulfill different tasks. They are coordinated by a central controller equipped with a business logic engine. An event-driven dialogue system (Fig. 2) is engaged for speech recognition and communication with other modules [12].

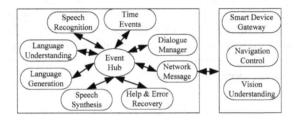

Fig. 2. FusionBot software architecture

Each module communicates with the others through TCP/IP socket messages. Meanwhile human-robot communication is achieved through the Abacus speech recognition system that is developed by Institute for Infocomm Research (I2R) and objects recognition is achieved through the vision understanding. Autonomous robot navigation is achieved by the navigation control. The communication between the coffee machine and the FusionBot is achieved through the smart device gateway. The detailed Dialog Management is described in [12]. This section mainly focuses on four modules; 1) speech recognition, 2) vision understanding module and 3) navigation control and 4) smart device gateway module.

3.1   Speech Recognition

The speech recognition module is one of the key components of the system. The tasks of the speech recognition module are to identify words and phrases in spoken language and convert them to a machine-readable format. The challenges are the noisy environment and different speaking pattern of different people.

To achieve robust speech recognition, we employed the ABACUS platform [13], a in-house developed multilingual and phonetic engine, which recognises continuous sentences or phrases. The engine supports both Large Vocabulary Continuous Speech Recognition (LVCSR) and Backus-Naur Form (BNF) grammar-based speech recognition. We employed the later approach as it helps to constrain the spoken language interaction to limited scope and task domain, while improving the speech recognition performance significantly.

ABACUS consists of two major components, the signal processing front-end and the recognition engine. The front-end uses liftered mel-scale cepstral coefficients to extract feature vectors in accordance with the ETSI standard. Meanwhile the voice activity detection is based on signal-Noise spectral distance in multiple Mel-frequency bands. The algorithm is performed in short term Fast-Fourier Transform (FFT) domain with a frame of 32ms and frame shift of 8ms. The noisy STFFT spectrum is smoothed by moving average and then is transformed in specific Mel-frequency bands. The speech recognizer is a frame-synchronous HMM-based recognizer, employing 3-state HMM to model context dependent tri-phones. Each of these states has a mixture of 16 Gaussians.

In addition to ABACUS, we integrated the microphone array system to further enhance the speech signal. The microphone array system consists of 2 main components; a) the beam-former, and b) the spectral subtraction stage. The beam-former and the spectral subtraction stage are implemented using the Griffith-Jim adaptive [14] and speech enhancement algorithm [15] technique respectively. These techniques significantly enhanced the speech recognition accuracy (e.g., babble: 47.9% and clean 84.3%).

3.2 Vision Understanding

The vision understanding module is one of the key components of the system. The tasks of the vision module are detection and tracking of a cup to guide the arm to grab the cup, and detection and tracking the coffee machine to guide the arm to put the cup in the correct position. The challenges are the varying lighting conditions, visual variations of the target objects viewed from different distances and viewing angles.

To achieve robust object detection from a mobile platform, we developed a novel method integrating bottom-up color-based segmentation and top-down appearance based detection. We employ a mixture of 3 Gaussians corresponding to different lighting conditions to characterize the color distributions of target objects. The Gaussian models are built in learning phase with dozens of training images captured in various lighting conditions and viewing angles. After the learning phase, the models are used for color-based object segmentation. For each incoming new image, each pixel is classified as object pixel or not according to the probability of the color belonging to one of the Gaussian models. A morphological operation is applied to extract the connected regions of cup colors. Noise and small regions are filtered out. For the remaining regions, a moment-based shape descriptor is applied to see if it is a cup.

To be robust in cluttered scenes, the HOG (Histogram of Oriented Gradients) object detector [16] is also employed in this vision system. In the learning phase, the HOG detector of target cup is built from the training images. During object detection, we scan the image with detection windows of multiple scales using sparse grid points. The overlapping detections are clustered according to strengths of the detection responses.

The detections from the bottom-up color-based segmentation and top-down appearance-based detection are fused. This vision system is robust to the possible failures of individual detectors. One example of detection of the cup is shown in Fig. 3.

Fig. 3. An example of cup detection. The images from the left to right are: color-based segmentation, HOG-based detection, final detection by fusion.

Similar approach is also used for detection and tracking the coffee machine. However, when the arm is close to the smart coffee machine, only part of coffee machine is within the view. Hence, the spout is exploited as the target object for detection and tracking when the arm is very close to the coffee machine.

3.3 Navigation Control

The navigational control main tasks are to acquire the FusionBot exact current position/location in the region of space, path planning and moving the robot to next designated by detecting and avoiding any obstacles in its path. To achieve these tasks, the navigation control employed a three-layered structured system - localization, path-planning and lastly obstacle avoidance.

To achieve robust navigation in a natural home environment, we employed the Adaptive Monte Carl Localization (AMCL) [17] and the Wavefront Propagation planer [18] techniques to localize and optimal path planning respectively. The AMCL algorithm uses a pre-defined map to compare against the perceived surroundings obtained by the laser-range finder and wheel odometry values. It then computes the instantaneous probabilistic location of the robot using adaptive particle filters that represent the possible locations. Similarly, the WP algorithm handles the path-planning portion using the pre-loaded map together with the localized position of the robot obtained from AMCL. Additionally, we employed the Nearest Diagram (ND) [19] technique to execute the instructions or control directions given by the WP planner.

The ND algorithm works as a local obstacle avoidance system whereby it constantly monitors the sensors and plans a local path that avoids both dynamic and static obstacles. It has the ability to deviate from the planned path that was computed by the WP algorithm in order avoid the obstacles. As the WP is constantly planning the optimal path, the robot will be able to reach its goal location eventually.

With the three algorithms tightly coupled together, the FusionBot was able to navigate autonomously to the planned location at an estimated error rate of roughly ±0.1m and ±5° heading while avoiding obstacles.

3.4 Smart Device Gateway

The smart device gateway module is a key module that interfaces with the smart coffee machine. The tasks of the device gateway module are sending and receiving wireless messages to and from the coffee machine. For instance, once the FusionBot placed a cup under the coffee machine spout, the robot sends a message (e.g., "♥CO71" for black coffee) using the smart device gateway to the smart coffee machine. Table 1 shows a sample message format (row 2) and actual dispense black

Table 1. Sample Message Format

1	Byte 0	Byte 1	Byte 2	Byte 3	Byte 4	Byte 5	Byte 6
2	Start	C1	C2	*1	*2	End	End
3	♥	C	0	7	1	-	-

coffee message (row 3), where, Byte 0 represents start byte, Byte 1 and 2 represents the type of coffee (i.e., ES = espresso, CO = black coffee, CA= cappuccino and MA = latte), Byte 3 & 4 represents input setting <0-9, 1 = low and 9 = high > for water and bean grind time respectively, and Byte 5 & 6 represents end byte.

Upon sending the message to the smart coffee machine, the smart device gateway receives various acknowledgement messages (e.g., ♥OK_COMMAND = correct message/message accepted, ♥NOK = incorrect message/message not accepted, ♥BUSY = coffee machine busy, ♥DISPENSE_OK= dispense completed, ♥OK_REMOVECUP = remove cup, ♥RESET_OK = ready to dispense coffee) and acts accordingly.

4 Experiment Setup

Since FusionBot is aimed at operation in a natural home environment, it was stationed in a miniature kitchen stall labeled "Coffee Corner" at the corner of the TechFest 2008 event hall for two days. The stall minimally resembled a home kitchen with two tables, a coffee machine and few paper cups.

The FusionBot experiments were carried out by five researchers: an experimental supervisor, two speech monitoring operators, a robot and coffee machine monitoring operator, and a vision monitoring operator. All operators were seated at the other end of the stall to minimize the presence of experimenters in the experiment area. Before conducting the experiments, subjects were given general instructions and simple introduction on how the robot system works. Subjects then performed experiment tasks (Fig. 4), i.e., ordering a cup of coffee. In the following we present a typical dialog of the coffee ordering scenario. (R: FusionBot and V: Visitor)

(a)   (b)

(c)   (d)

Fig. 4. Experiment tasks. Clockwise from upper left: a) getting a coffee order, b) grabbing the cup and moving towards the coffee machine, c) communicating with the coffee machine and dispensing the coffee, and d) fetching the cup of coffee to visitor.

R: "What would you like to drink?"
V: "I would like to have a cup of cappuccino."
R: "OK, I'll go and get you a cup of cappuccino."
[After getting a cup of cappuccino, the RobotBot places the cup on the table.]
R: "Here is your cup of cappuccino."

Thereafter, the subjects were asked to fill out the satisfaction questionnaires to find out their reaction and perception on the Fusionbot.

5 Results

Of just over 100 survey questionnaires handed out, sixty eight (68) valid responses (i.e. 68%) were received. Respondents were asked to complete a mix of five point Likert scales (i.e., range 1-5, higher = better) to indicate (or express) their agreement or disagreement towards a set of statements. Each degree of response agreement was given a numerical value from one to five (using 1 to signify highly negative, 2 fairly negative, 3 as neutral 4 as fairly positive and 5 as highly positive). Some Likert scales (i.e., positive and negative terms) in the questionnaires are reversed in consecutive attitude objects to ensure the respondent's attention does not waver when completing the questionnaires. Also, respondents were encouraged to list the negative and positive features of the FusionBot using open-ended questions. These questions helped to gather additional information and increased understanding of each respondent's general feeling about the FusionBot. Subjects' actual comments on the FusionBot are presented to complement the graphical analysis. No alterations are made on respondents' comments so as to avoid giving false or misleading information.

5.1 Respondent Characteristics

Of the sixty eight (68) respondents, 69 % were male (46) and 31% were female (21). One respondent did not fill up the gender question. One third (34%) of the respondents (23) were less than 30 years old, while more than half (61%) of the respondents (43) were 31 years or older. Two respondents did not fill up the age range question. More than half (67%) of the respondents (44) were research staff. The remaining 32% of respondents (21) were a mixture of students (5), and others (16). About 30% of the respondents (20) hold PhD degrees, 33% respondents (22) hold Master degrees, 35% respondents (24) hold bachelor degrees and 1% (1) respondents hold other degrees. Three respondents did not fill up the educational qualification question.

5.2 Robot Experience

Respondents were asked to complete a five point Likert scale (range 1-5, higher = better) on their robotic experience and whether they have learnt any robotics information from the demonstration of the FusionBot demonstration.

The pie chart (Fig. 5a) illustrates that almost half (38%) of the respondents (26) indicated (i.e., Likert scale values of '1' and '2') that they knew nothing about robotics, whilst more than one third (28%) of respondents (19) indicated (i.e., Likert scale values of '4' and '5') that they knew a lot about robotics. Two out of sixty eight respondents did not complete this question.

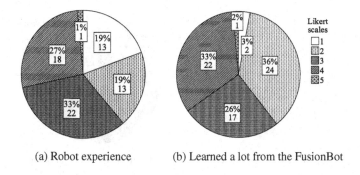

(a) Robot experience   (b) Learned a lot from the FusionBot

Fig. 5. a) Robotic experience and b) Learned from the FusionBot demonstration

The pie chart (Fig. 5b) illustrates that almost half (39%) of the respondents (26) indicated (i.e., Likert scale values of '1' and '2') that they did not learnt a lot about robotics from the FusionBot demonstration, whilst more than one third (35%) of respondents (23) indicated (i.e., Likert scale values of '4' and '5') that they learnt a lot about robotics from the demonstration. Two out of sixty eight respondents did not complete this question.

5.3 Satisfaction on Tasks

Respondents were asked to complete a five point Likert scale (range 1-5, higher = better) to indicate whether they were satisfied with what the FusionBot can do (i.e., to take a customer's order, to grab a cup, to move towards the coffee machine, to place the cup in front of the coffee machine dispenser, to communicate with the coffee machine to dispense coffee, and finally to bring and place the cup coffee on the table).

The pie chart (Fig. 6a) clearly illustrates that more than half (55%) of the respondents (37) indicated (i.e., Likert scale values of '4' and '5') that they were satisfied with what the FusionBot can do, whilst less than one quarter (8%) of respondents (6) indicated (i.e., Likert scale values of '4' and '5') that they were not satisfied with what the FusionBot can do. One out of sixty eight respondents did not complete this question.

Respondents were asked to complete a five point Likert scale to indicate whether the FusionBot was prone to technical difficulties and malfunctions. The pie chart (Fig. 6b) clearly illustrates that more than one third (36%) of the respondents (24) indicated (i.e., Likert scale values of '4' and '5') that the FusionBot was prone to technical difficulties and malfunctions, whilst less than one quarter (24%) of respondents (15) indicated (i.e., Likert scale values of '1' and '2') that the FusionBot was not prone to technical difficulties and malfunctions. One out of sixty eight respondents did not complete this question.

A question using a five point Likert scale was asked to respondents to indicate whether it was easy to communicate with the FusionBot. The pie chart (Fig. 6c) illustrates that more than one third (37%) of the respondents (25) indicated (i.e., Likert

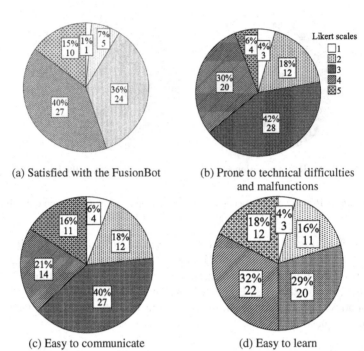

Fig. 6. Satisfaction on tasks: a) Satisfied with the FusionBot, b) Prone to technical difficulties and malfunctions, c) Easy to communicate, and d) Easy to learn

scale values of '4' and '5') that it was easy to communicate with the FusionBot, whilst nearly one quarter (24%) of respondents (16) indicated (i.e., Likert scale values of '1' and '2') that it was not easy to communicate with the FusionBot.

Respondents were asked to complete a five point Likert scale (range 1-5, higher = better) to indicate whether the FusionBot was easy to learn how to use it. The pie chart (Fig. 6d) illustrates that half (50%) of the respondents (34) indicated (i.e., Likert scale values of '4' and '5') that it was easy to learn how to use the FusionBot, whilst less than one quarter (20%) of respondents (14) indicated (i.e., Likert scale values of '1' and '2') that it was not easy to learn how to use it. One respondent did not complete this question.

5.4 Suggestion/Comments on the FusionBot

An open ended question was asked to respondents to list the things they like or dislike about the FusionBot. Table 2[3] shows the most frequently cited respondents' comments. NO alterations are made on the respondents' suggestion/comments so as to avoid giving false or misleading information.

An open ended question was asked to respondents what would they like a robot to help them in their home. Table 3 shows the most frequently cited respondents' feedbacks.

---

[3] Due to space limitations, for Table 2 and 3, only the most frequently cited comments are shown.

Table 2. Like and dislike about the FusionBot

Like		Dislike	
1.	If it is working, the efficiency.	1.	Appearance.
2.	Robot was able to accomplish requested task of serving coffee.	2.	Very mechanical looking.
		3.	No human touch in terms of appearance.
3.	His voice and wheels.	4.	Robot looks a lot raw as can see all mechanical parts.
4.	Voice recognition, precision.		
5.	Looks functional and well designed as prototyped.	5.	Accuracy on voice recognition needs improvement -> perhaps more training required.
6.	It is rather stable which I like.		
7.	It looks good, simple and sophisticated.	6.	The base of the robot can look better?
8.	Function OK.	7.	Too bulky & wires exposed.
		8.	Ergonomics may be considered later.
		9.	Not aesthetically appealing and need dressing up.
		10.	However the build (size) & speed could be better improved.
		11.	Too mechanical - should have & rounded outer cover.

Table 3. Feedbacks on the FusionBot

	Like
1.	Household chores like laundry, ironing cloths, vacuum, sweeping & mopping floors and serving coffee.
2.	Entertainment & communication; play music & simple tasks such as weather & news.
3.	To serve guest
4.	Locate items (keys, spectacles etc). Integrate with other tech e.g., Camera. To switch on the air-con when I'm on the way home.

6 Discussion and Conclusions

This paper described a service robot, named FusionBot that autonomously serves coffee to visitors on their request. The robot's main features are; 1) speech recognition, 2) objects recognition, 3) objects grabbing and fetching and 4) communication with a smart coffee machine. This paper also briefly presented the speech recognition, vision understanding, navigation control and smart device gateway (specifically aimed at interaction with the smart coffee machine). The robot has been tested successfully in an experimental environment that minimally resembled a home kitchen. Our experiments were successful in two main aspects; 1) the robot demonstrated the robustness of the hardware and software components in completing a challenging real-world task, and 2) like CARL[4], it provided some evidence towards the feasibility of using autonomous service robot and smart coffee machine to serve drink in a reception/home or acting as a host in an organization.

Over all, with regards to the FusionBot task satisfaction, more than half of respondents were satisfied with what robot can do. On the other hand, some respondents made some constructive comments, which include support of different languages, improve on speech and vision understanding, and improve on size and speed of robot. These could be due to; 1) the current robot was only designed to support English as spoken language, 2) the robot's speech module was initially trained with only male voice and it failed several times with female voice, 3) due to the safety reason, the speed of robot was programmed relatively slow, and finally, 4) the robot static appearance was not a major consideration in this project.

With regards to the technical difficulties and malfunctions, more than one third of the respondents indicated that the FusionBot was prone to technical difficulties and malfunctions. This could be due to two reasons; 1) the weak wireless network and 2) the coffee machine's hardware limitation. The communication between the robot and the smart coffee machine was lost several times due to the weak wireless network, resulting in the robot operator manually dispenses and serves coffee to visitors. According to the hardware specification, the coffee machine was designed to dispense around 20 cups daily. However during the experiment, we dispensed more than 100. Nearly one quarter of the respondents indicated that it was not easy to communicate with the robot. This could be due to occurrence of various background noises, which were falsely picked up by the robot as speech input from the visitor. Similarly, less than one quarter indicated that it was not easy to learn how to use the FusionBot. This could be due to the not knowing what to do with robot and not knowing what the robot does. However, due to the complex nature of the service robotics study, results reported in this paper are just one step in this direction. These results serve as a partial view of the phenomenon. More research needs to be done in order to support these findings, using larger samples, and if possible in a real-life scenario.

Few key lessons learned while developing this social robot, mainly; 1) static appearance is very important, 2) requires robust speech recognition and vision understanding, 3) requires comprehensive training on speech and vision modules with respective data, and finally 4) requires a robust computational architecture that seamlessly integrates various software and hardware modules to perform robotic tasks.

As ongoing project, FusionBot is continuously evolving. Currently, we are developing an open robotic software architecture, which would allow us to integrate and synchronize various distributed systems/modules and immense spectrum of different hardware/mechanical components to perform certain robotic tasks. In addition, we are also developing an attention directed dialog using the speech recognition and vision understanding technology.

References

1. Shiomi, M., et al.: Interactive Humanoid Robots for a Science Museum. IEEE Intelligent Systems 22(2), 25–32 (2007)
2. Thrun, S., et al.: MINERVA: a second-generation museum tour-guide robot. In: IEEE International Conference on Robotics and Automation, Detroit, Michigan, USA, pp. 1999–2005 (2005)
3. Gockley, R., et al.: Grace and George: Social Robots at AAAI. In: American Association for Artificial Intelligence (AAAI), San Jose, California, USA (2004)
4. Lopes, L.S., et al.: Towards a Personal Robot with Language Interface. In: 8th European Conference on Speech Communication and Technology, Geneva, Switzerland, pp. 2205–2208 (2003)

5. Spiliotopoulos, D., Androutsopoulos, I., Spyropoulos, C.D.: Human-robot interaction based on spoken natural language dialogue. In: European Workshop on Service and Humanoid Robots, Santoriri, Greece, pp. 25–27 (2001)
6. Montemerlo, M., et al.: Experiences with a mobile robotic guide for the elderly. In: 18th national conference on Artificial intelligence, Edmonton, Alberta, Canada, pp. 587–592 (2002)
7. Toptsis, I., et al.: Modality Integration and Dialog Management for a Robotic Assistant. In: 9th European Conf. on Speech Communication and Technology, Lisbon, Portugal, pp. 837–840 (2005)
8. Cook, D., et al.: MavHome: An Agent-Based Smart Home. In: 1st IEEE International Conference on Pervasive Computing and Communications, pp. 521–534. IEEE Computer Society, New Work (2003)
9. Cook, D., Das, S.: Smart Environments: Technologies, Protocols and Applications. Wiley Interscience, Hoboken (2004)
10. Kidd, C.D., et al.: The aware home: A living laboratory for ubiquitous computing research. In: 2nd International Workshop on Cooperative Buildings, Integrating Information, Organization, and Architecture, Pittsburgh, USA, pp. 191–198 (1999)
11. Intille, S.S., et al.: A Living Laboratory for the Design and Evaluation of Ubiquitous Computing Technologies. In: Human factors in computing systems, Portland, USA, pp. 1941–1944 (2005)
12. Jiang, R., et al.: Development of Event Driven Dialogue System for Social Mobile Robot. In: Global Congress on Intelligent Systems (GCIS), Xiamen, China (2009)
13. Haizhou, L., Bin, M., Chin-Hui, L.: A Vector Space Modeling Approach to Spoken Language Identification. IEEE Transactions on Audio, Speech and Language Processing 15(1) (2007)
14. Compernolle, D.V.: Switching Adaptive Filters for Enhancing Noisy and Reverberant Speech from Microphone Array Recordings. In: International Conference on Acoustics, Speech, and Signal Processing (ICASSP), Albuquerque, USA, pp. 833–836 (1990)
15. Cohen, I., Berdugo, B.: Speech enhancement for nonstationary noise environments. Signal Processing 81(11), 2403–2418 (2001)
16. Dalal, N., Triggs, B.: Histograms of Oriented Gradients for Human Detection. In: International Conference on Computer Vision and Pattern Recognition, San Diego, CA, USA, pp. 886–893 (2005)
17. Fox, D.: KLD-Sampling: Adaptive Particle Filters. In: Advances in Neural Information Processing Systems, pp. 713–720. MIT Press, Cambridge (2001)
18. Latombe, J.-C.L.: Robot Motion Planning. Springer, New York (1990)
19. Minguez, J., Osuna, J., Montano, L.: A Divide and Conquer" Strategy based on Situations to achieve Reactive Collision Avoidance in Troublesome Scenarios. In: EEE International Conference on Robotics and Automation (ICRA), New Orleans, USA, pp. 3855–3862 (2004)

Mutually Augmented Cognition

Florian Friesdorf[1], Dejan Pangercic[2], Heiner Bubb[1], and Michael Beetz[2]

[1] Institute of Ergonomics (LfE)
Mechanical Engineering
Technische Universitaet Muenchen
Garching/Munich, Germany
{friesdorf,bubb}@lfe.mw.tum.de
[2] Intelligent Autonomous Systems (IAS)
Computer Science IX
Technische Universitaet Muenchen
Garching/Munich, Germany
{pangerci,beetz}@in.tum.de

Abstract. In MAC, an ergonomic dialog-system and algorithms will be developed that enable human experts and companions to be integrated into knowledge gathering and decision making processes of highly complex cognitive systems (e.g. Assistive Household as manifested further in the paper). In this event we propose to join algorithms and methodologies coming from Ergonomics and Artificial Intelligence that: a) make cognitive systems more congenial for non-expert humans, b) facilitate their comprehension by utilizing a high-level expandable control code for human experts and c) augment representation of such cognitive system into "deep representation" obtained through an interaction with human companions.

1 Introduction

Next-generation technologies are developed and implemented, technologies that enable cognitive capabilities and behaviour, and allow machines to reprogram themselves (Sec. 4). Further, these machines are to closely interact with humans in assistive and collaborative tasks [1]. The resulting work systems are highly complex socio-technical systems, in which even the machines are not deterministic anymore, but show emergent behaviour. While the machines will reach a very high degree of autonomy in their planning, control and failure recovery processes, the human role is reduced to take influence on planning processes and to help the machine to recover from unknown/unmodelled situations. On the one hand, the human is to be assisted in order to reduce complexity to enable him/her to comprehend the complex systems' state and to make competent decisions. On the other hand, the machine should learn from the human to deal itself with similar planning and recovery situations in the future (Sec. 2). Therefore, on the one hand, algorithms and display modalities need to be researched that allow for reduction of complexity and assistance of humans in decision making scenarios, and, on the other hand, algorithms for gathering and deriving knowledge from interactions in such scenarios (Sec. 3).

J.-H. Kim et al. (Eds.): FIRA 2009, CCIS 44, pp. 152–161, 2009.

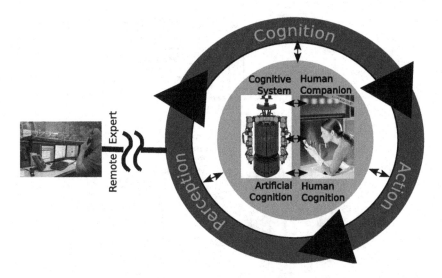

Fig. 1. *Mutually Augmented Cognition*: The system needs to be able to communicate its state to a human companion or remote operator, depending on the situation. Therefore, algorithms and display modalities need to be developed, that allow a robot to assess its state and situation, support humans with varying knowlegde levels in understanding the complex cognitive system's state, and to gain knowledge from humans.

The challenge is to benefit the human, where the machine is superior and to learn from the human, where the human is superior. Machine and human should mutually augment their cognition.

2   Challenge Scenarios

MAC is about the inclusion of humans into knowledge gathering and decision making processes. We distinguish between human companions, that are interacting with robots in a so-called *Assistive Household* (Fig. 2 and [1]), and human experts, that are remotely consulted in case situations cannot be resolved by help of companions (Fig. 1). The demonstrator shown in Fig. 2 is still in the evolutionary stage, yet very ubiquitous and complex. To deal with this complexity, a great amount of very specific knowledge is needed, which humans normally do not posses. In MAC we argue that using natural language-like expandable RPL control code in combination with the human mental model sufficiently reduces the system's complexity to be competence conducive for the human and to lead to higher acceptance. Situations are grouped into three categories: green, yellow, red; reflecting the likelihood and severity of a malfunction.

green, normal operation. The robot is not aware of any malfunction, a human companion is consulted because it is likely to gain needed knowledge. Examples:

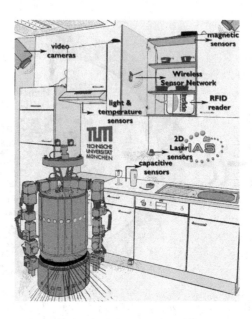

Fig. 2. *Assistive Household@IAS*

How many people are coming for dinner? Where can I find further plates? Human analogy: asking a colleague.

yellow, contradictory sensor information. There might be a malfunction or undetected environmental obstacles. Resolving the situation with help of a human companion is very likely to succeed. Example: robot wheels are turning, but robot is not moving - an object is stucked underneath. Human analogy: feeling sick, consulting a friend.

red, system malfunction. We are sure of a severe system malfunction. The situation can only be resolved by consulting a remote expert. Example: sensor/actor outage, which cannot be bypassed or compensated through human companion. Human analogy: being severly ill, consulting a doctor.

Due to the complex system and environment, situations may be wrongly assessed. To increase acceptance, we assume that it is desireable to solve as much as possible with human companions, willing to help, while not endangering them, and only to consult remote experts as a last resort.

3   Mutually Augmented Cognition

The goal is, to integrate humans into knowledge gathering and decision making processes of cognitive plan-based systems, especially in situations where a substantial implicit knowledge is required. Therefore, the machine needs to communicate its state, reduce complexity to enable the human to comprehend it, and to support the human in building a mental model of the situation to enable him/her to reach a competent decision. Based on this experience and through observation of the human's investigation and decision the machine is to learn in order to cope with similar future situations.

In Fig. 3 a schematic interaction sequences of a cognitive machine and a human is illustrated. Upon encountering an unforeseen event or a planning event, the machine presents the situation with its state to the human. The human is investigating the situation and builds up a mental model of the situation, it's quality is a prerequisite for the human to make a competent decision. In order to increase the efficiency, effectiveness and quality of building this mental model, the machine assists the human in the investigation, and hence augments the human's cognition. After coming to a conclusion, the human instructs the machine on how to proceed further. From this experience, the machine gathers knowledge and learns, being assisted by the human to increase the quality of the derived knowledge. Human and machine mutually augment their cognition, drawing benefit of their respective individual strengths. Derived from the interaction sequence (Fig. 3), three research goals can be identified and are described in the following subsections. In Sec. 3.4 MAC's nested perception-cognition-action loop is described.

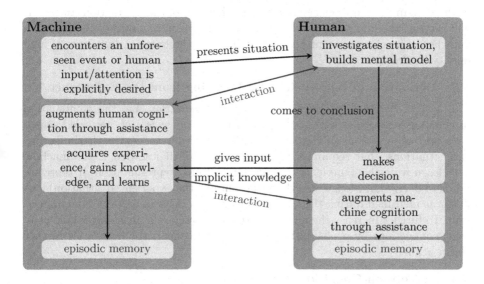

Fig. 3. Interaction sequence of cognitive machine and human: A machine decides to consult a human and presents the situation. The human investigates the situation and builds up a mental model of it, while being assisted by the machine. Upon coming to a conclusion, the machine receives instructions on how to proceed. Through this experience the cognitive machine generates knowledge and learns. Human and machine mutually augment their cognition to draw benefit from their respective individual strengths and gain episodic memory.

3.1 Presentation of Cognitive Systems' State

The plan-based cognitive systems described in Sec. 4.1 need to be able to perceive their own state through internal sensors and further, be able to estimate unmeasurable state variables. To pursue this, we will encapsulate MAC in the RPL which adherently allows for crude system's plan execution monitoring and failure detection, analysis, diagnosis and recovery. Possible malfunctional part of the system remains tractable thanks to the automatic high-to-low level code expansion techniques. MAC's objective will mind to extract relevant parts of the code, evaluate them in the situational context and provide algorithms and display modalities needed to allow and support a human to understand the cognitive system's comprehensive state.

3.2 Online Analysis and Optimization of Human-Machine Interaction

Based on the presented state, the human companion or operator will investigate the situationa and make inquiries to the machine to build up a mental model of the situation. In order to increase efficiency and effectiveness of this interaction the machine will perform an online analysis and optimization of the human-machine interaction. Patterns will be derived and used as a base for estimation of the human mental model's quality and efficiency of the interaction. In that it will support the human in building a high quality mental model of the situation and thereof reach competent decisions, which are the base for derivation of high quality knowledge.

3.3 Derivation of Knowledge and Learning from the Experience

The system detailed in [1] aims at *deep representation*, i.e. the combination of common sense knowledge obtainable from *world wide web*, recorded sensor data and observations of human activities. In order to deal with similar future situations, that require a high degree of implicit knowledge, the machine is to learn from the experience and afore mentioned representation. It therefore needs to derive knowledge, based on the context (the experienced situation), the patterns identified during the human-machine interaction, and the human decision. To further increase the quality of the knowledge, the machine will identify weak connections by probabilistic reasoning and consult the human for further clarification.

3.4 Perception-Cognition-Action Loop

For realisation of MAC an entire perception-cognition-action loop is proposed, it should be embedded as a nested loop within higher-level plan-based cognitive systems, like the *Assistive Household* [1].

Perception. The perception needs to be based on internal sensors for perception of the cognitive system's state; and external sensors for recognition of patterns in the human-machine interaction (HMI) during human investigation, and actions triggered by the human.

Cognition. Cognitive capabilities need to be developed to: (a) estimate un-measureable system state variables based on internal sensor information; (b) estimate the human mental model of the situation based on HMI patterns and previous experience; (c) decide when to involve a human to improve the esti-mation; (d) learn HMI patterns in context of the experienced situation; and (e) derive knowledge from the experienced situation and human interaction.

Action. The complex cognitive system takes action by presenting its state to a human; and by adaptation of the human-machine interaction based on HMI patterns, state of human mental model, state of cognitive system, and previous experience.

4   State of the Art

MAC is a joined approach of the institutes of ergonomics and intelligent au-tonomous systems, combining methods and technologies of system ergonomics and computer science. The state of the art is therefore presented with respect to those fields.

4.1   Computer Science / Plan-Based Control

Plan-based control, in particular of mobile robots has been first introduced by McDermott et al. in [2]. Plans are maps (symbolic descriptions) of the robot's future activity, which are analyzed, transformed and executed. Plans have two roles: (1) They are executable recipes that can be interpreted by the robot to accomplish its jobs; and (2) they are syntactic objects that can be synthesized and revised by the robot to meet its criterion of utility. Therefore, the plan is simply the *part* of the robot control program that the planner explicitly reasons about and transforms. The plans are implemented in RPL [2], a Reactive Plan Language based on Lisp. It's main data constructs are *fluents* and *designators*. The first are variables that change over time and signal alteration of e.g. bound control threads and the latter are partial objects' descriptions that get updated by the sensing routines.

Despite the high-level code and thus already shadowing part of the system's complexity, it remains comprehensive which makes understanding of its state extremely difficult for humans (compare Sec. 4.2).

The autonomous city explorer[3] integrates humans into knowledge gather-ing processes of *normal operation* to find its way in urban environments. It meanwhile facilitates a Bayesian state estimation and behaviour selection[4] so that uncertainty is kept under control and the likelihood of achieving goals is increased.

4.2   Human Factors and Ergonomics

In [5] the authors describe ten challenges for making machines play nicely in joint human-agent activities. These ideas are carried further in [6] on the way

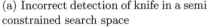

(a) Incorrect detection of knife in a semi-constrained search space

(b) Correct detection of knife in a narrowly constrained search space

Fig. 4. Demonstration of object search improvement upon human's intervention

to a "Theory of Complex and Cognitive Systems". According to Kiesler and Hinds [7] results from human computer interaction (HCI) are generally applicable to interaction with autonomous robots; software agents can be compared to robots in many ways. However, care must be taken as robots physically interact with humans and humans perceive autonomous and especially humanoid robots differently than computers: they feel less responsible [8], build up their mental model of a robots factual knowledge by extrapolating from their own as well as the robots origin and language [9]. Fong et al. conclude: "What are the minimal criteria for a robot to be social?", "How do we evaluate social robots?" [10] and based on Woods [11] recommend "It is not appropriate, or perhaps even necessary, for the robot to be as socially competent as possible. Rather, it is more important that the robot be compatible with the human's needs, that it be understandable and believable, and that it provide the interactional support the human expects".

These requirements are defined to a greater extent for human-machine dialog systems in ISO 9241-10 [12]: suitability for the task, suitability for learning, suitability for individualisation, error tolerance, conformity with user expectations, self descriptiveness, and controllability. Especially the latter three are in contrast to todays most advanced plan-based cognitive systems (see Sec. 4.1). In order to reach system ergonomic integration, we follow an approach of human, technology and organization (HTO/MTO, [13]). Based on human needs, technology and organization are adapted to achieve high efficiency, effectiveness and quality of the overall human-machine system. Our key method here is the analysis of information flow between human and machine as well as cognitive human modelling [14]. For evaluation of social acceptance we intend to adopt the methodology presented by Weiss et al. [15].

Last but not least, as technical systems gradually move towards highly complex socio-technical systems, work in other domains, like health care and especially regarding the hospital as a highly complex socio-technical system becomes relevant [16] in transferring principles of human interaction and assistance.

5   Preliminary Results

We demonstrate our approach by executing a "detect a table setting" scenario as thoroughly discussed by Pangercic et al in [17]. The objective therein is that the mobile robot identifies objects making up the table setting (cutlery and tableware, see Fig. 4 and subsequently cleans them up. The objects are perceived using CCD cameras and sought-after by applying a highly prolific state-of-the-art 3D CAD model-based vision algorithm [18]. Despite its outstanding performance there are still handful of issues that lead to erroneous matches e.g. heavy cutler, specular surfaces and rich textures. Fig. 4 left depicts a result of the search for 5 instances of the object knife on the kitchen table. One can see that thanks to the rich texture all of them are incorrect. In order to amend this, the algorithm's 3D search space has to be constrained and object's pose precisely estimated. In that event we propose a joint action between the robot and the accompanying human. After the robot has explored all onboard integrated solutions and still failed in finding the objects, it consults a fellow human via an audio channel to point into a direction of where the missing knife is located. Human's pointing gesture is identified by a multiple camera setup with the four ceiling-mounted CCD cameras. Gesture recognition is thereby only a subtle scenario in the scope of the markerless human motion tracking as presented in [1]. Final knife's position estimation is deduced from an intersection of the kitchen table plain and line extending human's pointing arm. The situation is depicted in Fig. 5.

Fig. 5. Estimation of object's(knife) position by pointing

6   Conclusion and Outlook

MAC will further enhance the machines' cognition, by augmenting it through human cognition, in situations where this is desireable. This will lead to faster recovery from failure situations and therefore to a higher availability of the cognitive systems. On the long run, cognitive machines will be able to recover from failure situations with help from even technical novices and human companions.

Further, MAC will allow the cognitive machines to better integrate and play nicely in socio-technical systems. The machines will become more controllable, conform to human expectations and support the human in increasing his/her competence. And finally, through feeling to be able to help the technically highly advanced machine, we suppose it will become more congenial to the human.

In order to achieve this, algorithms need to be developed that enable the self-reflection and self-presentation of cognitive plan-based systems and allow them to consult humans, similar to patients consulting a doctor. Evaluation within subject experiments are immanent and methods for assessing social acceptance are needed.

Acknowledgements

This work is funded by the German Research Fundation (DFG) as part of the excellence cluster Cognition for Technical Systems (CoTeSys). The authors would like to thank DFG and the board of CoTeSys.

References

1. Beetz, M., Stulp, F., Radig, B., Bandouch, J., Blodow, N., Dolha, M., Fedrizzi, A., Jain, D., Klank, Kress, I., Maldonado, A., Marton, Z., Mösenlechner, L., Ruiz, F., Rusu, R.B., Tenorth, M.: The assistive kitchen — a demonstration scenario for cognitive technical systems. In: IEEE 17th International Symposium on Robot and Human Interactive Communication (RO-MAN), Muenchen, Germany (2008); Invited paper
2. McDermott, D.: Robot planning. AI Magazine 13(2), 55–79 (1992)
3. Lidoris, G., Klasing, K., Bauer, A., Xu, T., Kühnlenz, K., Wollherr, D., Buss, M.: The autonomous city explorer project: Aims and system overview. In: Proceedings of the IEEE/RSJ International Conference on Intelligent Robots and Systems, San Diego, USA (October 2007)
4. Lidoris, G., Wollherr, D., Buss, M.: Bayesian state estimation and behaviour selection for autonomous robotic exploration in dynamic environments. In: Proceedings of the IEEE/RSJ International Conference on Intelligent Robots and Systems, San Diego, USA (October 2007)
5. Klein, G., Woods, D.D., Bradshaw, J.M., Hoffman, R.R., Feltovich, P.J.: Ten challenges for making automation a "team player" in joint human-agent activity. IEEE Intelligent Systems 19(6), 91–95 (2004)
6. Hoffman, R.R., Woods, D.D.: Toward a theory of complex and cognitive systems. IEEE Intelligent Systems 20(1), 76–79 (2005)
7. Kiesler, S., Hinds, P.: Introduction to this special issue on human-robot interaction. Human-Computer Interaction 19, 1–8 (2004)
8. Hinds, P., Roberts, T.L., Jones, H.: Whose job is it anyway? a study of human-robot interaction in a collaborative task. Human-Computer Interaction 19, 1–8 (2004)
9. Lee, S.l., man Lau, I.Y., Kiesler, S., Chiu, C.-Y.: Human mental models of humanoid robots. In: Proceedings of the 2005 IEEE International Conference on Robotics and Automation, 2005. ICRA 2005, April 2005, pp. 2767–2772 (2005)

10. Fong, T.W., Nourbakhsh, I., Dautenhahn, K.: A survey of socially interactive robots: Concepts, design, and applications. Tech. Rep. CMU-RI-TR-02-29, Robotics Institute, Pittsburgh, PA (December 2002)

11. Woods, D.D.: Decomposing automation: apparent simplicity, real complexity, pp. 3–19. Lawrence Erlbaum, Mahwah (1996)

12. International Organization for Standardization, Ergonomics of Human System Interaction. International Standard; ISO 9241:2006, Geneva, Switzerland: International Organization for Standardization (2006)

13. Strohm, O., Ulich, E.: Unternehmen arbeitspsychologisch bewerten. Ein Mehr-Ebenen-Ansatz unter besonderer Berücksichtigung von Mensch, Technik, Organisation. Zurich, Switzerland: vdf Hochschulverlag AG (1997)

14. Bubb, H.: Future Applications of DHM in Ergonomic Design. In: Digital Human Modeling, pp. 779–793. Springer, Heidelberg (2007)

15. Weiss, A., Bernhaupt, R., Tscheligi, M., Wollherr, D., Kuhnlenz, K., Buss, M.: A methodological variation for acceptance evaluation of human-robot interaction in public places. In: The 17th IEEE International Symposium on Robot and Human Interactive Communication. RO-MAN 2008, August 2008, pp. 713–718 (2008)

16. Carayon, P., Friesdorf, W.: Handbook of Human Factors and Ergonomics, 3rd edn. Human Factors and Ergonomics in Medicine, ch. 59, pp. 1517–1537. Wiley, Chichester (2006)

17. Pangercic, D., Tavcar, R., Tenorth, M., Beetz, M.: Visual scene detection and interpretation using encyclopedic knowledge and formal description logic. In: Proceedings of the International Conference on Advanced Robotics, ICAR (2009)

18. Wiedemann, C., Ulrich, M., Steger, C.: Recognition and tracking of 3D objects. In: Rigoll, G. (ed.) DAGM 2008. LNCS, vol. 5096, pp. 132–141. Springer, Heidelberg (2008)

19. Bandouch, J., Engstler, F., Beetz, M.: Evaluation of hierarchical sampling strategies in 3d human pose estimation. In: Proceedings of the 19th British Machine Vision Conference, BMVC (2008)

How Humans Optimize Their Interaction with the Environment: The Impact of Action Context on Human Perception

Agnieszka Wykowska[1], Alexis Maldonado[2], Michael Beetz[2], and Anna Schubö[1]

[1] Department of Experimental Psychology, Ludwig Maximilians Universität, München, Germany
[2] Computer Science Department, Chair IX, Technische Universität, München, Germany

Abstract. Humans have developed various mechanisms to optimize interaction with the environment. Optimization of action planning requires efficient selection of action-relevant features. Selection might also depend on the environmental context in which an action takes place. The present study investigated how action context influences perceptual processing in action planning. The experimental paradigm comprised two independent tasks: (1) a perceptual visual search task and (2) a grasping or a pointing movement. Reaction times in the visual search task were measured as a function of the movement type (grasping vs. pointing) and context complexity (context varying along one dimension vs. context varying along two dimensions). Results showed that action context influenced reaction times, which suggests a close bidirectional link between action and perception as well as an impact of environmental action context on perceptual selection in the course of action planning. Such findings are discussed in the context of application for robotics.

Keywords: Action Context, Action Planning, Human Perception.

1 Introduction

When humans perform a particular action such as, for example, reaching for a cup in a cupboard, they need not only to specify movement parameters (e.g., the correct width of the grip aperture) but also to select movement-related information from the perceptual environment (e.g., the size and orientation of the cups handle - but not its color - are relevant for grasping). Moreover, the context in which the grasping action is performed may also have an impact on both the action performance and the prevailing selection processes of the agent. If the cup is placed among other cups of different sizes and handle orientations, selection might be more difficult as compared to when the cup would be placed among plates. In the first case, the context varies along at least two dimensions that are relevant for grasping a cup (size and orientation of handles). In the second case, the cup is embedded in a homogeneous context also consisting of dimensions

J.-H. Kim et al. (Eds.): FIRA 2009, CCIS 44, pp. 162–172, 2009.

irrelevant for grasping a cup (breadth of plates). Therefore, the two environmental contexts might result in different processing speed of the environmental characteristics.

Several authors have investigated the influence of intended actions on perceptual processes. For example, Craighero and colleagues [2] showed that when agents were asked to grasp a tilted bar, onset latencies of their movement depended on the characteristics of a visually presented stimulus that signaled when the movement should begin (a so-called "go signal"). If the "go signal" shared action-related features with the to-be grasped object (e.g., was of the same orientation), the onset of the movement occurred faster as compared to the condition when the "go signal" differed from the to-be grasped object. These results support a close link between perception and action.

Similar results were reported by Tucker and Ellis [11]. The authors conducted a study in which participants were presented with natural objects (e.g., grape, cucumber) or manufactured objects (e.g., screw, hammer). The objects could be smaller (grape, screw) or larger (cucumber, hammer) implying either precision grip (small objects) or power grip (larger objects). The task was to categorize the objects as natural or manufactured. Half of the participants had to respond with a power grip to natural objects and precision grip to manufactured objects; the other half had opposite instructions. The results showed that although size of the objects was completely orthogonal to the categorization task, it influenced performance: Precision grips were faster to small objects relative to large objects and power grips were faster to large objects compared to small objects. This suggests that size was implicitly processed as an action-related feature of an object and, as such, had an impact on behavior.

More recent studies demonstrated that visual detection processes are highly dependent on intended action types [1] and that the perceptual system can bias action-relevant dimensions if they are congruent with the performed action [3]. Wykowska, Schubö and Hommel [14] conducted a series of experiments in which they observed action-related biases of visual perception already at early stages of processing. Importantly, these studies showed action-perception links in a situation where action and perception were completely unrelated and decoupled but had to be performed in close temporal order. Participants' task was to detect a visually presented target while they were preparing for a grasping or pointing movement. The target could be an object that deviated in size or in luminance from the other objects. Wykowska and colleagues found that performance in the visual search task was influenced by the intended movement although the movement was not executed but only planned. That is, detection of perceptual dimension (size or luminance) was better when accompanied by the preparation of a congruent movement (e.g., grasping for size and pointing for luminance) as compared to the preparation of an incongruent movement. These results indicate a close link between action and perception that merely coincide in time. Moreover, action preparation affected perceptual processing at the level of early mental representations.

These observations support the Theory of Event Coding [5] which postulates a common representational code for perception and action. We claim that such a common code implies bidirectional links that are used in such a way that action can have an impact on perception no less than perception can have an impact on action, for example, when the missing perceptual features of the cup are being specified for the grasping movement. It is obvious that action planning is informed and controlled by perceptual processes that provide information about the objects we act upon and the changes of our environment that we intend to achieve. Perhaps less obvious is the need for perceptual processes to be constrained and guided by action planning. Successful acting requires the focusing and selection of action-relevant information, so to adapt the action to the environmental conditions. That is, setting up a particular action plan may increase the weights for perceptual dimensions that are relevant for that action - intentional weighting in the sense of [5]. The aim of the present study was to investigate not only the impact of action planning on perceptual processes but also the impact of the context in which the action occurs. The action context might influence not only behavior and performance of an agent [12] but also the agents perceptual stages of processing.

2   Experimental Paradigm

A paradigm was used in which participants were asked to perform a perceptual task (an efficient visual search task) in close temporal order to a motor task. The visual search task consisted in detecting the presence of either a size or a luminance target (in separate blocks of trials) embedded in a set of other, distracting items. The motor task consisted in grasping or pointing to a specific object on the Movement Execution Device. The motor task should be planned before the visual search task but executed only after the search task was completed. In other words, while participants were performing the visual search task, the grasping or pointing movement was prepared but not yet executed. The action context consisted of objects of various sizes and surface luminance that were mounted on a movement execution device (MED), see Figure 2. The action context could either be simple in which the objects varied along only one dimension (luminance) or complex in which the objects varied along two different dimensions: size and luminance.

2.1   Participants

Nineteen paid volunteers (16 women, mean age: 23.3) took part. Participants were asked to perform a practice session on the day preceding the experiment. The practice session consisted of the motor task only, without the visual search task.

2.2   Stimuli and Apparatus

The search display contained 28 grey circles positioned on three circular arrays. The target could appear on one of four positions on the middle circular array

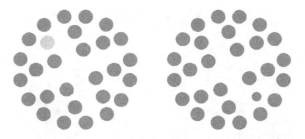

Fig. 1. Examples of visual search displays that contained a luminance target (left) or a size target (right)

at the upper left/right or lower left/right from the middle point. The target was defined either by luminance (lighter grey, cf. Figure 1, left) or by size (smaller circle, cf. Figure 1, right).

Below the computer screen, the movement execution device (MED) was positioned. It consisted of a box containing eight holes positioned also on a circular array. Round plastic items that could vary in luminance and size each covering a LED could be attached and detached from the box (cf. Figure 2). For the purpose of this experiment, we used the following combination of objects:

1. The simple action context (luminance dimension only) consisted of four grey, medium-sized objects and four objects that differed in luminance: two being darker and two being lighter than the standard elements (cf. Figure 2a).
2. The complex action context (luminance and size dimensions) consisted of four grey, medium-sized items and four grey items that differed in size, two being smaller and two being larger than the standard elements (cf. Figure 2b).

Fig. 2. Movement Execution Device (MED). Simple action context (A) consisting in objects that vary along the luminance dimension (LED lighting up and the objects on MED are all defined by luminance) and complex action context (B) consisting in objects that vary along the luminance dimension (LED lighting up) and size dimension (items on MED).

Fig. 3. Experimental setup. Participants were asked to perform the movement task (grasping or pointing) on the MED with their left hand. The movement ended when participants returned with their hand to the starting position (space bar on the keyboard). The search task was to be performed with the right hand on the computer mouse (target present: left mouse key; target absent: right).

2.3   Procedure

Participants were seated at 80 cm distance from the computer screen on which the visual search task was presented. The MED was positioned below the computer screen, in front of the participants (cf. Figure 3).

Each experimental trial began with a fixation cross displayed for 500 ms. Subsequently a movement cue (cf. Figure 4) appeared for 1000 ms. Participants were instructed to prepare for the movement but not execute it until a signal from the MED would appear. The cue was followed by a blank display (500 ms) and, subsequently, by the search display presented for 100 ms. A blank screen followed the search display and remained on the computer screen until participants responded to the search task. Upon response to the search task, one of the LEDs on the MED lit up for 300 ms. This event constituted the signal for participants to execute the prepared movement, i.e., to either point to or grasp the object that lit up. After the experimenter registered the participants movement with a mouse key, a new trial began (cf. Figure 4).

Participants were asked to be as fast and as accurate as possible in the search task. Speed was not stressed in the movement task. The experiment consisted of one practice block and 8 experimental blocks with 112 trials each. The two target types (size and luminance) were presented in separate blocks and also the configuration of objects on the MED was changed block-wise. The order of blocks was balanced across participants. The movement task was randomized within each block of trials.

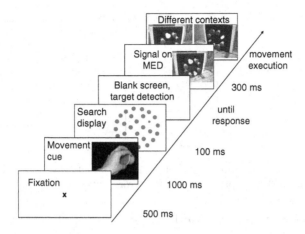

Fig. 4. A trial sequence. Each box represents an event of a trial (e.g., presentation of a fixation cross, of a movement cue). Presentation time for each of the event is shown on the right side of the figure.

2.4 Data Analysis

The analysis focused on the impact of context complexity on perceptual processing. Mean reaction times (RTs) and error rates in the search task were submitted to statistical analyses. Trials with incorrect movement execution and incorrect search trials were excluded from the analysis. From the remaining data, individual mean RTs in the search task were submitted to analysis of variance (ANOVA) with: target presence (target present vs. absent), target type (luminance vs. size), movement type (point vs. grasp), and action context (one vs. two dimensions) as within-subject factors.

If the environmental action context of a movement has an impact on perceptual processing of action-relevant characteristics, then the simpler context (varying along one dimension) should yield better performance in the search task as compared to the context consisting of two-dimensions.

3   Results

The statistical analysis showed that action context had a significant effect on search performance, $F(1, 18) = 5.5$, $p < .05$ indicating that the complex action context elicited slightly longer RTs in the search task ($M = 398$ ms, $SEM = 15$) as compared to the simple context ($M = 392$ ms, $SEM = 16$). Moreover, this effect was more pronounced for the grasping movement (simple context: $M = 390$ ms, $SEM = 15$; complex context: $M = 398$ ms, SEM $= 15$) as compared to the pointing movement (simple context: $M = 395$ ms, $SEM = 15$; complex context: $M = 398$ ms, $SEM = 16$), as indicated by the significant interaction of movement type and action context, $F(1, 18) = 5$, $p < .05$ (cf. Figure 5).

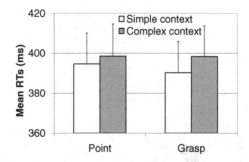

Fig. 5. Mean reaction times in the search task as a function of movement type and action context. Error bars represent standard errors of the mean (SEM).

Table 1. Mean error rates in the search task as a function of movement type (point or grasp) and action context

	Mean error rates (%) and SEM (in brackets)	
	Movement type	
	Point	Grasp
Action context		
Simple context	2.85 (0.5)	2.92 (0.7)
Complex context	2.86 (0.6)	2.96 (0.8)

Analogous analysis for error rates in the search task showed no significant effect, all $p > .5$. However, the small absolute differences were in line with the effects in RTs (cf. Table 1).

Moreover, the present results replicated previous findings [5] by observing the influence of intended movements on perceptual dimension: interaction of movement type and target type was significant, $F(1, 18) = 5$, $p < .05$ indicating again improvement in visual search performance for congruent movements as compared to incongruent movements: Size detection was faster when a grasping movement was prepared relative to the size-incongruent movement (pointing).

4   Discussion

The aim of the present study was to investigate the influence of action context on perceptual processes during movement preparation. Results showed that action context influences already early stages of processing, namely visual selection. Target detection in the visual search task was faster when the context in which the movements were executed consisted of only one dimension (luminance) as compared to two dimensions (size and luminance). Interestingly, the action context effect was more pronounced for the more complex movement (grasping) as compared to the simpler pointing movement. Importantly, the design of the present experiment made the movement execution (together with its context) completely decoupled from the perceptual task. Therefore, the influences of action context on performance cannot be attributed to late stages of the whole processing stream, i.e., as perceptual consequences of the movement execution. Rather, the present results show that action context influences already early perceptual processing. Such influence suggests that the perceptual system makes use of environmental hints when selecting action-relevant features of the environment. It is important to note that the present results were obtained in a rather restricted laboratory setting where all parameters were strictly controlled so that other factors such as prior knowledge or experience with the respective action context could have no influence on the results. This might have resulted in relatively small absolute differences in reaction times. These effects would probably become larger once they were examined in more natural settings, such as, for example, a real kitchen scenario.

The present findings can be interpreted in a broad, evolutionary context. Humans must have developed mechanisms that optimize their interaction with the environment. When planning actions, humans select specifically action-relevant information from environment. This might have been obtained by means of developing a common representational code [5] that allows for fast and efficient action-perception transition. Evidence for such codes comes from behavioral data [5] and from neurophysiology [9]. Moreover, the discovery of mirror neurons, e.g., [7] provided strong support for the coupling of action and perception. Although some authors postulate that the fast "how/where" pathway for action is separate from the slower "what" pathway for perception and conscious processing [4], recent evidence shows much closer interactions [8]. A prominent example of evidence for such a fast neuronal route linking perception with action was described by Humphreys and Riddoch [6]. The authors examined a patient MP who suffered from unilateral neglect. Following damage to the fronto-temporal-parietal regions of the right cerebral hemisphere, the patient was unable to direct his attention to objects presented in his left visual field: a symptom typical for unilateral neglect patients. When presented with a simplified visual search task, he was impaired in finding objects defined by their perceptual features when they were presented in his neglected field. For example, when asked to detect a red object, he was unable to detect a red cup. Importantly, the patient was not impaired on color discrimination in general (that is, patient MP was able to name the objects color when presented as a single item - in a condition when neglect symptoms usually do not occur). Interestingly, when the targets were defined by action-related features of the objects (e.g., look for something to drink from), the patients performance improved markedly also in the neglected field. This indicates that there might be a fast and efficient link between action and perception, a so-called pragmatic processing route that facilitates extraction of action-relevant features of the environment.

The present results suggest that the interactions between perception and action might be even more complex. Our results showed that not only intended actions play a role in perceptual selection processes but also environmental contexts in which the actions are performed. If the context varies along several dimensions, perceptual selection is presumably more difficult. When humans are preparing a movement, they try to optimize their performance through efficient selection of perceptual information that might be relevant for that movement. This optimization process might be especially important for complex movements as they are more difficult. Therefore, humans might benefit from the environmental setting (i.e., a simple action context) more in case of more difficult interaction (more complex movement) as compared to a movement that is simple enough not require such a level of optimization.

5   Implication for Robotics

As argued above, humans have developed certain mechanisms that allow for optimization of action control. When designing a robotic system, one might take into account similar solutions especially when dealing with fast real-time systems that produce very rich and fast data streams.

Due to the limited computing power available on current robots, it is advantageous to use the context information to choose which algorithms process the sensory stream. For example, in the case of the robots that serve as a demonstration scenario in our kitchen environment (cf. Figure 6), the high-level control system informs the camera-based perception system about the action that will be performed, and the needed update speed. For the task of cleaning the table, this activates a system that fits 3D CAD models of cups and other typical objects in the image stream, which is very expensive computationally. After the objects are detected, the processing power is devoted to the sensory information coming from the robotic arm and hand to correctly grasp the object. Both data

Fig. 6. Two robots for manipulation in human environments. Left: a B21r robot with Powercube robotic Arms. Right: Kuka LWR arm with a DLR-HIT hand. Both perform pick and place tasks in a kitchen at the Intelligent Autonomous Systems Laboratory, TUM. The high amount of sensory data and limited computing power on board the robots make it necessary to apply only the specific algorithms that deliver information needed for the current action.

processing systems can still run in parallel, with priority being given to the one that is most important for the task at hand.

Our current research draws inspiration from the findings described in this paper to design a system for grasping using the multi-finger DLR-HIT hand (cf. Fig. 6) that has paired control/perception models. A set of grasping controllers was developed, one for each type of basic grasp movement (e.g. power grasp, precision pinch, and others). We learn an associated perception model on the sensory stream from the fingers (joint position and torque), and train it using examples of correct grasps and incorrect ones. When executing a grasp, only the associated model is fed with the sensory stream coming from the hand, and the system can monitor the success of the action. This design also complies with the ideas of Wolpert and Kawato of multiple interconnected forward and inverse models [13].

Finally, it is important to note that successful human-robot interaction requires good temporal synchronization, especially when joint attention might prove beneficial. Imagine the following: a robot assists a human in the kitchen. The humans attention is captured by a falling cup behind the robots back. In order to rescue the cup, not only must the robot make use of human-specific attention directing hints (such as, for example, gaze direction) [10] but also its processing system needs to be temporally synchronized (within the milliseconds range) with the humans system. Moreover, in order optimize the movement the robot needs to select the most relevant information from the environment for efficient fast processing and action programming. Therefore, implementing mechanisms analogous to those governing human perception and action planning into artificial systems might prove indispensable for human-robot interaction.

6   Conclusions

The present experiment showed that not only action planning but also the environmental context in which an action takes place affects perceptual processing. The present results have been interpreted in a broad evolutionary context postulating that humans have developed means of optimizing action planning by efficient selection of relevant information from the environment. Selection depends on the complexity of the task at hand and the congruency between the intended action and the given environmental setting. Artificial systems designed to interact with humans need to take such interactions and constraints into account to allow for temporal compatibility and fluency between human and robot movements in joint action tasks.

Acknowledgements

This study was supported by the Deutsche Forschungsgemeinschaft (Excellence Cluster Cognition for Technical Systems (CoTeSys), Project # 301).

References

1. Bekkering, H., Neggers, S.F.W.: Visual search is modulated by action intentions. Psychological Science 13, 370–374 (2002)
2. Craighero, L., Fadiga, L., Rizzolatti, G., Umiltà, C.A.: Action for perception: a motor-visual attentional effect. Journal of Experimental Psychology: Human Perception and Performance 25, 1673–1692 (1999)
3. Fagioli, S., Hommel, B., Schubotz, R.I.: Intentional control of attention: Action planning primes action related stimulus dimensions. Psychological Research 71, 22–29 (2007)
4. Goodale, M.A., Milner, A.D.: Separate visual pathways for perception and action. Trends in Neurosciences 15, 20–25 (1992)
5. Hommel, B., Müsseler, J., Aschersleben, G., Prinz, W.: The theory of event coding (TEC): A framework for perception and action planning. Behavioral and Brain Sciences 24, 849–937 (2001)
6. Humphreys, G.W., Riddoch, M.J.: Detection by action: neuropsychological evidence for action-defined templates in search. Nature Neuroscience 4, 84–89 (2001)
7. Rizzolatti, G., Craighero, L.: The mirror-neuron system. Annual Review of Neuroscience 27, 169–192 (2004)
8. Rossetti, Y., Pisella, L., Vighetto, A.: Optic ataxia revisited: visually guided action versus immediate visuomotor control. Experimental Brain Research 153, 171–179 (2003)
9. Schubotz, R.I., von Cramon, D.Y.: Predicting perceptual events activates corresponding motor schemes in lateral premotor cortex: An fMRI study. Neuroimage 15, 787–796 (2002)
10. Sebanz, N., Bekkering, H., Knoblich, G.: Joint action: bodies and minds moving together. Trends in Cognitive Sciences 10, 70–76 (2006)
11. Tucker, M., Ellis, R.: The potentiation of grasp types during visual object categorization. Visual Cognition 8, 769–800 (2001)
12. Turner, R.M.: The context-mediated behavior for intelligent agents. International Journal of Human-Computer Studies 48, 307–330 (1998)
13. Wolpert, D.M., Kawato, M.: Multiple paired forward and inverse models for motor control. Neural Networks 11, 1317–1329 (1998)
14. Wykowska, A., Schubö, A., Hommel, B.: How you move is what you see: Action planning biases selection in visual search. Journal of Experimental Psychology: Human Perception and Performance (in press)

Development of a Virtual Presence Sharing System Using a Telework Chair

Yutaka Ishii[1] and Tomio Watanabe[2,3]

[1] Information Science and Technology Center, Kobe University, 1-1 Rokkodai, Nada, Kobe,
Hyogo 657-8501, Japan
ishii@kobe-u.ac.jp
[2] Faculty of Computer Science and System Engineering, Okayama Prefectural University,
111 Kuboki, Soja, Okayama 719-1197, Japan
[3] CREST of Japan Science and Technology Agency
watanabe@cse.oka-pu.ac.jp

Abstract. There has been much discussion on remote communication support for a telework that will enable employees to work at remote offices. We have already developed a remote communication support system via embodied avatars based on users' behaviors. However, there are various problems associated with an avatar-mediated interaction, particularly with regard to the relation between users and their avatars. In this study, we propose the concept of a presence sharing system Ghatcha [GHost Avatar on a Telework CHAir] in which the users' embodiment is not indicated by the avatars but by the chairs that suggest the presence of avatars. This system provides the same communication space for the users' embodiment, thus creating a feeling of working alongside remote workers. Moreover, the effectiveness of the prototype system is confirmed in the experiment.

Keywords: Embodied Interaction, Avatar, Remote Communication, Telework, Remote Operating Chair.

1 Introduction

A telework increases productivity and operational efficiency by offering employees the flexibility to work from their home offices. A telework would gain popularity as it can be utilized in different ways. However, the quality or efficiency of work might deteriorate as a result of a telework as it leads to a sense of isolation or a lack of concentration. Thus, it is important to examine remote collaboration support in detail. Remote collaboration has various purposes and applications, and it is expected to support for each situation. The subjects of this research are not remote users performing a group task but individual users performing their own specific tasks wherein all their co-workers also perform tasks with the same aim such as a job of a home-based worker or individually pursuing online distance learning. For example, these include software developments, data inputs, and assembling parts together. There has been an intensive discussion on and a remarkable improvement in the remote collaboration

J.-H. Kim et al. (Eds.): FIRA 2009, CCIS 44, pp. 173–178, 2009.
© Springer-Verlag Berlin Heidelberg 2009

support systems for the group task [1],[2]. These systems are constructed by a video image, voice, or virtual reality technique, and these techniques are quite effective for remote collaboration or realistic communication.

In the case of tasks that are not synchronized, however, the video image might contradict our expectations. In order to solve the problem, Honda et al. proposed a virtual office system "Valentine" using an awareness space and provided a work support environment for home-based workers [3]. However, when a user's own avatar is used as a communication media for an embodied interaction, many issues arise with regard to the relation between the users and their avatars. For example, if a human-type avatar is used, the correspondence of the user's motion and that of the avatar's would be hindered by input devices. Otherwise, the appearance of an avatar cannot appropriately represent a user's embodiment. We have developed an avatar mediated communication system for remote users using a human type avatar called a "VirtualActor" The importance of the relation between the users' behavior and that of their avatars' has been confirmed by the communication experiment [4].

In this research, we propose a new communication system using not the explicit virtual avatar but a chair in which a user's embodiment is represented. The chair motions indicate the presence of the remote users with the implicit avatar in the same communication space. Wesugi et al. have so far developed a chair communication system called "Lazy Susan" as a motion sharing system for remote users [5]. Their system introduces the presence of remote users by sharing the mutual chair motions of the users, which is linked to the partner's chair motion. However, the system was not evaluated from the viewpoint of the interaction with the avatar in the same virtual communication space. In human interaction, both users would interpret the chair motions in a different manner on the assumption that the partner behaves in the same way within the same interface. The present study aims to recreate the environment in which remote users interact with their co-workers in the same virtual office, and enhance their motivation in performing their tasks. In this paper, a prototype of the system using a virtual environment is developed, and the effectiveness of this system is demonstrated by an evaluation experiment.

2 Avatar-Mediated Interaction

The schema of a remote embodied interaction system is shown in Fig. 1. In avatar-mediated interaction, the relation between users and avatars is developed from avatar information, environment information, and the input devices that connect a user to the system. It would be necessary to analyze and synthesize these factors systematically for the development of a human-oriented interaction support system. In addition, the recognition model of interaction awareness via communicative avatars is shown in Fig. 2. User A identifies avatar A as "himself/herself" in the virtual communication space through the recognition of the correspondence between his/her behavior and that of the avatar's. Avatar A, which visualizes the communicative behaviors of user A and avatar B as the substitute of user B, can have embodied interaction in the same space. The embodied relation of a user and his/her own avatar is very important when designing the interface as it enables effective interaction awareness in the shared communication space.

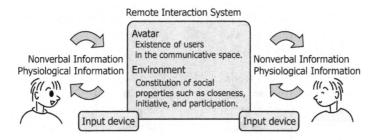

Fig. 1. Embodied avatar-mediated interaction

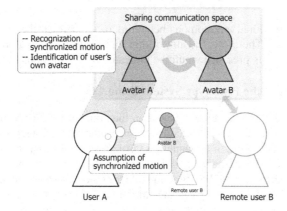

Fig. 2. Recognition model of interaction awareness via communicative avatars

3 Presence Sharing System for a Telework: Ghatcha

3.1 Concept of the System

We have developed an avatar-mediated communication system using a human type avatar called "VirtualActor" for the embodied interaction in the previous chapter, and the effectiveness of the system has been confirmed [4]. Moreover, we have investigated the importance of a mutually shared embodiment by the communication experiments using the system. However, as shown in Fig 1, the embodiment is not always indicative of the avatar information in the input/output of the embodied interaction system. It would be useful to integrate the embodiment with the environment information for the development of an effective interaction support system. Hence, this paper proposes a new presence sharing system called Ghatcha: GHost Avatar on a Telework CHAir. The Ghatcha system is based on the embodiment of the environment information of the chair motion rather than that of the avatar.

The concept model of the system is illustrated in Fig 3. The user is able to identify the existence of someone from the motions of the chair, which responds to the user in the same space. In the case of voice speech, the voice of the partner leads the user to assume that the partner is present. According to the study of the intentional motion of

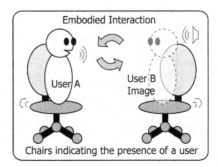

Fig. 3. Concept of the system

the chair as the intentional autonomous behaviors of an artifact, the chair evinces the existence of the user. Therefore, the aim of this study is to develop a presence sharing system wherein a user's presence would be indicated by the motions of the chair.

3.2 Development of the Prototype System Using CG

The prototype system was developed based on the concept mentioned in the previous section. In this system, the chair motions are measured by various sensors such as a gyroscope, an accelerometer, or a magnetic sensor. The virtual chair motions are represented based on the measurements, and is shared on the network. The mutual motions of each user are transmitted to the office model from the shared communication space. This collaborative system determines the third interaction space with the chairs for each remote co-worker. The prototype system is generated by an HP workstation xw4200, OS: Windows XP Professional SP1, and DirectX9.0b. The frame rate is 30 fps. The chair motions are measured by a laser sensor mouse (Logicool MX Air) attached under the chair. The communication scene using the system is shown in Fig. 4. This example displays only the user's human type avatar.

Fig. 4. Communication scene using the prototype system

4   Evaluation of the Experiment

4.1   Experimental Setup

The system evaluation experiment was performed by the prototype system using CG in the previous chapter. The subjects consisted of 10 pairs, and they worked on a simple task wherein they made paper cranes by folding pieces of paper. The task was repeated twice using two scenes: one where the chair system was connected with the motions of the user and another where they were not connected. The subjects were ordered to fold the papers as much as they could. After the task was finished in each scene, the user's behavior was observed during a waiting period of 3 minutes. The only information that was shared through the system was the motion of the chair. Only the user's human type avatar was displayed in addition to both the users' chairs. Thus, the user makes his/her presence felt not as the chair but as the avatar. The partner's avatar was not represented in the virtual space. The subjects answered the questionnaire after the task in each scene. They were provided an explanation of the conditions and the setting of the experiment, and they agreed to the experiment before the experiment started. The time taken to conduct the experiment was about 40 minutes on average including the waiting time and the time taken to answer the questionnaire.

4.2 Sensory Evaluation

The two scenes were evaluated on a seven-point bipolar rating scale ranging from –3 (lowest) to 3 (highest), in which 0 denotes a moderate score. For the sake of convenience, the results of the means and the standard deviations are shown in Fig. 5. The questionnaire consisted of eight categories: four categories on the impression of the work and the other four categories on the evaluation of media communication. In most of the categories, the significant difference between the two scenes was obtained by administering the Wilcoxon's rank sum test; a significance level of 0.1% for the items of "Do you feel like sharing the same space with a partner?", and "Do you feel like working together with a partner?" A significance level of 1% was obtained for the items "Do you enjoy your task?" "Do you believe that you could associate yourself with the character?" and "Do you recognize a partner's motion?" The effectiveness of the prototype system is evinced by the positive evaluation of each category in the scene that the chair motions were connected. The scene that the chair motions weren't connected has a negative evaluation that is significant at the 5% level for the item "Are you bored by your task?"

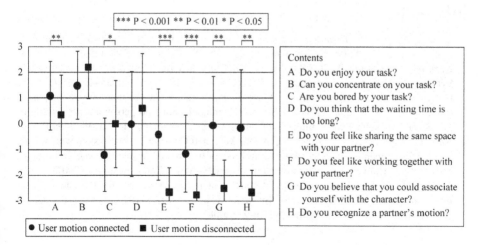

Fig. 5. Results of the questionnaire

5 Consideration

The task in the evaluation experiment of the prototype system was a simple one where participants had to fold a piece of paper into a figure of a crane. The waiting time for carrying out the task was arranged. This experimental setup was prepared for evaluating the shared interaction awareness during the time when the users worked so as to reduce their sense of isolation or lack of concentration. The experimental time was about 40 minutes on average. Therefore, the essential effectiveness of this system may not be determined for the purpose of long-term support. These studies should be further investigated. The users in the positive group of the possession to the avatar were effectively evaluated for the item on the recognition of the partners' motion. In other words, if a user cannot place himself/herself in the communication media, the interaction awareness would be obstructed by the indifference of the partner. By "self media-izing" a user can assume the role of the avatar in the media space, and this would lead the user to identify with the avatar in the interaction space. Accordingly, these features are important for the development of an effective interaction system. A more direct interaction design would be required using the user's embodiment for interaction awareness.

6 Conclusion

In this paper, the concept of the presence sharing system using a chair for telework was proposed, and the prototype of the system using a virtual environment was developed. Moreover, the evaluation experiment was performed using the prototype of the system, and the effectiveness of the system was demonstrated by a sensory evaluation.

The distinctive feature of this system lies in the usage of the chair as environment information based on a user's embodiment instead of the avatar as a substitute in the communication space. A cooperative work environment can be effectively created by this system because users can freely arrange a human-type or abstract character as their co-workers' avatars on each chair and perceive the relation among the co-workers.

References

1. Ishii, H., Miyake, N.: Toward an Open Shared Workspace: Computer and Video Fusion Approach of TeamWorkStation. Communications of the ACM 34(12), 37–50 (1991)
2. Kuzuoka, H., Kosuge, T., Tanaka, M.: GestureCam: A Video Communication System for Sympathetic Remote Collaboration. In: Proceedings of CSCW 1994, pp. 35–43 (1994)
3. Honda, S., Tomioka, H., Kimura, T., Oosawa, T., Okada, K., Matsushita, Y.: A company-office system "Valentine" providing informal communication and personal space based on 3D virtual space and avatars. Information and Software Technology 41, 383–397 (1999)
4. Ishii, Y., Watanabe, T.: An Embodied Avatar Mediated Communication System with VirtualActor for Human Interaction Analysis. In: Proc. of the 16th IEEE International Conference on Robot and Human Interactive Communication (RO-MAN 2007), pp. 37–42 (2007)
5. Wesugi, S., Miwa, Y.: "Lazy Susan" Chair Communication System for Remote Whole-Body Interaction and Connectedness. In: Proc. of the Third IASTED International Conference Human-Computer Interaction, pp. 93–99 (2008)

PLEXIL-DL: Language and Runtime for Context-Aware Robot Behaviour

Herwig Moser[1], Toni Reichelt[2], Norbert Oswald[3], and Stefan Förster[3]

[1] University of Stuttgart, Germany
herwig.moser@ipvs.uni-stuttgart.de
[2] Chemnitz University of Technology, Germany
tonr@hrz.tu-chemnitz.de
[3] EADS Military Air Systems, Germany
{norbert.oswald,stefan.foerster}@eads.com

Abstract. Faced with the growing complexity of application scenarios social robots are involved with, the perception of environmental circumstances and the sentient reactions are becoming more and more important abilities. Rather than regarding both abilities in isolation, the entire transformation process, from context-awareness to purposive behaviour, forms a robot's adaptivity. While attaining context-awareness has received much attention in literature so far, translating it into appropriate actions still lacks a comprehensive approach. In this paper, we present PLEXIL-DL, an expressive language allowing complex context expressions as an integral part of constructs that define sophisticated behavioural reactions. Our approach extends NASA's PLEXIL language by Description Logic queries, both in syntax and formal semantics. A prototypical implementation of a PLEXIL-DL interpreter shows the basic mechanisms facilitating the robot's adaptivity through context-awareness.

Keywords: Context-awareness, Description Logic, PLEXIL, Robotics.

1 Introduction

Coping with the complexities of application scenarios social robots are faced with, requires an understanding of the environmental circumstances and acting in response to that knowledge. This connection between *Context-Awareness* (*CA*) and responsive action results in *adaptive behaviour*, respecting an ever changing environment.

The last decade has seen a surge in research in the area of CA, to a great extent coming from the mobile devices and ubiquitous computing sector as reported by [1,2]. While robotics in general is able to benefit from those research results, actions taken by mobile devices are limited when compared to the physical embodiment of a robot. Lacking the need for complex responses, CA research for mobile devices often falls short when it comes to translating awareness into action using a comprehensive *uniform* approach.

J.-H. Kim et al. (Eds.): FIRA 2009, CCIS 44, pp. 179–186, 2009.

Even though there are CA frameworks [3,4,5,6] dealing with responsive action, they view CA and application in isolation. In principle, these frameworks provide context-to-action maps, where the *action* is only a *placeholder* for some *external* application functionality, in form of a function (Gaia [3]), method (RCSM [7] and the framework in [6]), service (SOCAM [8]) or "behaviour" (CASS [5]). Application code is thus developed conventionally, the ensuing adaptivity provided by an intermediary, hiding the interaction of context and functionality. In contrast, CA languages, such as COP [9], Olympus [10] and PLUE [11], attempt to make the association of context with functionality *explicit*. In [9], the author presents the Python-based *Context-Oriented Programming (COP)* framework. The COP idea is to have code skeletons "filled" at run-time with the implementation most pertinent to the current context, but lacks means to express reactions to changing context. The "Olympus" [10] approach provides a collection of C++ classes, handling context-sensitive behaviour based on the Gaia framework [3]. Context and CA are not first-class entities of the language but merely represented by plain C++ features. The Java-based "PLUE" [11] language allows for the specification of "states" which specify consistency and action conditions, contingent on context. States can evolve via context-triggered "transitions." In PLUE, context is modelled using an informal semantic network, precluding context reasoning through logical inferences.

In this paper, we present a *Description Logic (DL)*-based extension to NASA's concise but computationally complete [12] *Plan Execution Interchange Language (PLEXIL)* [13], such that CA, achieved through *DL-reasoning*, forms an *integral* part of the language. Based on the formal framework in [12], we integrated DL queries [14] into the language syntax and semantics and provide a software run-time capable of executing the extended language, called *PLEXIL-DL*. Representing a unified approach, PLEXIL-DL allows expressive context-conditions specified in a language powerful enough to formulate the complex responses necessary to drive adaptive robot behaviour.

2   Description Logic

Description Logic denotes a family of expressive knowledge representation formalisms. It has gained in attention in recent years, due to its use as the basis for the *Web Ontology Language (OWL)* [15] but also has found use in CA research as a way to represent context and exploit the power of DL reasoning [16].

DL describes domain knowledge in terms of a *Knowledge Base (KB)* $\mathcal{K}$, which consists of a *TBox* $\mathcal{T}$ and an *ABox* $\mathcal{A}$. While $\mathcal{T}$ describes general properties of domains, defining the relations between *concepts*, $\mathcal{A}$ contains statements about concrete instances of concepts, called *individuals*. Note that the assertions in $\mathcal{A}$ may change over time, reflecting newly gained knowledge. Example 1 presents a small KB, where $I_{\mathcal{K}}$ denotes the set of individuals present in $\mathcal{K}$.

Example 1 (Knowledge Base). *Let $\mathcal{K}$ consist of the tuple $(\mathcal{T}, \mathcal{A})$, with*

$$\mathcal{T} = \{Obstacle \equiv PhysicalObject \sqcap \exists crosses.Path\} \ and$$
$$\mathcal{A} = \{PhysicalObject(o), Path(p), crosses(o, p)\}.$$

```
AvoidObstacle:
{
    StartCondition: LookupOnChange(ObstacleSensor_obstacleInPath) == True;
    PreCondition: LookupNow(FuelGauge) - evasionCost > fuelThreshold;
    Command: evade();
}
```

Listing 1. Example of a PLEXIL node

$\mathcal{T}$ *defines the Obstacle concept to be any PhysicalObject that crosses some Path. $\mathcal{A}$ asserts that the individual o is a PhysicalObject and that it crosses p which is a Path. Accordingly, $I_{\mathcal{K}} = \{o, p\}$.*

We will briefly present the notion of *conjunctive ABox queries*, see [14] for details. A conjunctive ABox query Q consists of conjunctions of DL concept and role membership expressions, of the form $C(a)$ and $R(a, b)$, where a, b can either be individual names or existentially quantified variables. Query satisfaction by $\mathcal{K}$ is denoted as $\mathcal{K} \models Q$. The expression $\langle x_1, \ldots, x_n \rangle \leftarrow Q$ denotes that the *distinguished* variables $x_1, \ldots, x_n$ occurring in Q must be bound to individual names, forming the query answer. The *answer set* of a query Q is the set of n-ary tuples such that $\{\langle a_1, \ldots, a_n \rangle \in I_{\mathcal{K}}^n \mid \mathcal{K} \models Q[x_1/a_1, \ldots, x_n/a_n]\}$, where $Q[x_i/a_i]$ represents the query Q, where every occurrence of distinguished variable x_i is substituted by the individual name a_i. We will denote the answer set of Q w.r.t. $\mathcal{K}$ as answer$_{\mathcal{K}}(Q)$.

Example 2 (Query). *Let $\mathcal{K}$ be the KB of Example 1. Let $Q = Obstacle(x)$ be a query asking for Obstacles, where x is a distinguished variable such that $\langle x \rangle \leftarrow Q$. Then answer$_{\mathcal{K}}(Q) = \{\langle o \rangle\}$ because $\mathcal{K} \models Q[x/o]$.*

3   PLEXIL

PLEXIL is a "high-level plan execution language" [12], where plans represent the output of an automated or mixed-initiative planning process [13]. See [17] for the full language specification.

The basic building block of a PLEXIL plan is a *node*. Node execution can be made contingent on changes in external world state. Monitoring such changes, through so called *lookup* operations, can be incorporated in *node conditions*, causing nodes to respond to environmental circumstances. PLEXIL differentiates between two kinds of lookup operations, LookupNow and LookupOnChange. The former performs an immediate lookup, while the latter performs a LookupNow operation and if the returned value does not satisfy the overall condition, passively waits for the environment to change.

Listing 1 shows a PLEXIL node which becomes active as soon as an obstacle is in the path, under the precondition that there is enough fuel left.

3.1   Semantics

In this section we present an excerpt of a formal semantics of PLEXIL, described in [12], being relevant to our DL extension.

External World State. Let Σ represent the external world state, constituted as a set of associations of type $\langle \mathbf{X} = v \rangle$, where $\mathbf{X}$ is an external state variable name and v its value. A plan holds a local copy of Σ, represented by Γ, which is updated from Σ upon an external event and thus may not contain the same *values* as Σ in-between updates. The sets Σ and Γ are called *environments* and assumed to be functional, i.e., if $\langle \mathbf{X} = v \rangle$ and $\langle \mathbf{X} = w \rangle$ are both in the same environment then $x = w$. Lookup operations access the environment exclusively via Γ to guarantee consistency due to potential changes in Σ during an execution cycle. PLEXIL supports a limited number of possible value domains for external variables: integers, floats, ternary truth values and character strings [17].

Execution Semantics. A plan executes in discrete time steps referred to as *macro steps*, which are triggered upon external events. Incoming events are processed in order of their reception and only one event is taken care off at a time. The processing of an event and *all its cascading effects* proceeds until *quiescence*, i.e., no more node state transitions are enabled [17]. This is referred to as *run-to-completion* semantics [12].

Macro steps are modelled using a so called *macro relation*. We present only that part of the macro relation definition which is relevant for our extension (cf. [12]). The occurrence of an external event is indicated by the abstract predicate $\texttt{event?}(\Sigma_i, \Gamma, \pi)$, where π is the internal state of the PLEXIL plan and Σ_i the external environment before the i-th macro step iteration. In case of an event, the macro relation is defined to transfer the values of Σ_i to the local environment Γ, denoted Γ' after the update, and trigger a quiescence cycle. Equation (1) defines this transfer process.

$$\Gamma' = \begin{cases} \Sigma_i & \text{if } \texttt{event?}(\Sigma_i, \Gamma, \pi) \\ \Gamma & \text{otherwise} \end{cases} \tag{1}$$

4   PLEXIL-DL

We extend the PLEXIL syntax and semantics by allowing DL queries to appear in lookup expressions, making node execution contingent on query results. The extensions of the environment Σ and its transfer to Γ forms the core of PLEXIL-DL semantics. Whereas in [12], Σ is defined as a set of native PLEXIL values and their associated identifier, we extend the set of identifiers to also represent DL queries and the values to be corresponding query results. Since we preserve PLEXIL operator semantics, results of query expressions ultimately need to yield a value compatible with the PLEXIL type system.

4.1   Semantics

Extended External World State. Let Σ_Q represent that part of the external environment, such that for all associations $\langle \chi = \nu \rangle \in \Sigma_Q$, χ is interpreted as a DL query and ν its answer set, i.e., $\nu = \texttt{answer}_{\mathcal{K}}(\chi)$ w.r.t. $\mathcal{K}$ (see Section 2).

Changes in $\mathcal{K}$ which affect $\mathsf{answer}_\mathcal{K}(\chi)$ are assumed to be directly reflected in Σ_Q. Our formalism does not rely on any concrete update mechanism but leaves it up to an implementation how these changes are propagated.

We define $\Sigma_{\mathsf{DL}} = \Sigma \cup \Sigma_Q$ to represent the extended environment, thus including the native environment Σ. Figure 1 illustrates the extended environment Σ_{DL} and also shows the connection between the n-tuples of individuals in $I_\mathcal{K}$, depicted using rectangles, grouped as a set of query results, depicted using circles, with the queries of Σ_Q.

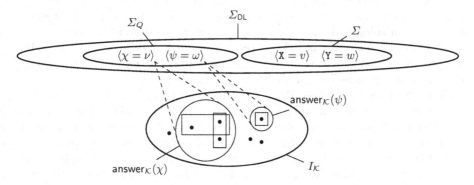

Fig. 1. Illustration of Σ_{DL} and the connection of query answers with Σ_Q

Extended Execution Semantics. We have extended Equation (1), such that query results are projected to a value of one of the native PLEXIL type domains. The projection is performed by a function $\mathsf{project}_\chi(\nu)$, which projects the query result ν of query χ to a value of a native PLEXIL type. The definition of $\mathsf{project}_\chi(\nu)$, for concrete χs, is to be done by plan creators. Equation (2) defines the replacement of Equation (1) extended by the projection of query results.

$$\Gamma' = \begin{cases} \Sigma_i \cup \{\langle \chi = \mathsf{project}_\chi(\nu)\rangle | \langle \chi = \nu \rangle \in \Sigma_{Q_i}\} & \text{if } \mathsf{event?}(\Sigma_{\mathsf{DL}i}, \Gamma, \pi) \\ \Gamma & \text{otherwise} \end{cases} \quad (2)$$

5   Runtime

PLEXIL plans can be executed by the so called *Universal Executive* (UE)[1]. We have developed a runtime which interfaces with the UE to facilitate execution of PLEXIL-DL plans. The runtime consists of a preprocessor, an execution layer and an auxiliary DL-based *Publish/Subscribe* (P/S) system called *Information Management System* (IMS) [18]. Internally, the IMS administrates a DL KB with integrated reasoner, such that incoming publications represent KB updates which cause subscriptions to be re-evaluated for potential matches. Subscriptions are formulated as DL queries and any changes in the query answer is propagated to subscribers. Additionally, the IMS can be queried using a standard request/response protocol.

---

[1] Available from http://plexil.sourceforge.net

```
AvoidObstacle:
{
    StartCondition: LookupOnChange(@[query = (?x, rdf:type, Obstacle), proj = card]@) > 0;
    PreCondition: LookupNow(FuelGauge) - evasionCost > fuelThreshold;
    Command: evade();
}
```

Listing 2. Exemplary PLEXIL-DL node

Extension of Lookup Operations. PLEXIL's syntactical means to define external state variable identifiers are insufficient to provide the expressiveness required by our extensions. We thus permit to specify an escaped expression using @[as opening and]@ as closing marker, in place of an identifier.

We represent queries using *Resource Description Framework (RDF)* triples and provide a predefined set of projection functions with reserved names, such as card, which returns the cardinality of the answer set.

Listing 2 shows the adapted PLEXIL node from Listing 1, having an extended LookupOnChange expression. The RDF expression corresponds to the DL query from Example 2 and query results are projected using the previously mentioned card function, yielding the current number of obstacles and thus prompting an action as soon as at least one obstacle exists. Given $\mathcal{K}$ of Example 1, the environment corresponds to $\Sigma_{\mathsf{DL}} = \{\langle \mathtt{FuelGauge} = v \rangle, \langle Obstacle(x) = \{\langle o \rangle\} \rangle\}$.

In order to be able to execute such a plan using the standard UE implementation, the query statement of the extended lookup needs to be transformed into a valid state variable identifier. Let $\mathsf{stateVar}(\chi) = \delta$ be an *invertible* function, which maps a query χ to a valid PLEXIL state identifier δ, applied by the aforementioned preprocessor. Let $\mathsf{stateVar}^{-1}(\delta) = \chi$ be the corresponding inverse function.

System Architecture. Based on Figure 2, which contains elements of the examples and code of previous sections, we now present the overall system architecture. The topmost element on the left shows a PLEXIL-DL plan file, containing the query of Listing 2, abbreviated as qryOb, which references an OWL Ontology, shown at the top right. Before being able to execute the plan, the preprocessor, shown underneath the plan file, transforms PLEXIL-DL queries into valid PLEXIL identifiers. The preprocessor keeps track of the original expressions, exporting mapping information used later on to evaluate queries and project their results.

The UE is now able to execute the plan, which has been reduced to the native PLEXIL format. Note that the PLEXIL-DL wrapper, shown in the middle, Registers the state variable it maintains. During execution, the UE invokes callbacks for each Lookup operation it encounters, requesting the associated value from state maintainers. By using $\mathsf{stateVar}^{-1}$, the wrapper is able to retrieve the original query. Depending on whether the UE requests an immediate lookup or a later notification, the wrapper submits a Query or Subscribes at the IMS. Any answer by the IMS, via an immediate Result or by Notifying the wrapper, is projected to the PLEXIL domain and propagated to the UE. In case of a subscription notification, an external Event is triggered by the wrapper.

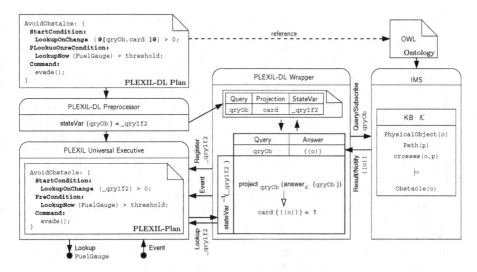

Fig. 2. System Architecture

6   Conclusions

In this paper we have presented work on a unified approach of specifying context-awareness and corresponding reactions within the same language, allowing us to express context contingencies intrinsic to the action specifications of a robot. We have chosen NASA's PLEXIL language to describe robot actions and make use of Description Logic for context modelling, exploiting DL queries to access the current context held in a DL Knowledge Base. Accordingly, we have extended PLEXIL such that these DL queries form an integral part of those language features that express conditional actions, unifying CA and sentient reactions to form adaptive robot behaviour.

The required syntactical supplements to PLEXIL have a formal grounding in the execution semantics of the plan language. Based on the freely available PLEXIL interpreter *Universal Executive* (*UE*), we have implemented a PLEXIL-DL wrapper which acts in cooperation with the so called *Information Management System* (*IMS*), a Publish/Subscribe system backed by a DL Knowledge Base, to execute PLEXIL-DL plans. Future work will exploit the formal semantics to allow plan validation with respect to CA.

References

1. Baldauf, M., Dustdar, S., Rosenberg, F.: A survey on context-aware systems. Int. Journal of Ad Hoc and Ubiquitous Computing 2(4) (2007)
2. Kjær, K.E.: A survey of context-aware middleware. In: SE 2007: Proc. of the 25th conference on IASTED Int. Multi-Conf., ACTA Press (2007)
3. Ranganathan, A., Campbell, R.H.: An infrastructure for context-awareness based on first order logic. Personal Ubiquitous Computing 7(6) (2003)

4. Gu, T., Pung, H.K., Zhang, D.Q.: A middleware for building context-aware mobile services. In: Proceedings of IEEE Vehicular Technology Conference (2004)
5. Fahy, P., Clarke, S.: Cass - middleware for mobile context-aware applications. In: Proc. of Mobisys Workshop on Context Awareness (2004)
6. Korpipää, P.: Blackboard-based software framework and tool for mobile device context awareness. PhD thesis, University of Oulu (2005)
7. Yau, S.S., Karim, F., Wang, Y., Wang, B., Gupta, S.K.: Reconfigurable context-sensitive middleware for pervasive computing. IEEE Pervasive Comp. 1(3) (2002)
8. Gu, T., Pung, H.K., Zhang, D.Q.: A service-oriented middleware for building context-aware services. Journal of Network and Comp. Applications 28(1) (2005)
9. Keays, R.: Context-oriented programming. BEng thesis, Univ. Queensland (2002)
10. Ranganathan, A., Chetan, S., Al-Muhtadi, J., Campbell, R.H., Mickunas, M.D.: Olympus: A high-level programming model for pervasive computing environments. In: Proc. of the Int. Conf. on Pervasive Comp. and Communications. IEEE, Los Alamitos (2005)
11. Cho, E.-S., Lee, K.-W., Kim, M.-Y., Kim, H.: Scenario-based programming for ubiquitous applications. In: Youn, H.Y., Kim, M., Morikawa, H. (eds.) UCS 2006. LNCS, vol. 4239, pp. 286–299. Springer, Heidelberg (2006)
12. Dowek, G., Muñoz, C., Pâsâreanu, C.S.: A formal analysis framework for plexil. In: 3rd Workshop on Planning and Plan Exec. for Real-World Sys. (2007)
13. Verma, V., Jónsson, A., Passareanu, C., Iatauro, M.: Universal executive and PLEXIL: Engine and language for robust spacecraft control and operations. In: Proc. of AIAA Space (2006)
14. Tessaris, S.: Questions and Answers: Reasoning and Querying in Description Logic. PhD thesis, Univ. of Manchester (2001)
15. Baader, F., Calvanese, D., McGuinness, D., Nardi, D., Patel-Schneider, P. (eds.): The Description Logic Handbook, 2nd edn. Cambridge University Press, Cambridge (2007)
16. Krummenacher, R., Strang, T.: Ontology-based context modeling. In: Proc. of the 3rd Workshop on Context Awareness for Proactive Systems (2007)
17. NASA: PLEXIL Reference Manual. NASA (2008)
18. Moser, H., Reichelt, T., Oswald, N., Förster, S.: Information management for unmanned systems: Combining DL-reasoning with publish/subscribe. In: Proc. of SGAI 2008. Springer, Heidelberg (2008)

Ambient Intelligence in a
Smart Home for Energy Efficiency and Eldercare

Liyanage C. De Silva[1,2], M. Iskandar Petra[1], and G. Amal Punchihewa[2]

[1] Faculty of Science, University of Brunei Darussalam, Brunei Darussalam
liyanagecd@yahoo.co.nz, merce5964@yahoo.com
[2] School of Engineering and Advanced Technology (SEAT), Massey University,
Palmerston North, New Zealand
L.desilva@massey.ac.nz, g.a.punchihewa@massey.ac.nz

Abstract. In this paper we present our research results related to smart monitoring, control and communication with the main objective of energy efficiency and eldercare in mind. One of the main objectives of this research work is to use multitude of different sensors to monitor activities in a smart home and use the results to control the home environment to meet the objectives of energy efficiency and eldercare. Here we present the application of the smart monitoring to a prototype system.

Keywords: Ambient Intelligence, Energy Efficiency, Eldercare, Smart Homes, Sensor Integration, Environment Monitoring.

1 Introduction

It is observed that people in many countries including Brunei, America, Japan, Singapore and New Zealand are now living longer and living well for longer periods of time. This has created a relatively new and growing area of health care and provider services, known as elder care. Elder care encompasses a wide variety of issues, including choosing a safe environment for the elderly person to live happily and safely and other related areas. In addition, what if elderly couples choose to live on their own or if elderly persons are living alone? For effective elder care, the traditional healthcare services need re-thinking. Instead of focusing on providing services to cure illnesses, it would be more effective to provide constant healthcare monitoring and raise an alert if something amiss is suspected. This not only lowers the cost of the healthcare as early treatment is less costly and very likely the elderly patient is still able to visit the nearby clinic or hospital, but also the quality of life of the elderly would be better as nobody likes to be sick or bedridden. Without proper care, the will to live could be impacted. Constant monitoring is also needed in order to detect cases that require urgent attention which otherwise could be life threatening, such as in cases of serious household accidents including fall, burn etc.

In our research project we will also look at the use of different modalities such as video and audio sensors for home monitoring for eldercare and energy efficiency. In this project, we used multi-modal sensing to model and analyze humans and their

J.-H. Kim et al. (Eds.): FIRA 2009, CCIS 44, pp. 187–194, 2009.

behaviour patterns with the aim of understanding better the well being of humans through constant monitoring and also help reduce their energy bills by central monitoring and control to make the society healthy and energy efficient.

Mainly we developed technologies for energy efficient homes and also homes that can support elderly people, disabled people using smart technologies. Alternatively these techniques can serve as a mode of protecting homes/factories and their contents from theft. Moreover this can be a remote monitoring and automatic alerting facility for a control centre operated by a single person. In this project stationary audio/video sensors, mobile sensors, floor sensors, and other sensing devices are installed in a model home/office and connected to a stationary and a mobile processing unit. The processing units will then determine the activities, usage of energy and other events by integrating the multitude of sensing devices to a main control center and subsequently it will act accordingly to reduce the energy usage and alert the authorities if any abnormalities are detected.

2 Energy Efficient Smart Home Technologies

There are a growing number of new research proposals and findings in related to new and alternative energy technologies. However there are many easy and cheap ways to reduce energy use at our homes by efficient energy management. Most of which simply require a change in behaviour. However we did not see much effort in this direction of research where by one can reduce the energy usage by monitoring and automatic control to make a home energy efficient. In the paper 1 we have proposed such a system to reduce the energy usage of a typical home using WIFI technology enabled smart switches. This is a prototype system intended to change the energy usage pattern of people. In this project we are looking ahead to enhance this technology by adding various different types of sensors to enhance the monitoring and control of the environment.

Energy plays an important role in many of our homes. We use it for many purposes including keeping cool during the day and providing light to our homes in the night, refrigerating and cooking our food and boiling our water. Apart from reducing the energy usage it may also require to find out ways of increasing the renewable energy input to the home. In the paper 2 the authors propose the use of solar cells as renewable energy source.

There are some other approaches in which researchers have devised artificial intelligence based techniques to build energy efficient systems 3. This paper presents and overview of commonly used methodologies based on the artificial intelligence approach with a special emphasis on neural networks, fuzzy logic, and genetic algorithms. A description of selected applications to building energy systems of AI approaches is also outlined. In particular, methods using the artificial intelligence approach for the following applications are discussed: Prediction energy use for one building or a set of buildings (served by one utility), Modelling of building envelope heat transfer, Controlling central plants in buildings, and Fault detection and diagnostics for building energy systems.

In Australia there are many organizations that promote the energy efficient homes. Sustainable Energy Development Office is one such organization 4. They provide methodology and required skills for the new home builders and people who intend to renovate their existing homes so that the finished home is energy efficient. In recent days there is a growing demand for intelligent homes and there are a number of professional service providers for such homes 5. They have commercial level light controllers, curtain controllers and other type of sensors and controllers.

In Singapore there is a model smart home known as STAR home 6. There they conduct research in relation to intelligent homes and among the themes they investigate entertainment, health care, security and power efficiency has been given high priority. In Japan they often call smart home a Ubiquitous home 7. In many smart home installations it was quite interesting to see the integration of multitude of sensors to harmonize the activities in a home with the help of modern technology. In the Unites States some companies even mass produce smart homes 8. It is a fine example of how fast the smart or intelligent homes are going to invade the communities who can afford the new technologies.

However these commercial level devices/systems developed in developed countries are not particularly applicable for countries like Brunei as it is required to understand the climate and the living patterns of Brunei people and their culture. Hence in our paper we first tried to acquire some of these commercially available sensors and install them in a prototype system in Brunei and monitor the patterns.

3 Experiment Setup

The preliminary data collection for this project has been performed at a model room in the Institute of Infocom Research, Singapore. However, in order to get realistic results it is necessary to acquire further data and analyze them for different scenarios and different sensor inputs suitable for Brunei climate and culture. The authors have demonstrated the use of smart home technologies to reduce energy consumption in an average house in their research work partly presented in the papers [1]&[2].

In this research we design, implement and monitor with the aim of future energy efficient and eldercare enabled home. The following two figures show the execution scenario in pictorial form.

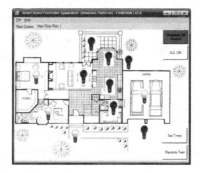

Fig. 1. Smart Home Control System for Energy Efficiency – Device Map (as of [2])

Fig. 2. Smart Home Control System for Energy Efficiency – GUI (as of [2])

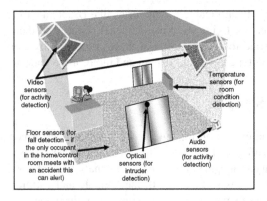

Fig. 3. Smart Home with various sensors

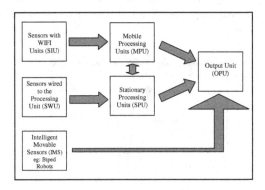

Fig. 4. Example Scenario of the Operation

The first task of this project is to install a single SWU (Sensor wired to the Processing Unit) connected to a SPU (Stationary Processing Unit). For the SWU a high resolution video camera will be used. For the SPU a high end desktop computer with a capture card is used. Then the video data will be acquired and stored in the hard disk of the desktop for offline processing. The video data captured contained various different human activities, such as walking, crying, shouting, talking on the phone, falling etc. Subsequently the video data are analysed and activities are extracted using image and video processing algorithms. Subsequently the offline processing will be upgraded to real-time processing when the processing algorithms are fully optimized.

Then a testing with a single SIU (Sensors with WIFI Units) and a MPU (Mobile Processing Unit) will be carried out. For the SIU we will initially use a WIFI capable mobile phone with a built in camera. Subsequently the data are analysed and activities are extracted.

In the next phase of the project other kinds of sensors such as audio sensors, optical sensor etc. are tested. The detected activities will then be sent to an OPU (Output Unit) for alerting the necessary personnel if there is amiss inside the house or inside the control room. Once all the necessary components are investigated the sensors and processing units will be installed in a real model house for real time testing with real occupants.

4 Video Sensor Based Event Recognition

We used a state-based approach to recognize events and actions. The state diagram in Fig. 5 shows the transitions between states defined for a tracked human in the image sequence.

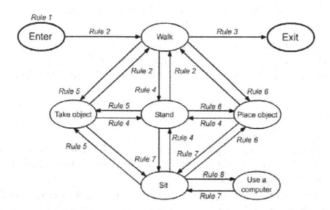

Fig. 5. State Transitions for Video Sensor Based Activity Detection

5 Audio Sensor Based Event Detection

Video sensor based event detection approach has some short falls like it fails to cover the entire room and also event non-detection due to occlusion. Hence in this section we

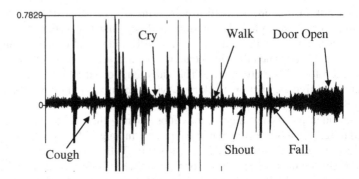

Fig. 6. A sample long audio file used for audio based event detection and segmentation

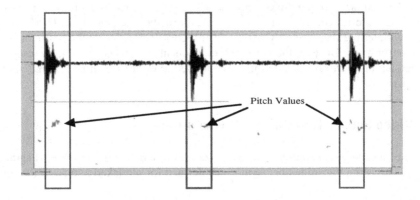

Fig. 7. Pitch values detected at coughing events

consider audio sensor based actions. Here we adopted a combined pitch and intensity based audio event detection to classify cough, walk, cry, door open, fall, and shout audio events into their respective group. Example audio file are shown in the Fig. 6-7.

6 Results

Forty image sequences containing different actions and events were used to evaluate the accuracy of action and event recognition. Table 1 shows the accuracy of recognition of events in our proposed system.

Then we used audio data for event detection. The separation of cough, cry and shout (vocal track generated sounds) from walk, door-open and fall was done using pitch contour detection. This is due to the fact that vocal track generated audio signals consists of its inherent formant frequency component. Then cough, cry and shout were further separated using the intensity contours. Cry had a constant intensity profile, while cough and shout had some abruptly increased intensity values. However cough and shout were easily separated by using the energy of the audio pulses. Walk has its inherent property of gradual increase of the intensity profile till the steps are

Table 1. Video based event detection

Video events (no of events in the video clips tested)	Classified correctly	Classified incorrectly	Not classified	Average Detection Accuracy %
Enter (20)	17	3	0	85.0
Walk (35)	31	4	0	88.5
Exit (20)	19	1	0	95.0
Stand (15)	13	0	2	86.7
Sit (15)	14	0	1	93.3
Use PC (12)	9	3	0	75.0
Take object (5)	5	0	0	100.0
Place object (5)	5	0	0	100.0
Unusual event (9)	8	0	1	88.9
Overall average accuracy				**90.3**

Table 2. Audio based event detection

Audio Events (no of events in the audio clips tested)	Classified correctly	Classified incorrectly	Not classified	Average Detection Accuracy %
Cough(16)	15	1	0	93.8
Walk (25)	24	1	0	96.0
Cry (12)	12	0	0	100.0
Door open (19) (Enter or Exit)	16	2	1	84.2
Fall (14)	12	2	0	85.7
Shout (10)	9	0	1	90.0
Overall average accuracy				**91.6**

getting close to the microphone and then gradual decrease when walks past the microphone. The following table (Table 2) shows the results of the audio monitoring sensors we have obtained.

7 Conclusions

In this paper we have presented our research results related to smart monitoring, control and communication with the main objective of energy efficiency and eldercare in mind. Our video based analysis has given us a comprehensive set of results to understand the human actions in an enclosed room or in a home environment with the possible detection of Entering, Walking, Exiting, Standing, Sitting, Using a PC, Taking an Object, Placing an Object and any other unusual event including falling. These activities can be used to control the room lighting, air-conditioning etc. to reduce the total energy usage of the house.

Then the introduction of the audio event detection increased the possible types of actions that can be detected like cough, cry and fall which may be hard to detect just by video only. These short duration and scattered events may occur outside the coverage area of the video camera system in the house and hence may go undetected if only a video based system was employed. These audio based sensors are vital in homes aimed at automated eldercare to reduce the privacy problems that may occur due to video based sensors.

Currently we are working in the prototype implementation of other sensors such as ultrasound and temperature sensors to increase the knowhow of the ambient intelligence to provide the features of a smart room facility constructed in one of our research facilities to obtain real life data and their analysis.

References

1. De Silva, L.C., Mathew, S.: Energy Efficient Smart Homes. In: Published in the proceedings of the 1st International Conference of Institution of Engineering and Technology Brunei Darussalam Network (IETBIC2008) held in The Rizqun International Hotel, Brunei Darussalam, May 26-28 (2008)
2. Lach, C., Punchihewa, A., De Silva, L.C., Mercer, K.: Smart Home System Operating Remotely Via 802.11b/g Wireless Technology. In: Published in the proceedings of the 4th International Conference Computational Intelligence and Robotics and Autonomous Systems (CIRAS2007), held in Palmerston North, New Zealand, November 28-30 (2007)
3. Malik, Q., Ming, L.C., Sheng, T.K.: The effect of temperature on the power output of photovoltaic solar cells. In: Proceedings of the World Renewable Energy Congress, Paper No. 14-RTPV10. Elsevier, Amsterdam (2006)
4. Krarti, M.: An overview of artificial intelligence-based methods for building energy systems. Journal of Solar Energy Engineering 125(3), 331–342 (2003)
5. http://www.sedo.energy.wa.gov.au/index.asp (accessed on March 1, 2009)
6. http://www.hometouch.com.hk/newok/index.html (accessed on March 1, 2009)
7. http://starhome.i2r.a-star.edu.sg/ (accessed on March 1, 2009)
8. http://www.nict.go.jp/ (accessed on March 1, 2009)
9. http://www.msnbc.msn.com/id/12253119/ (accessed on March 1, 2009)

Intelligent Technologies for Edutainment Using Multiple Robots

Naoyuki Kubota, Yuki Wagatsuma, and Shinya Ozawa

Tokyo Metropolitan University, Japan
6-6 Asahigaoka, Hino, Tokyo 191-0065, Japan
kubota@tmu.ac.jp,
wagatsuma-yuki@sd.tmu.ac.jp,
ozawa-shinya@sd.tmu.ac.jp

Abstract. This paper aims to realize the next generation of edutainment using multiple robots based on the integration of network technology, intelligent technology, and robot technology. First, we explain the hardware specification and control mechanism of tele-operated mobile robots. Next, we explain the user interface based on GUI for the tele-operation. Finally, we show several experimental results of the developed multiple robots in the educational fields.

Keywords: Edutainment, Tele-operation, Multiple Robots, Human Interface, Intelligent Technologies.

1 Introduction

Recently, the emerging synthesis of information technology (IT), network technology (NT), and robot technology (RT) is one of the most promising approaches to realize a safe, secure, and comfortable society for the next generation [1-3]. Furthermore, various types of robots such as surveillance robots, rescue robots, and partner robots have been developed for the next generation society. As the development of cheap and small sensor devices, these have been easily incorporated into such a robot. Furthermore, NT can provide the robot with computational capabilities based on various types of information outside of robots. Actually the robot directly receives the environmental information through a local area network without the measurement by the robot itself. As the development of ubiquitous computing and sensor network, we should discuss the intelligence technology in the whole system composed of robots and environmental systems. Here intelligent technologies related with measurement, transmission, modeling, and control of environmental information are called as ambient intelligence. Information resources and the accessibility within an environment are essential for people and for robots. The environment surrounding people and robots should have a structured platform for gathering, storing, transforming, and providing information. Such an environment is called inforamtionally structured space [10].

In our previous study, we proposed the cooperated system of a partner robot and environmental system based on a sensor network [4]. Furthermore, we discussed the human interface for the monitoring system based on a tele-operated mobile robot [5].

J.-H. Kim et al. (Eds.): FIRA 2009, CCIS 44, pp. 195–203, 2009.

And also, we discussed the role of robots in edutainment from the viewpoint of project-based learning [9].

In this paper, we discuss the applicability of the multiple tele-operated mobile robots based on sensor network to the field of edutainment. First, we explain the developed tele-operated mobile robots and environmental system based on sensor network. Next, we explain the GUI environment for the tele-operation based on touch panel and joystick and control mechanism of the robot. Finally, we show two examples of multiple tele-operated mobile robots.

2   Multiple Robots for Edutainment

2.1   Robot Edutainment

Various types of robots have been applied to the fields of education [9,11,12]. Basically, there are three different aims in robot edutainment. One is to develop knowledge and skill of students through the project-based learning by the development of robots (Learning on Robots). Students can learn basic knowledge on robotics itself by the development of a robot. The next one is to learn the interdisciplinary knowledge on mechanics, electronics, dynamics, biology, and informatics by using robots (Learning through Robots). The last is to apply human-friendly robots instead of personal computers for computer assisted instruction (Learning with Robots). A student learns (together) with a robot. In addition to this, such a robot can be used for supporting teachers by the teaching to students and the monitoring of the learning states of students. An educational partner robot can teach something through interaction with students in daily situation. Furthermore, the robot can observe the state of friendship among students. This is very useful information for teachers, because it is very difficult for a teacher to extract such information from the daily communication with students.

Figure 1 shows an example of education by using multiple robots. In this example, the teacher controls three robots through a host computer, and shows synchronized behaviours of three robots. The teacher asks students about the difficulty of synchronized behaviours. Basically, it is very difficult for people to realize synchronized behaviours, but it is very easy for robots by the wireless network. In this way, students can consider how to realize various behaviours among people.

Fig. 1. Edutainment by multiple robots

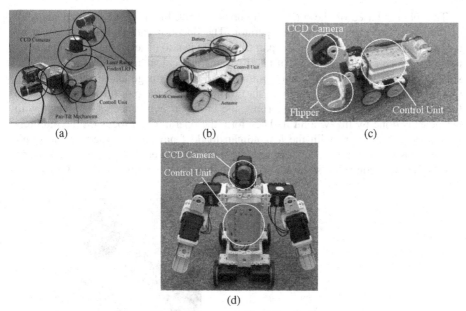

(a)   (b)   (c)

(d)

Fig. 2. Tele-operated mobile robots

2.2 Mobile Robots

We developed various types of mobile robots using Bioloid [6]. The robot can perform the wireless communication by ZigBee with host computers or other robots. Figure 2 shows four types of developed mobile robots in this study. The first mobile robot shown in Fig.2 (a) equips a laser range finder and two CCD cameras. The size of this robot is 260 x 100 x 230 [mm]. The front CCD camera is equipped on the pan-tilt mechanism. The moving ranges of the pan and tilt are -90° to 90° and -30° to 90°, respectively. The size of the robot in Fig.2 (b) is 180 x 115 x 120 [mm]. We used wireless small CMOS cameras in order to reduce the size of the robot. The diameter and height of the camera are 19.5 and 11.5 [mm]. The diameter of the lens is 0.8 [mm]. The communication range of the camera is 30 [m]. A human operator sends a control command to the robot from the host computer based on images received directly from the CCD cameras. Figure 2 (c) shows a soccer robot with a small flipper at the front of the robot and with a wireless CCD camera at the head part of the robot. Furthermore, we developed two more types of soccer robots with a wireless camera at different position on the robot in order to discuss the operability of the robot by using a joystick. Figure 2 (d) shows a mobile robot with a human upper body and a wireless CCD camera. This robot is developed to discuss the gesture communication of robots between two people.

2.3 Human Interface

We developed a human interface for the tele-operation based on graphical user interface (GUI) from the host PC. The measured data of wireless sensor devices are transmitted to the host computer, and the operator can monitor those data (Fig.3 (a)). We used five wireless acceleration sensor nodes for the environmental system. The sensor device equipped with a door can measure the acceleration of the X axis and illuminance.

The robot is based on the semi-autonomous control. Figure 3 (b) shows a snapshot of the control of the robot by the operator. The operator can navigate the robot by using a joystick and can control the pan-tilt mechanism to observe the environment by using the touch panel. The angle of pan-tilt camera is calculated automatically by the simple relationship between the position touched on the image and its corresponding direction of the camera. If the robots start to move, then the pan-tilt is going back to the normal position automatically. Here, the collision avoidance of the robot with obstacles is done by fuzzy control. Therefore, the operator can focus on the monitoring of the environment with the low human load of navigation based on the semi-autonomous control mode. Furthermore, the operator can easily choose the operating mode of manual control or semi-autonomous control by the button of the joystick.

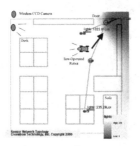   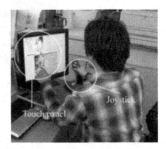

(a)The environment of remote control robot   (b) Control by touch panel and by joystick

Fig. 3. Remote control of robots based on sensor networks

2.4   Intelligent Control of Mobile Robots

We apply simplified fuzzy inference to control the mobile robot [7,8], because fuzzy rules are easily designed. In general, a fuzzy if-then rule is described as follows,

> **If** x_1 is $A_{i,1}$ and ... and x_m is $A_{i,m}$
> **Then** y_1 is $w_{i,1}$ and ... and y_n is $w_{i,n}$

where $A_{i,j}$ and $w_{i,k}$ are the Gaussian membership function for the jth input and the singleton for the kth output of the ith rule; m and n are the numbers of inputs and outputs, respectively. Fuzzy inference is described by,

$$\mu_{A_{i,j}}(x_j) = \exp\left(-\frac{(x_j - a_{i,j})^2}{b_{i,j}^2}\right) \tag{1}$$

$$\mu_i = \prod_{j=1}^{m} \mu_{A_{i,j}}(x_j) \tag{2}$$

$$y_k = \frac{\sum_{i=1}^{R} \mu_i w_{i,k}}{\sum_{i=1}^{R} \mu_j} \tag{3}$$

where $a_{i,j}$ and $b_{i,j}$ are the central value and the width of the membership function $A_{i,j}$; R is the number of rules. Outputs of the robot are motor output levels. Fuzzy controller is used for vision-based target tracing behavior. The fuzzy controller is applied to collision avoidance and target tracing behaviors.

3 Experimental Results

3.1 Tele-operation of Soccer Robots

We developed various types of soccer robots using Bioloid. In this experiment, we used two types of soccer robots shown in Fig.2.(c). The size of this robot is 230 x 100 x 230 [mm]. The task is for the left robot to shoot a ball through one pass from the right robot (Fig.4 (a)). The size of soccer court is 3 [m] x 1.5 [m]. There are five goals of different colors in the end of court. The score of the central goal is the highest. After the score is obtained, two robots and the ball are brought back to their initial potions. The playing time is two minutes. Each student can control the soccer robot by watching the remote image sent from the wireless CCD camera equipped with the robot (Fig.4 (b)). The robot can kick a ball by using a flipper according to the command from the student. We used the manual control mode for the soccer. It is very difficult for each student to understand the global position of the ball, the goal, and the robot itself, because the angle of view of the equipped camera is very limited and narrow. Therefore, the students surrounding the soccer court can tell the operating student the relative angle and distance to the ball or goal, but it is inhibited that the operating student directly watches the court.

Figure 5 shows snapshots of en experiment of the remote control by two students. The task was successfully done by the control of two robots, and the surrounding students also told to the operating students about the rotating directions and suitable actions. The communication skill for soccer of students was gradually improved. As a result, the students learned the importance of communication in the cooperative task, and how to tell about the suitable actions.

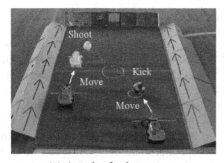

(a) A task of robot soccer   (b) Remoto control of the soccer robot

Fig. 4. An experiment of robot soccer

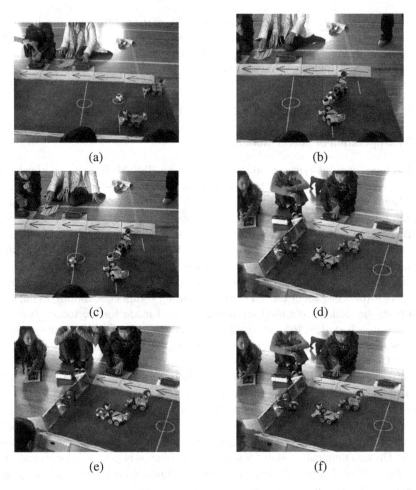

(a)

(b)

(c)

(d)

(e)

(f)

Fig. 5. Snapshots of the robot soccer

3.2 Tele-operation of Human-Like Robots

We discuss the cooperation of two robots in this subsection. The human-like robot composed of two arms where the degree of freedom of each arm is 2. The size of this robot is 230 x 100 x 230 [mm]. This is used for body language with other robots. The operator can control the robot by watching image sent from the wireless CCD camera equipped with the robot. However, it is very difficult for the operator to perceive the environment surrounding the robot, since the angle of view of the camera is very narrow. Therefore, we developed the follower mobile robot that sends the image from the backward of the human-like robot to the operator. The maximal speed of each robot is 170 mm/sec. Figure 6 (a) shows a snapshot of two robots. The follower mobile robot has a sensor unit at the front of the robot. The sensor unit equips with three infrared sensors and illuminance sensor. The follower mobile robot measures the distance information up to 500 [mm] by swinging the sensor unit including infrared distance sensor. The number of directions measured by the swinging motion is 9. By using the distance information, the follower mobile robot traces the human-like robot by the fuzzy controller while keeping the predefined distance between two robots.

We conducted preliminary experiments on gesture communication. The human-like robot shows the turning sign as gesture communication when the robot turns right or left. Figures 6 (b) ~ (d) show the experimental results of gesture communication when the robot turns right.Figure 6 (b) shows a camera image from the environmental system. This view is easy to understand, but it is difficult for the operator to perceive the environment by the image from the human-like robot (e.g., Fig.6 (d)). Actually, the operator cannot see its foot. Therefore, the camera image of the follower mobile robot is required. Figure 6 (c) shows the view from the follower mobile robot. From this image, the operator can perceive the environment surrounding the robot. Figure 6 (d) shows the gesture for the right turn from the camera image of the other robot. It is very easy for the operator to understand the meaning of this kind of gesture. Figure 7 (a) shows the right and left motor output (maximum 50,minimum -50), and Fig.7 (b) shows the measured sensor data with the range of (0, 255). The value becomes

(a) A bird view of the robots

(b) The overview from the environmental camera

(c)The view from the follower

(d) The view from the other robot

Fig. 6. An experimental result of gesture communication by robots

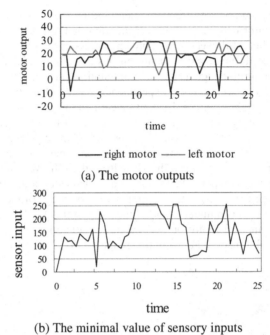

(a) The motor outputs

(b) The minimal value of sensory inputs

Fig. 7. The control of robots

low as the robot approaches to objects. There are two types of couplings among the robots in this experiment. One is the tight coupling between the leader and follower. If this combination is broken, the control of the robot becomes difficult. The other is the loose coupling between two human-like robots. In this coupling, it is important to send and receive the intention of the operator through the gestures of the robots, but the detailed actions are not so important. In this way, student can discuss the cooperation and synchronization of robots from various points of view.

4 Summary

In this paper, we showed a system of multiple tele-operated mobile robot based on sensor networks. A mobile robot can observe the local environmental information, while the environmental system based on the sensor network can obtain the global environmental information. Based on the global and local environmental information, the operator can control the robot flexibly and adaptively. Furthermore, we discussed the applicability of multiple robots in the filed of education. The experimental results show that students can consider and discuss the cooperation, synchronization, and communication among robots from various points of view.

As a future work, we intend to develop an edutainment textbook based on the co-operation and synchronization of multiple robots, and conduct the experiments in elementary schools.

This instruction file for Word users (there is a separate instruction file for LaTeX users) may be used as a template. Kindly send the final and checked Word and PDF files of your paper to the Contact Volume Editor. This is usually one of the organizers of the conference. You should make sure that the Word and the PDF files are identical and correct and that only one version of your paper is sent. It is not possible to update files at a later stage. Please note that we do not need the printed paper.

We would like to draw your attention to the fact that it is not possible to modify a paper in any way, once it has been published. This applies to both the printed book and the online version of the publication. Every detail, including the order of the names of the authors, should be checked before the paper is sent to the Volume Editors.

Reference

[1] Kubota, N., Shimomura, Y.: Human-Friendly Networked Partner Robots toward Sophisticated Services for A Community. In: Proc. of SICE-ICCAS 2006, pp. 4861–4866 (2006)

[2] Khemapech, I., Duncan, I., Miller, A.: A survey of wireless "sensor networks technology". In: PGNET, Proc. the 6th Annual Post Graduate Symposium on the Convergence of Telecommunications, Networking and Broadcasting, EPSRC (2005)

[3] Kubota, N., Nishida, K.: Cooperative Perceptual Systems for Partner Robots Based on Sensor Network. International Journal of Computer Science and Network Security (IJCSNS) 6(11), 19–28 (2006)

[4] Kubota, N., Koudu, D., Kamijima, S., Taniguchi, K., Nogawa, Y.: Vision-based Teleoperation of A Mobile Robot with Visual Assistance. Intelligent Autonomous Systems 9, 365–371 (2006)

[5] Kubota, N., Ozawa, S.: Tele-operated Robots for Monitoring Based on Sensor Networks. In: Proc. of SICE Annual Conference 2008, Chofu, Tokyo, Japan, August 20-22, pp. 3355–3360 (2008)

[6] http://www.robotis.com/zbxe/main

[7] Jang, J.-S.R., Sun, C.-T., Mizutani, E.: Neuro-Fuzzy and Soft Computing. Prentice-Hall, Inc., Englewood Cliffs (1997)

[8] Fukuda, T., Kubota, N.: An Intelligent Robotic System Based on a Fuzzy Approach. Proceedings of IEEE 87(9), 1448–1470 (1999)

[9] Kubota, N., Tomioka, Y., Ozawa, S.: Intelligent Systems for Robot Edutainment. In: Proc. of 4th International Symposium on Autonomous Minirobots for Research and Edutainment (2007)

[10] Satomi, M., Masuta, H., Kubota, N.: Hierarchical Growing Neural Gas for Information Structure Space. In: IEEE Symposium Series on Computational Intelligence 2009 (2009)

[11] Kim, H.: Veltman, Edutainment, Technotainment and Culture, Veltman, K.H., Città Annual Report (2003)

[12] Mizuko, I.: Engineering play: Children's software and the cultural politics of edutainment. Discourse 27(2), 139–160 (2004)

Remote Education Based on Robot Edutainment

Akihiro Yorita[1], Takuya Hashimoto[2], Hiroshi Kobayashi[2], and Naoyuki Kubota[1]

[1] Tokyo Metropolitan University, Graduate School of System Design,
6-6, Asahigaoka, Hino, Tokyo, Japan
yorita-akihiro@sd.tmu.ac.jp,
kubota@tmu.ac.jp
[2] Tokyo University of Science, Graduate School of Mechanical Engineering,
1-14-6, Kudankita, Chiyoda-ku, Tokyo, Japan
{tak,hiroshi}@kobalab.com

Abstract. This paper discusses the role of robots in remote education. There are three different aims of robot edutainment, i.e., Learning on Robots, Learning through Robots, and Learning with Robots. The last is to apply human-friendly robots instead of personal computers for computer-assisted instruction. Especially, natural communication capability is required to educational robots in the learning with robots. In this paper, we apply human-friendly robots to remote education and discuss the requirements and specifications of robots for the remote education.

Keywords: Robot Edutainment, Human-Robot Interactions, Remote Control and Monitoring, Computational Intelligence.

1 Introduction

Learning is one of the fundamental rights of all people. Recently, although the population size of children in developed countries, the educational expenses per child are increasing, and the need to high quality of education is increasing much more. Furthermore, open and distance learning is fast becoming an accepted and indispensable part of the main stream of educational system, as the development of information and communication technologies [1]. The open and distance learning realized the individualized learning and teaching style in the education. Furthermore, online lecture or presentation of teaching materials becomes accessible to many learners. As a result, group learning and teaching through the Internet has also been one stream of open and distance learning. On the other hand, the remote education is one of important and efficient approaches in order to realize high quality of education in underpopulated area where the number of teachers is not enough for the education. Therefore, in this paper, we discuss the role of robots in the remote education.

Various types of robots have been applied to the fields of education [8,9]. Basically, there are three different aims in robot edutainment. One is to develop knowledge and skill of students through the project-based learning by the development of robots (Learning on Robots). Students can learn basic knowledge on robotics itself by the development of a robot. The next one is to learn the interdisciplinary knowledge

J.-H. Kim et al. (Eds.): FIRA 2009, CCIS 44, pp. 204–213, 2009.
© Springer-Verlag Berlin Heidelberg 2009

on mechanics, electronics, dynamics, biology, and informatics by using robots (Learning through Robots). The last is to apply human-friendly robots instead of personal computers for computer assisted instruction (Learning with Robots). A student learns (together) with a robot. In addition to this, such a robot can be used for supporting teachers by the teaching to students and the monitoring of the learning states of students. An educational partner robot can teach something through interaction with students in daily situation. Furthermore, the robot can observe the state of friendship among students. This is very useful information for teachers, because it is very difficult for a teacher to extract such information from the daily communication with students. We showed the effectiveness of the learning with robots in the previous works [3-5]. A partner robot in educational fields is not the replacement of a human teacher, but the replacement of a personal computer. A student seldom shows physical reactions to a personal computer in the computer-assisted instruction (CAI), because the student is immersed into 2-dimensional world inside of the monitor. However, a student aggressively tries physical interactions to a robot, because the robot can express its intention through physical reactions. Of course, the robot should play the role of a personal computer.

In this paper, we discuss the applicability of robots in the remote education. First, we explain the robots used in remote education and the remote education system. Next, we discuss the roles of robots in the remote education. Finally, we discuss the future vision toward the realization of educational partner robotics.

2 Robots Used for Edutainment

2.1 Android Receptionist Robot: SAYA

We aim for realization of human-like natural behaviors with android receptionist robot SAYA in which the Face Robot [6,14] is used as shown in Fig.1.

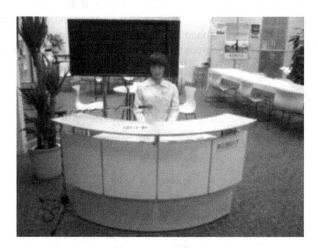

Fig. 1. Receptionist robot "SAYA"

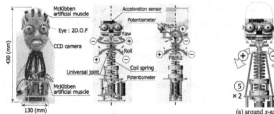
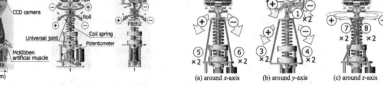

Fig. 2. Internal structure of SAYA   **Fig. 3.** Actuator distribution for head rotations

Figure 2 shows internal structure of the Face Robot. McKibben pneumatic actuators [7] are used for controlling displacements of facial skin. They are put in the internal frame. McKibben pneumatic actuator generates contraction force by sending compressed air. Since it is small, light and flexible, it can be distributed to carved surface of the skull like human muscles. In addition, its viscoelastic property is similar to human muscle. There is an oculomotor mechanism which controls both pitch and yaw rotation of eyeballs by 2 DC motors. Two eyeballs move together since they are linked to each other. A CCD camera is mounted in the left-side eyeball. Since coil springs can move flexibly like a human's neck, we adopted the coil spring for the head motion mechanism. Referring to anatomical knowledge, we decided movable positions as shown in Fig. 2. In human, forward and backward motions of the head are flexed by combination of a head rotation and a bending of the neck. Therefore we set the center of rotation for the pitch rotation ("Pitch1"), and we also set the center of rotation for the yaw rotation in the base of head. We form the facial skin of the face robot with soft urethane resin to realize the texture like a human facial skin. There is 2-axis acceleration sensor for measuring roll and pitch rotations of the head. A potentiometer is attached in the root of the head and the neck in order to measure a relative angle of the head to the neck in the pitch rotation. In addition, there is a potentiometer in the bottom of the neck for measuring the yaw rotation of the head.

McKibben pneumatic actuators are also used for head rotations. Each rotation is driven by two antagonistic McKibben pneumatic actuators as shown in Fig. 3. For example, the head pitch rotation is controlled by differential pressure between actuator 1 and 2.

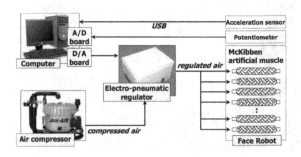

Fig. 4. System configuration

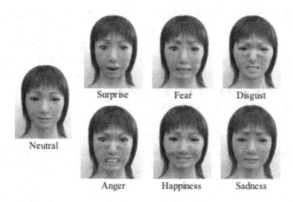

Fig. 5. 6 typical facial expressions

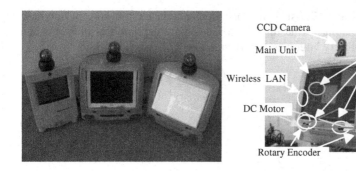

Fig. 6. Human-friendly Partner robots; MOBiMac

The composition of the control system is shown in Fig.4. 2-axis acceleration sensor detects gravity acceleration and sends it to a host PC. Two potentiometers also detect rotation angles and send voltage signal to an A/D board in the PC. Voltage signals are calculated in the PC and they are sent to an electro-pneumatic regulator through a D/A conversion board. Here, the electro-pneumatic regulator is a regulator to control air pressure in McKibben pneumatic actuators.

We have been able to express more minutely all six basic facial expressions (Surprise, Fear, Disgust, Anger, Happiness, and Sadness) on the face robot. Fig.5 shows seven facial expressions including "Neutral".

2.2 Partner Robots: MOBiMac

We developed human-friendly partner robots a mobile PC called MOBiMac (Fig. 6) in order to realize human-friendly communication. This robot has two CPUs and many sensors such as CCD camera, microphone, and ultrasonic sensors. Furthermore, the information perceived by a robot is shared with other robot by the wireless communication. Therefore, the robots can easily perform formation behaviors.

We have applied steady-state genetic algorithm (SSGA), spiking neural networks (SNN), self-organizing map (SOM), and others for human detection, motion extraction, gesture recognition, and shape recognition based on image processing [2,10-12]. Furthermore, the robot can learn the relationship between the numerical information as a result of image processing and the symbolic information as a result of voice recognition [13]. MOBiMac can be also used as a personal computer and its development cost is much cheaper than that of humanoid robots.

We have also applied fuzzy inference systems to represent behavior rules of mobile robots, because the behavioral rules can be designed easily and intuitively by human linguistic representations. A behavior of the robot can be represented using fuzzy rules based on simplified fuzzy inference [18]. In general, a fuzzy if-then rule is described as follows,

If x_1 is $A_{i,1}$ and ... and x_M is $A_{i,M}$
Then y_1 is $w_{i,1}$ and ... and y_N is $w_{i,N}$

where $A_{i,j}$ and $w_{i,k}$ are the Gaussian membership function for the jth input and the singleton for the kth output of the ith rule; M and N are the numbers of inputs and outputs, respectively. Fuzzy inference is performed by,

$$\mu_{A_{i,j}}(x_j) = \exp\left(-\frac{(x_j - a_{i,j})^2}{b_{i,j}^2}\right) \tag{1}$$

$$\mu_i = \prod_{j=1}^{M}\mu_{A_{i,j}}(x_j) \tag{2}$$

$$y_k = \frac{\sum_{i=1}^{R}\mu_i w_{i,k}}{\sum_{i=1}^{R}\mu_j} \tag{3}$$

where $a_{i,j}$ and $b_{i,j}$ are the central value and the width of the membership function $A_{i,j}$; R is the number of rules. Outputs of the robot are output levels of the left and right motors ($N=2$). Fuzzy controller is used for collision avoidance and target tracing behaviors. The inputs to the fuzzy controller for collision avoidance are the measured distance to the obstacle by ultrasonic sensors ($M_c=8$). The inputs to the fuzzy controller for target tracing are the estimated distance to the target point and the relative angle to the target point from the moving direction ($M_t=2$).

3 Remote Education System

3.1 Remote Control System

Figure 7 shows the total architecture of the remote control system for robots and the remote education. Basically, this system is composed of two rooms of a class room and operation room.

There are a host computer, robots, and monitoring system in the class room. The host computer is connected with a monitoring system of the class room and robots by wireless communication, and has the data-base on the personal information of students, and educational environment. Furthermore, the host computer can send the state of students obtained from the robots to the teacher in the remote operation room. The camera equipped in the monitoring system takes an overall image of the class room, while the camera equipped with the robots takes a local image on the table and the state of students.

There are a remote control computer, a main monitor, and monitoring system in the operation room. The teacher gives a talk to the student. Here, if the teacher calls a student A, then the operator shows the state of the student A to the main monitor. Furthermore, the operator sends the basic control commands to the robots in the class room, and performs the monitoring of students in the class room (Fig.8, the view from MOBiMac). The robots are controlled by semi-autonomous control. The operator selects the meta-level control mode of (1) standard lecture mode, (2) robot instructor lecture mode, and (3) interaction mode. In the the standard lecture mode, each robot makes scenario-based utterances according to the content of a lecture or experiment. The operator moves ahead with the scenario according to the talk of the teacher. Furthermore, the robot instructor lecture mode, one of the robots plays the role of a teacher. According to the command from the remote operator, the robot moves ahead with the scenario instead of a teacher. On the other hand, in the interaction mode, the robot autonomously performs conversation and interaction with students according to the perceptual information received from the host computer, and the results of the image processing and voice recognition.

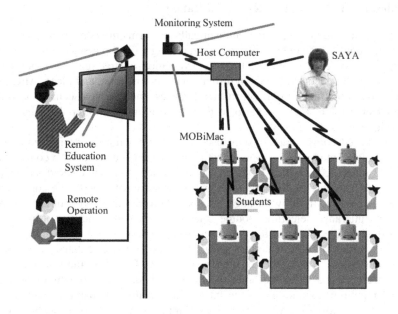

Fig. 7. A total architecture of remote education system

Fig. 8. The view from MOBiMac on the table

3.2 Lecture Mode for the Remote Education

In both of (1) standard lecture mode and (2) robot instructor lecture mode, a scenario is used for a lecture or experiment. The scenario is composed of (1) main utterances, (2) assisting utterances, and (3) interaction utterances. The leading teacher or robot reads the main utterances including the important explanation of topics according to the time schedule of the lecture or experiment. The assisting utterances are used for other robots in the table. The robots in the table make the assisting utterances for the students in order to explain the topics in detail. The interaction utterances are used to stimulate or encourage the students to focus on the study.

3.3 Interaction Mode in the Remote Education

In the interaction mode, the robot automatically performs the conversation with students. The contents of conversation are (1) greetings, (2) topic selection utterances, and (3) scenario-based utterance. In the conversation of greeting, the robot selects suitable sentences according to the information of season, time, and place. After short greetings, the robot makes topic selection utterances, and selects a scenario according to the selected topic. Each scenario in this mode is much shorter than that in the lecture mode.

The proposed conversation system used in the interaction mode is composed of three interrelated modules; (1) topic selection modules, (2) conversation control module, and (3) utterance selection module. The topic selection module decides the global flow of conversation based on the selection probabilities of topics. The conversation with people controls the flow of utterances based on transition probabilities of utterances. The utterance selection module selects the next utterance according to the internal states of the robot and the responses from the person.

The conversation control module selects a scenario and calculates the transition probabilities based on the order of utterances (Fig.9). Basically, the transition probabilities are designed when the order of utterances is decided. Sometimes, owing to the local repetition of utterances, the backward transition probabilities are also included in the system. Some of utterances are skipped if the selection probability of the topic in the scenario selection is low. In this way, the flow of conversation is controlled according to the internal state and the response from the person.

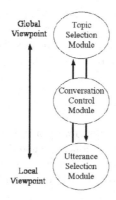

Fig. 9. Control of conversation

4 Experimental Results

We show preliminary experiments of remote education. The experiment was done on November 2007. Figure 10 shows a preliminary experiment of the robot instructor lecture mode. SAYA plays the role of teacher, and reads the main utterances in the scenario sequentially according to the command from the remote operator. The other robots of MOBiMac share the information of scenario, and reads the assisting utterances after SAYA reads the main utterances. In the interaction mode, each MOBiMac performs image processing to extract people and objects, and performs the conversation with the students. Figure 10 (b) shows a photo of SAYA interacting

(a)   (b)

(c)

Fig. 10. A preliminary experiment of the robot instructor lecture mode

with SAYA. Although SAYA is kind and gentle to students, but SAYA gets angry if a student touches on the face of SAYA. Figure 10 (c) shows the photos of MOBiMac interacting with students. The experimental results show that the students can interact with robots with pleasure and interest.

5 Summary

In this paper, we discussed the applicability of robots in remote education. First, we explained the robots used in remote education and the remote education system. Next, we discussed the roles of robots in the remote education. Finally, we showed preliminary experimental results. We will conduct experiments in several elementary schools in May, 2009.

As future work, we will develop a method for cognitive development of robots through the learning with students, and furthermore, we intend to develop the monitoring system of the learning state of students.

Acknowledgments. We thank Takeru Mori, Shiho Wakisaka, Aiko Yaguchi, Yuki Wagatsuma, and Rikako Komatsu, Kaori Tajima for their support to conduct experiments in the elementary schools.

References

1. Unesco, Open and Distance Learning-Trends, Policy and Strategy Considerations (2002)
2. Kubota, N., Nishida, K.: Cooperative Perceptual Systems for Partner Robots Based on Sensor Network. International Journal of Computer Science and Network Security (IJCSNS) 6(11), 19–28 (2006)
3. Kubota, N., Ozawa, S.: Tele-operated Robots for Monitoring Based on Sensor Networks. In: Proc. of SICE Annual Conference 2008, pp. 3355–3360 (2008)
4. Kubota, N., Tomioka, Y., Ozawa, S.: Intelligent Systems for Robot Edutainment. In: Proc. of 4th International Symposium on Autonomous Minirobots for Research and Edutainment (2007)
5. http://www.robotis.com/zbxe/main
6. Hashimoto, T., Hiramatsu, S., Tsuji, T., Kobayashi, H.: Development of the Face Robot SAYA for Rich Facial Expressions. In: Proceeding of SICE-ICASE International Joint Conference 2006, pp. 5423–5428 (2006)
7. Schulte, H.F.: The characteristics of the McKibben artificial muscle. In: The Application of External Power in Prosthetics and Orthotics, National Academy of Sciences-National Research Council, Publication 874, pp. 94–115 (1961)
8. Veltman, K.H.: Edutainment, Technotainment and Culture, Veltman, K.H., Civita, Annual Report (2003)
9. Ito, M.: Engineering play: Children's software and the cultural politics of edutainment. Discourse 27(2), 139–160 (2004)
10. Kubota, N.: Visual Perception and Reproduction for Imitative Learning of A Partner Robot. WSEAS Transaction on Signal Procesing 2(5), 726–731 (2006)
11. Kubota, N.: Computational Intelligence for Structured Learning of A Partner Robot Based on Imitation. Information Science 171, 403–429 (2005)

12. Kubota, N., Nojima, Y., Kojima, F., Fukuda, T.: Multiple Fuzzy State-Value Functions for Human Evaluation through Interactive Trajectory Planning of a Partner Robot. Soft Computing 10(10), 891–901 (2006)
13. Kubota, N., Yorita, A.: Structured Learning for Partner Robots based on Natural Communication. In: Proc. (CD-ROM) of 2008 IEEE Conference on Soft Computing in Industrial Applications (SMCia), pp. 303–308 (2008)
14. Hashimoto, T., Hiramatsu, S., Tuji, T., Kobayashi, H.: Realization and Evaluation of Realistic Nod with Receptionist Robot SAYA. In: 16th IEEE RO-MAN International Conference on Robot & Human Interacive Communication, pp. 326–331 (2007)
15. Kubota, N., Shimomura, Y.: Human-Friendly Networked Partner Robots toward Sophisticated Services for A Community. In: Proc. of SICE-ICCAS 2006, pp. 4861–4866 (2006)
16. Khemapech, I., Duncan, I., Miller, A.: A survey of wireless sensor networks technology. In: PGNET, Proc. the 6th Annual Post Graduate Symposium on the Convergence of Telecommunications, Networking and Broadcasting, EPSRC (2005)
17. Jang, J.-S.R., Sun, C.-T., Mizutani, E.: Neuro-Fuzzy and Soft Computing. Prentice-Hall, Inc., Englewood Cliffs (1997)
18. Fukuda, T., Kubota, N.: An Intelligent Robotic System Based on a Fuzzy Approach. Proceedings of IEEE 87(9), 1448–1470 (1999)
19. Kubota, N., Aizawa, N.: Intelligent Cooperative Behavior Control of Multiple Partner Robots. In: Proc. (CD-ROM) of IEEE/RSJ International Conference on Intelligent Robots and Systems, pp. 2783–2788 (2008)

Not Just "Teaching Robotics" but "Teaching through Robotics"

Andrew W. Eliasz

First Technology Transfer Ltd.
awe@ftt.co.uk

Abstract. This paper explores strategies for teaching robotics not simply as a subject in its own right, but, using robotics in the teaching environment as an opportunity to stimulate creative thinking and generating an interest in science and technology as creative endeavours. The spirit is very much that espoused by C.P.Snow in his attempts to bridge "the two cultures" i.e. that of the arts on the one hand and that of science and technology on the other.

Keywords: Cross curriculum teaching, Arduino, PicoCricket, Lego Mindstorms, VEX, Autistic, Scratch.

1 Introduction

It is often stated by politicians that a "healthy economy" depends very much on the presence of a sufficiently large pool of well educated scientists and technologists. Yet, at the same time science and scientists are perceived by many to be somewhat strange and possibly dangerous types.Robots are also perceived by many as a potential threat, and by others as a source of great wealth and power and control.

The effective use of robotics, in the classroom requires that many teachers, not only science and maths teachers, are confident in using and adapting the various technologies in the classroom, and also that there is a much greater degree of "cross disciplinary" teaching. For this to happen it is necessary to provide teachers with the means to "handle the complexity" inherent in using robotics. Partly this must come from adding these subjects into the teacher training curriculum, partly by providing courses and workshops for teachers, and, most importantly by providing suitable tools with user friendly interfaces and abstractions to make the use of these technologies "feel natural".

2 Robots and Robotics - A Brief History

The history of robots a robotics goes back a long way, as a brief look at the relevant Wikipedia [1] will show. The term "robot" originated in a 20th century play which addressed the theme of "robot rights" in the context of serfdom of Androids working in an industrial world. In this play the "robots" are "the workers".

J.-H. Kim et al. (Eds.): FIRA 2009, CCIS 44, pp. 214–223, 2009.

The history of Robots is closely related with developments in digital process control and artificial intelligence. The "Mega Giant Robotics" web site has a "short history of robotics" [2] web page surveys these developments. Associated with the development of robotics systems there have been extensive developments in software and processor hardware image processing and pattern recognition.

It does not require much imagination to see the possibilities of projects, teaching modules and research activities based on material associated with the history of robots and robotics.

Robotics needs to be considered together with Artificial Intelligence and Simulation and Computer Game Programming as there are extensive areas of overlap between these various disciplines. Teaching through robotics should therefore also include these topics in its ambit.

All of the above are complex subjects. However, the spirit underlying "teaching through robotics" is not oriented towards mastering such technologies in detail, but in being able to use them effectively, and get students to use them effectively and to think about their implications e.g.

How might you motivate your listener to appreciate the importance of search strategies ?

What if you were talking about search in the context of a biology lesson that was teaching about how honey bees discover nectar bearing flowers?

- what about constructing a "little simulation" using Logo? e.g. programming a little "robot bee" that moved around on the floor and where nectar bearing flowers were represented by e.g. red disks of a certain size ?
- how might the "artificial bee" communicate where the flowers were to its fellow bees ?, what if the bees and hives were part of a computer game simulation package constructed along the lines of "Sim City" ?, how might different search strategies be investigated ?

3   The UK National Curriculum - An Example Curriculum

An application of "teaching through robotics" must be appropriate to the level of educational development of the students and also contextually appropriate to the subject being taught. Very often it will involve "cross disciplinary" teaching, indeed this is one of its greatest virtues. The UK National curriculum is broken up into several Key Stages covering different age ranges and, in total spanning the age range from 5 to 16. A *Key Stage* is a stage of the state education system in the UK.

It specifies the educational knowledge expected of students at various ages.

Within this curriculum there are provisions for "Cross Curricular Dimensions". [A full account can be downloaded from the QCA website [3].

The Cross-curriculum dimensions include:

- identity and cultural diversity
- global dimension and sustainable development
- technology and the media
- creativity and critical thinking

There are potentially many opportunities here for introducing robotics as a topic that can be used to explore various issues e.g.

- artistic and technical applications of robotics [technology and media, community participation, creativity and critical thinking ...]
- robotics competitions involving creativity and design skills as, for example, in a robot dance competition (in the sense of robots dancing both by themselves and with humans) [5] [technology and media, community participation, creativity and critical thinking ...]

In addition the National curriculum includes a Personal Learning and Thinking Skills (PLTS) framework that is made up of six groups of skills beyond the functional skills of English, mathematics and ICT

- independent enquirers
- creative thinkers
- reflective learners
- team workers
- self-managers
- effective participators

There is a huge scope for "teaching through robotics" here, ranging from discussion of rights and obligations e.g. using the metaphors raised by portrayal of robots in films and plays through to the use of analogy as for example in developing a robot playing soccer team.

The QCA documents contain various examples of cross-curriculum teaching. The example chosen here shows the skills and imagination needed to make this a success, and to it I have added some thoughts and conjectures as to where "teaching through robotics" might fit in, and the extra skills and experience that a teacher might need to do this successfully. The case study is called "Where do sounds belong" [6]

The project story is as follows:

A teacher wanted the class to link sounds with their sources and to distinguish between a sound and its source. She began by asking the mixed year 1 and 2 class to sit quietly on the carpet and listen carefully to the sounds the could hear around them. This led to a spider diagram (shown below)

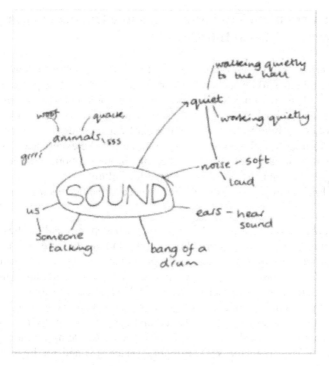

Fig. 1. Spider Diagram of SOUND Concepts

Following on from this was a reading of the 'The Sound Collector' , a poem by Roger McGough. The ensuing discussion led to a realisation that a sound has a source e.g. "the sizzle needed to go with the bacon in the pan, the crying needed to go with the baby" and so on. Further class discussion led to a deeper exploration of the difference between sounds and their sources and explained carefully that these were objects that made sounds e.g. class then talked about, described and imitated the sounds that 'crisps' and 'my cat' make.

Now, imagine, the teacher was confident in working with robots that could respond to sound, generate sound, distinguish between different sounds and record ('memorise') sound and could program various behaviour patterns into the "classroom" robot. What kinds of further themes might have emerged ?

Who can tell, but we might imagine questions such as

- remembering sounds
- being able to reproduce sounds from memory
- knowing where a sound came from
- being able to pair a sound with an object
- forgetting (imagine what would happen if you could not forget anything ?)

What would it take for a teacher to be able to use "robotic" resources in this context ?

4 Cross-Curriculum Teaching - Valuable Opportunity or Yet Another Set of Good Intentions ?

In many countries teaching curricula are very full . Apart from preparing and teaching lessons much time has to be spent on administrative work such as producing detailed formal course plans, marking homework, writing reports and assessments, meetings with parents and school governors. A key issue, as regards fostering "teaching through robotics" is how to motivate and reward teachers, especially those teaching non-technical subjects. Robotics and programming must be experienced as something enjoyable and creative. Approaching "teaching through robotics" as if it were yet another "imposed curricular chore" is a "recipe for failure".

For "teaching through robotics" to be a success factors need to be considered include

-- helping teachers overcome a whole range of misconceptions and hurdles
-- programming (at least with a suitable programming language) is not as difficult as it seems - teachers by nature are good communicators and hence skilful users of language - they should therefore be good at using a programming language if it enables them to express themselves in a natural and effective way
-- mastering the technology underlying robotics - conveying the message and experience that the technology can be mastered and can be fun e.g. by means of classes and workshops such as those run by the Robotics Academy at Carnegie Mellon University (CMU) both for VEX Robotics systems[7] and for Lego Mindstorms systems [8]
-- similar courses have been developed in the UK [9] and can also be accessed via the UK Robotics Education Foundation web site [10]' and, in Canada the University of Alberta has also developed a variety of LEGO Mindstorms oriented resources for use by teachers [11]
-- newer , more accessible technologies that are being developed, key examples include
-- Arduino [12], Scratch [13] for interactive programming, and Python ACT-R [14] (for modelling human cognition)
[Arduino and Scratch workshops for teachers are not yet widely available, but are starting to appear [15]]
-- the HEXBUG "insectoids" from Innovation - used well offer a vast range of opportunities for teachers [16]

5 Attempts and Developments to Make Robotics, Computing, Modelling and Computer Game Programming and AI More Accessible - Cricket Logo, Arduino, Scratch

One of the earliest graphical programming environments was the Cricket Logo programming environment developed for use with the Handy Board embedded computer [17], developed by Fred Martin and colleagues at MIT [19]

The Cricket project itself has evolved further, independently of Lego Mindstorms.

One such evolutionary path, based on the Scratch IDE is the PicoCricket, a tiny "Cricket like" computer developed and marketed by "The Playful Invention Company" [20] that can plug into a whole range of devices including motors, light, sound and touch sensors, beamers and sound boxes. and grew out of work involved with the Playful Invention and Exploration Network [21].

The interesting thing about this combination of a sensor board and a graphical application development environment is that it is possible to alter the appearance and behaviour of sprites in response to various "sensor events" , e.g. changing the appearance of a sprite whenever there is a loud sound.

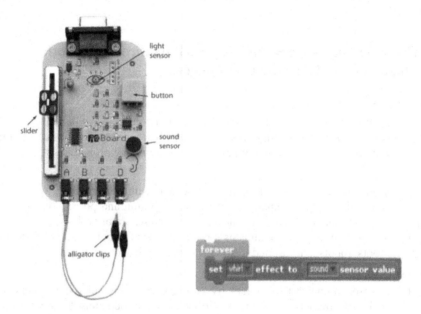

Fig. 2. Scratch Pico Board and "Scratch code fragment"

The Arduino project also tries to make the building of interactive systems and devices (including robots) more accessible to non professional programmers. It grew out of a Java based IDE called "Processing" and an underlying programming language called "Wiring". It represents an attempt to develop an open-source electronics prototyping platform characterised by its flexible and simple to use hardware as well as a relatively easy to use C like programming language.

As with the Lego Mindstorms developer and builder community the Arduino community is very active with many enthusiasts. The use of Arduino in Arts colleges to develop interactive "art forms" is very encouraging and suggests that some of the approaches can be adapted to use in junior and middle school teaching, providing the necessary workshops and funding for teachers to attend them are made available.

Use of AI and Computer Games as part of teaching through robotics is still at an early stage of development. Artificial Intelligence and Computer Games programming encompass some of the most advanced and difficult algorithms and programming techniques currently known. If AI is to be explored by non-specialists then, as shown with the Cricket, Mindstorms, Scratch and Arduino projects the underlying principles and functionality must be somehow encapsulated and exposed through simple to use programming interfaces and development environments. There are several candidate technologies that show potential and that are being explored. These include Rule Based programming languages such as CLIPS and Jess, Prolog, and reasoners such as Fact++ that can be used in connection with OWL based Ontology development environments. There are also cognitive modelling frameworks such as ACT-R. These are complex systems and their incorporation in "teaching through" robotics would, initially, be largely through demonstrations of various projects and applications that have used these tools.

6   Examples and Suggestions for Teaching through Robotics Opportunities in a Range of Subjects

Mathematics
- geometry of position estimation and collision / obstacle avoidance
- working out the best way to get from A to B when there are obstacles in the way, maze following, searching, planning, estimating probabilities

Science
- mechanics and physics problems associated with constructing, driving and controlling robots - equations of motion , friction and power, sensors and measurement (includes chemical sensors and laboratory robots)
- models of biological organisms e.g. behaviour of simple insects, swimming robots, robots that respond to stimuli such as light, sound, touch

Art and Design
- representational art vs. interactive art, incorporating microcontrollers into designed artefacts e.g. clothes with sensors and actuators built into them, adaptive surroundings in interior designs

Design and Technology - there is a degree of overlap with Art and Design - though here the emphasis is more on working and functional appliances and systems
- robotics and assisted living, smart sensors
- control of machines and motors, mechanical actuators
- object and image recognition

Geography
- data collection - using static and mobile sensors
- traffic control - e.g. adhoc sensor networks in cars
- ground surveillance, environmental monitoring , exploration (e.g. researching the geography of other planets)

History
- history of the development of robot technology
- robots and war

Citizenship

- robots and surveillance, robots and terrorism, robots and civil liberties
- ethical and moral issues concerned with robots, duties of owners of robots

Music and Dance

- can robots recognise tunes, can robots keep a beat
- can robots dance in response to different kinds of music

Media studies

- robots in plays, robots in films, robots in marketing and advertising

Information and Communication Technology

- communication between man and robot
- surveillance data collected by robots - data protection issues
- how much should robots know ?

Modern Languages

- language recognition, language translation, robots acting as interpreters

Physical Education

- balance and two legged robots, robot fitness instructors, table tennis playing robots

7 Robots and Teaching Those with Special Needs - Using Autism as an Example

It is now possible to build robots with quite sophisticated gesture, face and speech recognition capabilities. These robots tend to be quite expensive and the software running on them very complex and relatively unique to each individual research group. However as parts and software become standardised the cost will fall.

Autistic children and adults typically have difficulty with social interaction and maintaining social relationships and also may exhibit strange compulsions and obsessions.

It has been observed that people with autism often interact 'naturally' with computer technology and can use it quite creatively. It is not surprising therefore that attempts have been made to use humanoid robots in helping children with autism. The high costs of developing and building humanoid robots has led some researchers to explore the possibilities of working with cheaper "robotic toy dolls", for example, the Aurora project explored how non-humanoid mobile robots can be used in an environment in which autistic children can explore and discover interaction skills instead of simply being taught them e.g. by playing chasing games with a mobile robot. This work was later extended to use a humanoid robot doll, called "Robota" which was based on a commercially available doll which was engineered to have movable microcontroller controlled legs and arms and head, and was connected by a serial link to a PC which contained speech synthesis and video image processing software. The system was capable of tracking up and down arm movements of the user child when the child was facing the camera. The robot had rudimentary touch responses by being able to detect its limbs being moved via potentiometers. Trials with Robota proved to be encouraging and demonstrated the potential of this approach. [22]

The Interaction Lab which , part of the Center for Robotics and Embedded Systems at USC, directed by Maja Mataria, have studied the therapeutic potential for robot - human interaction in autism, and have developed a humanoid robot called LabBandit (actually now LabBandit2) and that is built of relatively inexpensive standard parts. [23] and,also, have had some promising results.

It is not necessary to use especially realistic robots in work with autistic children. BeatBots,forexample, has developed a very simple looking robot called Keepon that has been used in research on social development and interpersonal coordination as well as in therapeutic practice for children with developmental disorders such as autism. [24]

References

1. Wikipedia entry for Robots, http://en.wikipedia.org/wiki/Robot
2. Mega Giant Robotics - "a brief history of robotics",
   http://robotics.megagiant.com/history.html
3. UK Cross Curriculum planning guide for schools (2009),
   http://curriculum.qca.org.uk/uploads/
   Crosscurriculumdimensionsplanningguideforschoolspublication_
   tcm8-14464.pdf
4. UK Cross Curriculum Dimensions,
   http://curriculum.qca.org.uk/key-stages-3-and-4/
   cross-curriculum-dimensions
5. Flight of the Phantom Phoenixes - RobocupJunior Dance International Champions (2006),
   http://www.ictamber.org.uk/
6. Where do sounds Belong - Cross curricular teaching and creativity,
   http://curriculum.qca.org.uk/uploads/
   4-where-do-sounds-belong_tcm8-12082.pdf
7. CMU Robotics Academy VEX Robotics courses for teachers,
   http://www.education.rec.ri.cmu.edu/content/vex/index.htm
8. CMU Robotics Academy LEGO Robotics courses for teachers,
   http://www.education.rec.ri.cmu.edu/content/lego/index.htm
9. FTT - First Technology Transfer - robotics courses for teachers,
   http://www.ftt.co.uk/roboticsteachers.php
10. UK Robotics Education Foundation - courses for teachers,
    http://www.ukref.org.uk/Teaching_Thru_Robotics.html
11. University of Alberta - Faculty of Education - Teaching and Learning with LEGO Robotics, http://www.quasar.ualberta.ca/legorobots/index.htm
12. Arduino, http://arduino.cc/en/Guide/HomePage
13. Scratch, http://scratch.mit.edu/
14. Python ACT-R, http://www.carleton.ca/ics/ccmlab/actr/index.html
15. Arduino and Scratch workshops being developed by FTT,
    http://www.ftt.co.uk/arduino_teachers.php
16. Innovation First – HEXBUG, http://www.hexbug.com/
17. Handyboard, Cricket Logo, http://handyboard.com/cricket/
18. Martin, F., Mikhak, B., Silverman, B.: MetaCricket A Designer's Kit for Making Computational Devices, http://www.research.ibm.com/journal/sj/393/part2
19. Martin, F.G.: Robotic Explorations, A Hands-on Introduction to Engineering. Prentice-Hall, Englewood Cliffs (2001)

20. The Playful Invention Company : URL to PicoCricket, http://picocricket.com/
21. The Playful Invention and Exploration Network homepage,
    http://www.pienetwork.org/about/
22. Dautenhahn, K., Billard, A.: In: Keates, S., Langdon, P.M., Clarkson, P.J., Robinson, P. (eds.) Proc. 1st Cambridge Workshop on Universal Access and Assistive Technology (CWUAAT), pp. 179–190. Springer, Heidelberg
23. Using Robots for the Education of Children with Autism in the Classroom,
    http://robotics.usc.edu/interaction
24. Research into social and interpersonal development using Keepon,
    http://beatbots.org/research/

A Proposal of Autonomous Robotic Systems Educative Environment

Jorge Ierache, Ramón Garcia-Martinez, and Armando De Giusti

Computer Science PhD Program, Computer Sc. School, La Plata National University
Instituto de Sistemas Inteligentes y Enseñanza Experimental de la Robótica FICCTE
Universidad de Morón
Intelligent Systems Laboratory, Engineering School, University of Buenos Aires,
Instituto de Investigación en Informática LIDI, Facultad de Informática, UNLP
jierache@unimoron.edu.ar, rgarciamar@fi.uba.ar,
degiusti@lidi.info.unlp.edu.ar

Abstract. This work presents our experiences in the implementation of a laboratory of autonomous robotic systems applied to the training of beginner and advanced students doing a degree course in Computer Engineering., taking into account the specific technologies, robots, autonomous toys, and programming languages. They provide a strategic opportunity for human resources formation by involving different aspects which range from the specification elaboration, modeling, software development and implementation and testing of an autonomous robotic system.

Keywords: Robotic, Autonomous Systems, Technologies in Education.

1 Introduction

The development of the technologies applied to education contributes to the learning process; particularly the application of an Autonomous Robotic Systems Laboratory (ARSL) collaborates with different areas in the training process of Information Technology students, from the interpretation of requirements to the autonomous system implementation, enhancing student's creativity as regards physical construction, software optimization and sensor integration, as well as the development of cooperative and competitive environment among robots. Watching how a turtle moves around in our monitor, while avoiding virtual obstacles to reach its goal in the corner of the monitor, does not have the same emotional impact on a student as observing how an Autonomous Robotic System (ARS) can avoid obstacles to achieve its goal in the corner of a room, and interacts with us by means of our mobile phone. We consider that the Computer Engineering, especially those associated with the Autonomous Robot Laboratories become an aid to learning processes in the case of beginner and advanced students; in the contextual framework [1], stated in figure 1, it is considered: [a] Paradigm under which the student carries out his/her work,[b] Methodology that is applied under the selected paradigm, [c] Techniques that facilitate the development of the phases and stages of the methodology applied, [d] Tools on which the techniques are applied, [e] Programming languages, [f] Robots. In this order, for

J.-H. Kim et al. (Eds.): FIRA 2009, CCIS 44, pp. 224–231, 2009.

beginners we can consider the imperative paradigm, that of objects, their methodologies and techniques, such as the Nassi-Shneiderman Diagrams [2], UML [3], programming languages like C, particularly NQC [4], and JAVA, in particular LeJOS Java [5], for the development of software running on RCX [6], NXT [7] robots. For advanced students, we consider the multiagent paradigm, with methodologies like MaSE [8], [9], techniques like Agent-UML [10].

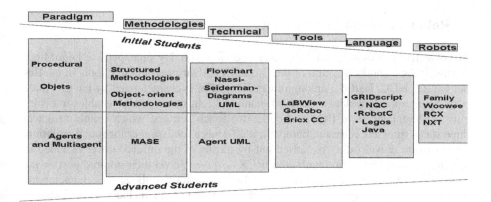

Fig. 1. Contextual Framework Learning

Meanwhile, the ARSL makes it easier to state explicitly the requirements under the IEEE 830 Standards [11], their validation and contextual framework the testing with the application of techniques such as Complexity Menasure [12]. It also contributes to improve the construction processes in a teamwork environment, where students are highly motivated for the development of their robots or pets. Many projects involve a centralized control, the computer instructs motor 1 to start, to turn in a clockwise direction, at half power under a planned action sequence, but the same robot agent can be applied to explore decentralized systems and those of self-organized behaviors [13]. For instance, if we consider an agent that wanders around its habitat, which has lit areas and dark areas, our agent has two rules, one indicating it to move forward when lit areas are detected and the other indicating to move backward when dark areas are detected; the agent wanders about until it reaches a shaded spot, so it moves back until it comes out of that spot and moves forward again; it goes on oscillating at the edge of the shade; in this case, we can consider our robot agent as a creature that detects edges; this capacity is not explicitly stated in its two rulers, in fact it is a group behavior which emerges from the interaction between the two rulers, similarly to the way in which a flock behavior emerges from the interaction between birds [14]. At different moments, students tend to consider their creatures at different levels; they sometimes see them at a mechanistic level, when analyzing how a piece of the mechanism moves another one. At times, they see them at an information level, and they explore how information is transmitted from the computer to the motors and sensors. On another occasion, they see their creatures at a psychological level, by attributing them a certain purpose or personality. One creature wants to go towards light, another one prefers darkness, another one is afraid of noise. Students go quickly through these levels and learn according to the context situation what level is better; they think about systems in terms of levels [14]. The idea of learning through design is one aspect of what Seymour Papert [15] called "constructionist approach to learning and education". The human beings build their knowledge in a particularly efficient way when they participate in the construction of products they are emotionally involved in.

2 Robots, Languages and Tools

The objective of this section is to give an overview of today's inexpensive LegoMindstorms, RCX Robotic Kits and of the latest NXT, their programming tools in the Robot C [16], NQC, LeJOS , among others. The RCX is characterized by having: three ports for motors, five slots to keep programs, a Light sensor, which enables to distinguish different levels of light and dark, two touch sensors, which enable to detect three states (pressed, released, bumped); it also has a loudspeaker for sound emission. The program downloading is carried out by means of the infrared tower included in the kit. The communication with other RCXs is possible via their infrared port on the front. The NXT is the new generation of Legomindstorms robots; it is characterized by having higher computing power than the RCX. The NXT includes some functions to test the sensors, to personalize the sounds it may reproduce, three ports for motors, four ports for sensors. It is equipped with a light sensor, a sound sensor, two touch sensors, which enable to detect three states (pressed, released, bumped), an ultrasound sensor functioning as a radar, thus enabling the detection of object, which may be set to detect close or distant objects; it detects objects at a distance from 0 to 255 centimeters with a precision of +-3 centimeters; it also has a high-fi loudspeaker, improved, and three servo motors, that have been improved as regards the RCX version. The servo motors have built-in rotation sensors which enable precise and controlled movements and a perfect motor synchronization. The NXT has a USB port, intended for program downloading. It supports Bluetooth wireless communication, thus enabling both program downloading and interaction with cell phones, PCs and laptops, etc. The communication with other NXTs is also carried out via bluetooth. These robots can be programmable in a native graphical environment, in the case of RCX [17] and LabView [18] in the case of NXT. Regarding LabView, it is worth mentioning that it was developed by National Instruments and used by the NASA to monitor and control Sojourner Rover robot, during the mission to explore the surface of Mars [19]. These environments use blocks which assemble with one another to form a complete program. These blocks include: motor control (forward, reverse, on and off), repetitive cycles (while, repeat), control structures (if else), data collection from the sensors, variable use, constants and timers. In addition to these graphical environments, there exists a series of programs which enable their programming in more traditional codes, such as Java. That is the case of the LeJOS API for the RCX [20] and the LeJOS NXJ for the NXT [21], iCommand is a Java package to control the communications over a Bluetooth connection [22]. One of the mostly used programs, which highly increases programming possibilities is the NQC [23], [17], developed by Dave Baum and used to program the RCX in a language similar to the C one. For the

NXT there exists a program called RobotC [16], which is much more complete than the NQC for the RCX, and includes its own firmware that makes it very powerful. Here follow the most important features of the main programming tools of Lego Mindstorms. Among the languages, the ones that can be mentioned are: NQC for the RCX and Robot C, similar to NQC, for the RCX, however it is much more powerful and enables robot programming in limited C. It includes a firmware, support for Bluetooth communication. This is one of the new languages existing nowadays to develop with NXT Lego Mindstorms. For JAVA NXT programming it can be mentioned the Lejos Java. Its alternative firmware for the NXT is characterized by: [a] enabling program development in JAVA in order to monitor NXT robots, [b]functioning under Windows and Linux, [c] communicating with the NXT via USB, [d] supporting Bluetooth communication; NXT firmware enables a Master/Slave-type set up for Bluetooth communication. Up to three NXTs can communicate via Bluetooth. The new JAVA API for the NXT is called iCommand, that includes, among other features, Webcam Support and Electronic Compass Support.

The Multimodule Robots are introduced as Bioloid Robot kit [24], is characterized by having a total of 18 servomotors, infrared sensors in the head to communicate with other robots and sensors to detect proximity forward and towards its sides, microphone and loudspeaker. Bioloid Comprehensive is the modular robotic platform kit suitable for building advanced robots having up to 18 degrees of freedom like humanoids. It is suitable for learning, hobby, research and competition. The kit is like an upgraded version of Meccano and is made up with many constructive blocks the user may assemble with screws. Its programming language is C.

3 Autonomous Toys, Programming Languages and Tools

Regarding Autonomous Toys, here follow the most relevant features of "Robosapien" (humanoid robot) and "Robopet, Robotail, Roboraptor" (quadruped robots), Roboquead (hexapod robots) from the Woowee family [25].Moreover, communication interfaces and programming tools are considered, particularly GoRobo. Although they are sold as toys, they offer so advanced features that they become an excellent way of experimenting on robotics. The humanoids robots have stereo sound sensors, infrared vision, and touch sensors to detect obstacles and several degrees of freedom. We can find in this category: [a] Robosapien V1 is a version with less features concerning sensors than the V2, it does not incorporate vision, the displacement capacities are similar in their functionality, though the RS-V1, being smaller, has a better displacement, [b] Robosapien V2, apart from the above-mentioned characteristics, it includes touch sensors in the gloves, and in palms of its hands, thus enabling to take objects. It also has a camera which lets it recognize colors. [c] Robosapien Multimedia increases even more the capacities of the RS-V2; it includes as an important characteristic a mini SD memory, in which it can be directly programmed, by means of a graphical-type code editor, existing only in this version. It has 4 personalities by default, which can be modified by the user. It can also record videos and mp3-format sounds, take pictures, and then reproduce them all on its Liquid Crystal Display (LCD).

The Quadrupeds robots [25] are also equipped with infrared vision, stereo sound sensors and motors. The most relevant ones are: [a] Robopet: apart from the above-mentioned characteristics, it is able to interact with Robosapien, and also detect edges, for instance table edges. [b] Robotail has a touch sensor on its back, which by being pressed makes the robot have a different behavior. Moreover, when it is "hungry", it becomes very "aggressive" and can only be calmed down by "finding food". [c] Roboraptor is the roboreptile that is able to interact with Robosapien. [d] the arthropods are introduced as Roboquad, which has four legs with a chassis designed to move in any direction at three different speeds, has as a special feature that of identifying motion at a distance of about three meters; once identified, the robot can follow the object movement. It has edge sensors which allow to detect doorframes, table and chair edges.

A "GoRobo" programming environment allows to control most of the above-mentioned robots from the WowWee family (Roboraptor, Robopet, Roboreptile, RSV2 and RS Multimedia). The programming language used is called GRIDscript (Go-Robo ID script) [26]. It uses a simple and consistent programming syntax, based on modern commercial practices of programming products (Visual Basic, C++, etc). GRIDscript uses a basic programming syntax (While/EndWhile, For, If/Else/Endif, Repeat/EndRepeat), for the creation of procedures and the use of variables. The beginners can use this language to define simple procedures which may be later combined to create more complex ones. Moreover, the robot can be programmed to interact with each other, since the software allows the simultaneous control of six of them. The commands are transmitted via an infrared tower that, as an interpreter, sends them to each robot, by identifying the type through an infrared tower, such as USB-UIRT [27] and RedRat3 [28]. Here follow some actions to be carried out with GoRobo: use of conditional and instruction repetition blocks, use of events conditioned by timers, possibility of introduction of random code execution. This language was designed to be suitable for every age and to be used in both an educational and a professional context, where there exists an interaction of classical and formal programming languages with the natural commands of the robots that are used. It also includes a scene editor which can be upgraded with sound. Other programming options, for the Robosapien and the Robopet, receive their commands via IR through a remote control; in this way there exist those that have performed a mapping to hexadecimal of said commands [29], thus making it possible the Robosapien´s programming by downloading the code with the Lego Mindstorms´ IR Tower. The problem is that there is a constraint in the quantity of instructions that can be received by the RS, up to twenty. Another more radical option is the brain transplant to the Robosapien; sometimes it has been decided to replace the Robosapien V1 head with a Palm [30], in this way the problem of the quantity of instructions that can be sent to RS is eliminated, greatly improving the Robosapien V1´s ability to do calculations.

4   An Application Case between Robots and Autonomous Toys

This case simulates the behaviour of a herbivore, wandering along its environment developed with the NXT robot, which in this case, was a carpet outlined by a wooden structure, with green papers distributed at random representing food. So this herbivore (the prey) wandered easily until it found a food area. At this point it stopped to "feed" and it was also able to detect the borders of the habitat, thanks to its touch sensor and avoid them. At the moment the prey, with its sound sensor, detected the sound of a predator, represented by the Robotail (autonomous toy) or something got the backwards position (detected by the ultrasound sensor) it "scared" failing to eat and beginning to run away at high speed. In this case, you can clearly see two types of behavior, the first one that looks for food and the other one, that flees, both depending on the NXT interaction with its environment and with Robotail (figure 2). The NXT (robot) which represents the prey was programmed with RobotC and the Robotail (toys) which represents the prey was programmed in a GoRobot environment.

Fig. 2. Robot NXT (prey) interacting with the environment and Robotail (predator) TE&ET 07

5 Autonomous Robot System Development Laboratory

The Autonomous Robot System Laboratory (ARSL) presents an opportunity for students´ learning, particularly in the context of programming robots that work in dynamic and cooperative environments and require the creation of strategies aimed at reaching their goals to confront their opponents, without the action of external supervision. The Autonomous Robot System control programs cannot define explicitly every possible action in view of all the possible situations that may arise in its environment. The robot must not be fully pre-programmed; it must have a cognitive architecture that enables to establish a relationship between its sensory input and its actions on the environment [31]. It must have the ability of generating its autonomous sensorization map–actions to survive and achieve its goals. An Autonomous robot System Laboratory (ARSL) also offers a favorable scenario for the development of applications centered on context where the participation of robot and human players may be of interest in an interactive environment through cell phones, Internet, etc. The initial communication strategy to support the interaction between autonomous agents and human beings is based on the use of the possibilities provided by the wireless Bluetooth communication among agents. The advanced students are also interested in the methodologies in multiagent context, tools, intelligent autonomous systems, Artificial Intelligence concepts, vision and distributed processing [32]. Moreover, an ARSL can include global information from the environment by means of the integration of a vision system that allows the detection and localization of objects and autonomous robots in the scenario; in this case, the complexity level is even higher, thus enabling that, besides processing the information given by its own sensors, the robot may have information of everything happening around it and be dynamically adapted, interacting among them and with the environment, as well as develop capabilities to facilitate sharing of knowledge between systems of autonomous robots [33], [34].

6 Conclusions and Future Research Lines

The use of robotic technology proposed on the present paper helps the development of different educational experiences such as robot soccer, rescue, navigation and so on. the experiences developed by the students within the robotics laboratory context turn out to be stimulating for them as they can see the result of their work through the action performed by their robots while strengthening the learning process. Furthermore the present paper has been developed on the last five years´ experience with a participation of an average of twenty initial level students per semester, working in teams for the construction of robots, scenarios, software development and tests. On the advanced level an average of eight students worked for two semesters, they developed final works where robots were integrated with the application of intelligent systems techniques and multiagent methodologies.Future research lines are aiming to the development of a framework where different robots are integrated, to the development of interoperating simulation capability between virtual and real worlds in order to support the robots learning scenario.

Acknowledgements

This research is supported by PID A01-007- FICCTE-UM.

References

1. Ierache, J., Bruno, M., Dittler, M., Mazza, M.: Robots y juguetes autónomos, una oportunidad en el contexto de las nuevas tecnologías en educación. In: VIII Ibero-American Symposium on Software Engineering, pp. 371–379 (2008)
2. Nassi, I., Shneiderman, B.: Flowchart techniques for structured programming, SIGPLAN Notices XII (August 1973)
3. UML, http://www.uml.org/
4. NQC – Not Quite C, http://bricxcc.sourceforge.net/nqc/index.html
5. Lejos, Java for Legomindstorms. SourceForge, http://lejos.sourceforge.net/
6. Lego Mindstorms RCX,
   http://www.lego.com/eng/education/mindstorms/
7. Lego Mindstorms NXT, http://mindstorms.lego.com/
8. DeLoach, S.: Analysis and Design using MaSE and agent Tool. In: Proceedings of the 12th Midwest Artificial Intelligence and Cognitive Science Conference, MAICS (2001)
9. Ierache, J.: Elaboración de una Aproximación Metodológica para el desarrollo de Software Orientado a Sistemas Multiagentes (2003),
   http://www.fi.uba.ar/materias/7570/index.htm
10. Bauer, B., Muller, J.P., Odell, J.: Agent UML: A Formalism for Specifying Multiagent Software Systems. In: Proc. ICSE 2000 Workshop on AOSE 2000, Limerick (2000)

11. IEEE recommended practice for software requirements specifications -IEEE Std 1028-1988, IEEE Standard for Software Reviews and Audits (ANSI) Software Requirements Specifications. IEEE. Std 830-1
12. Mc Cabe, T.: A Software Complexity Menasure. IEEE Transactions on Software Engineering 2(4), 309–320 (1976)
13. Resnick, M.: Tortugas, Termitas y Atascos de Tráfico, Gedisa, Barcelona (2001)
14. Morrollon, M., Segoviano, A.: 1, 2, 3... Logo (Ideas e Imaginación). Centro de Orientación de Sociología y Psicología Aplicada. Cospa, Madrid (1985)
15. Papert, S.: Situating constructionism, en I. Harel y S. Papert (comps.), Constructionism. Abel Publishing, Norwood (1991)
16. Quick start guide, Robotics Academy, Carnegie Mellon University, http://www.robotc.net/
17. Baum, D., Hansen, J.: NQC, http://bricxcc.sourceforge.net/nqc/doc/NQC_Guide
18. National Instruments. LabVIEW, http://www.ni.com/academic/mindstorms/
19. National Instruments LabVIEW Software Monitors Health of Mars Pathfinder Sojourner Rover (1997), http://findarticles.com/p/articles/mi_m0EIN/is_1997_July_18/ai_19593795
20. Lejos RCX, http://lejos.sourceforge.net/p_technologies/rcx/lejos.php
21. Lejos NXJ, http://lejos.sourceforge.net/p_technologies/nxt/nxj/nxj.php
22. Icommand.NXT, http://lejos.sourceforge.net/p_technologies/nxt/icommand/icommand.php
23. Baum, D.: NQC Manual, http://bricxcc.sourceforge.net/nqc/doc/NQC_Manual
24. Bioloid Constructive Kid, http://www.tribotix.com/Products/Robotis/Bioloid/Bioloid_info1.htm
25. WowWee, http://www.woowee.com
26. Go-Robo, http://www.q4tecnologies.com/
27. USB-UIRT, http://www.usbuirt.com/
28. RedRat3, USB Universal Remote Control, http://www.redrat.co.uk/RedRat3/index.html
29. Lego IR-Tower.Trondheim-Bratislava, http://www.robotika.sk/maine.php
30. Sven, B., et al.: Using Handheld Computers to Control Humanoid Robots Proceedings Dextrous Autonomous Robots and Humanoids (2005)
31. García Martínez, R., Borrajo, D.: An Integrated Approach of Learning, Planning and Executing. Journal of Intelligent and Robotic Systems 29, 47–78 (2000)
32. Wooldrige, M., Jennings, N.: Agent Theories, Architectures and Languages: a Survey in Eds. Intelligence Agents 1(22) (1995)
33. Ierache, J., Naiouf, M., García Martínez, R., De Giusti, A.: A Un modelo de arquitectura para el aprendizaje y compartición de conocimiento entre sistemas inteligentes autónomos distribuidos. In: VII Ibero-American Symposium on Software Engineering pp. 179–187 (2007)
34. Ierache, J., García-Martínez, R., De Giusti, A.: Learning Life-Cycle in Autonomous Intelligent Systems. World Computer Congress. In: Bramer, M. (ed.) Artificial Intelligence in Theory and Practice II, pp. 451–455. Springer, Boston (2008)

Mechatronics Education: From Paper Design to Product Prototype Using LEGO NXT Parts

Daniel M. Lofaro, Tony Truong Giang Le, and Paul Oh

Drexel Autonomous Systems Lab (DASL)
Department of Electrical and Computer Engineering
Bossone Research Center
3120-40 Market Street
Philadelphia, PA 19104-2875 United States of America
dml46@drexel.edu

Abstract. The industrial design cycle starts with design then simulation, proto-typing, and testing. When the tests do not match the design requirements the design process is started over again. It is important for students to experience this process before they leave their academic institution. The high cost of the prototype phase, due to CNC/Rapid Prototype machine costs, makes hands on study of this process expensive for students and the academic institutions. This document shows that the commercially available LEGO NXT Robot kit is a viable low cost surrogate to the expensive industrial CNC/Rapid Prototype portion of the industrial design cycle.

Keywords: Control, Robotics, WhIP, Wheeled Inverted Pendulum, Design Cycle, Design Process, LEGO NXT.

1 Introduction

The inverted pendulum has long been considered a classic controls problem and has thus become one of the *industry standards* for control design examples. The basic nature of the inverted pendulum, i.e. rotation about a central pivot point, makes the system a prime example for linearization and linear control. Methods such as Proportional Integral Derivative (PID), State Variable Feedback (SVF) and non-linear control such as Sliding Mode Control (SMC).

In recent years it has been common place to add wheels around the pivot point of the inverted pendulum resulting in control inputs of velocity or position with respect to the ground. Feedback from these wheeled inverted pendulums now includes the angle of the inverted pendulum, just like in a simple inverted pendulum, and the desired position or velocity of the wheeled inverted pendulum. A good example of a commercially available wheeled inverted pendulum is the Segway®.

Currently there are companies, such as Quanser[1], that are mainstays in university control lab courses. They are high quality turnkey systems that typically cost around 10,000 USD. Such costs often limit the number of units a university can procure.

---

[1] www.quanser.com

J.-H. Kim et al. (Eds.): FIRA 2009, CCIS 44, pp. 232–239, 2009.
© Springer-Verlag Berlin Heidelberg 2009

Consequently this limits how many students can actually have hands-on experiences using such devices. The use of the latter systems gives students experience implementing different forms of control in the real world. This experience allows students to learn more quickly about how a simulated solution to a control problem will compare with the real world solution. The methods described in this paper will expand upon the idea of having students implement control in the physical world through *hands on* exercises which have a low monetary cost. The authors envision a system that costs about the price of typical control systems textbook ($100 to $200 USD per unit). Each student will experience the design process through, modeling, simulation, and implementation. A trade study is also presented which supports the desired methods.

2 Proposed Course Model

The proposed model for a *hands on* controls design course will guide students through an abbreviated version of the entire design process starting with meeting design requirements thru real world implementation. The proposed course is designed to be completed in about 10 to 20 weeks, that is, a 1 or 2 quarter terms in a typically 3-credit course. The overall objective of the proposed course is to have the students design, build, and apply closed loop control to their own wheeled inverted pendulum (WhIP). Each student will design and build their WhIP utilizing the commercially available LEGO® NXT robotics kit. The sequence of topics for the course are described in the following sections.

2.1 Design

In industry there are always design requirements that a given control system must meet. Typical design requirements include performance specifications, such as rise time, settling time, power consumption, noise immunity, size, weight, and power. Design requirements also include monetary cost constraints because a real world control system has to function properly while having a reasonable cost. Because of this each student will be required to make a bill of parts for their implemented design. The design requirements for this course are to make the following:

Physical Constraints. The WhIP must fit inside of a 0.3m x 0.3m x 0.3m box. No more than two wheels may be touching the ground at any given time. No other part of the WhIP may touch any external surface. It must be powered by the battery that comes with the LEGO NXT kit and it may only use the parts included in the LEGO NXT kit and the rate gyro.

Free Standing WhIP. The WhIP must be able to stand upright in the same location for an extended period of time regardless of the slope of the ground. The slope will range from -35° to 35°. The completion of this will demonstrate the students' ability to use position control to control the WhIP in a stable manor.

Robust Moving WhIP. The end product must be able to traverse uneven terrain both in forward and in reverse at a constant linear speed. The completion of this will demonstrate students' ability to implement a form velocity control while keeping the WhIP standing upright.

Navigate Obstacle Course. The WhIP must be able to navigate through an 8x8 segment maze where each segment is a box with one or more sides measuring 0.5m in width and 0.5m in length. Each wall in the segment will be 0.5m tall. This maze will be similar to that of the maze in the popular Micromouse[2]competitions. This will show that the students are able to control the direction of the WhIP while implementing obstacle avoidance/path planning algorithms and keeping it standing upright.

Low Cost WhIP. Each student will be limited to only the parts that they need to make the WhIP functional by putting a restriction on how much each WhIP can cost. This will force students to re-evaluate the design requirements and think about what is absolutely needed to complete their objectives. Due to the fact that each device is made out of LEGOs, a monetary cost, which is given in the *LEGO price catalog,* will be given to each piece used. The LEGO price catalog enables students to order a wide range of parts. The price for each part enables one to access the total fabrication cost of realizing one's WhIP. This enables students to economically assess their realization and possibly identify parts to eliminate or exchange to reduce costs.

Limited Sensor Input. The only sensor inputs available to the students will be a rate gyro and an ultra-sonic range finder. By limiting the information that the system feeds back to the controller, the students will have to work with constraints. These constraints are representative of those found in industry, where cost can limit the number of sensors a system can posses and thus affect the number of observable states.

2.2 Simulation

When the design phase is completed, a 3D model of the WhIP will be created in a CAD program, such as SolidWorks[3]. SolidWorks was chosen because it is one of the most popular CAD packages in industry and is often taught in undergraduate engineering courses. SolidWorks also contains information on the material properties of each piece used. This is important because SolidWorks can then calculate the Center-of-Mass and Moment-of-Inertia of the resulting WhIP. These calculations then feed into simulation packages like LabVIEW[4] and Simulink[5].

2.3 Control

The motivation and purpose of creating a model and simulation is to create an effective control algorithm. When designing the control algorithm multiple methods can be used. If it is desired to teach the students about linearization and linear control, the wheeled inverted pendulum can be linearized around its vertical point. Then linear

---

[2] http://en.wikipedia.org/wiki/Micromouse
[3] www.SolidWorks.com
[4] www.ni.com
[5] www.mathworks.com

control methods, such as PID and SVF, can be applied to it to stabilize the system[1]. If the desire is to learn more about non-linear control, then the system can be taken as is and non-linear control methods such as Sliding Mode Control can be applied to stabilize the system[2][3].

2.4 Prototype

After the control algorithm of choice has been implemented and tested in the simulated environment, the control can then be ported to the real world. At this point, a prototype will be made. The beauty of the WhIP is that instead of making the prototype using a CNC or a costly rapid prototype machine, each student will construct their kit out of LEGO parts that they listed in their bill-of-materials. If the students require more parts for their prototype than is supplied in their kit then an extra cost will be added to their bill-of-materials which would include loss of time and money.

2.5 Testing and Evaluation

During this phase, the control algorithm will be implemented on the prototype. If the control is unsuccessful in the real world but is successful in the simulation, the design must be modified by changing the simulation properties or creating a new design. The process is then repeated until, a control algorithm is found that works with the new model of the system in simulation and also works on the real world system.

It is important to note that the main difference between the *Industry Design Cycle* and the *Proposed Design Cycle* is that the CNC/Rapid Prototype phase of the *Industry Design Cycle* has been replaced by the LEGO NXT Kit in the *Proposed Design Cycle*. LEGO provides a means to physically realize one's design. This is important because designs performed in CAD do not always translate well in real-world fabrication. For example parts, that seem to mate well in CAD, may not provide proper tolerance. By constructing with LEGO, the students will better recognize such oversight.

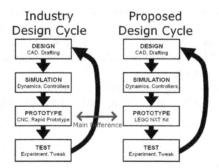

Fig. 1. The block diagram to the bottom left shows the *Industry Design Cycle*. The block diagram on the right shows the *Proposed Design Cycle*. Each design cycle is identical with the exception of the prototype phase. The *Industry Design Cycle* uses CNC and Rapid Prototype machines to create the prototypes while the *Proposed Design Cycle* uses the inexpensive and commercially available LEGO NXT Kits to create prototypes[8].

After multiple iterations of the design cycle there will be an accurate model of the inverted pendulum, and a real world functional wheeled inverted pendulum.

3   Case Study

A case study was conducted to demonstrate the feasibility of the *Proposed Design Cycle* as described earlier. The system used in the trade study is the Matlab® and Simulink® Embedded Coder Robot NXT software[4]. This software acts as a control interface for the LEGO NXT kit. The design that was chosen was based off of the NXTway-GS, a two-wheeled balancing robot[5]. This model was chosen because of the simple and proven design. *Figure 2* below shows the design. The design is called the LEGO NXT Wheeled Inverted Pendulum or the NXT-WhIP for short.

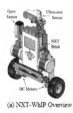 

(a) NXT-WhIP Overview   (b) Simulink® Virtual Reality Toolbox

Fig. 2. Shown below in (a) is the wheeled inverted pendulum which was based off of the NXTway-GS. The design has been modified to suit the needs of the proposed course and is now called the NXT-WhIP[6]. Shown below in (b) is the SolidWorks® model of the NXT-WhIP in the Simulink® Virtual Reality Toolbox running over a smooth bump in a stable manner. The simulation shown below shows a NXT-WhIP with larger wheels than that shown in (a). This is because the simulation below was from a revision of the NXT-WhIP which occurred during implantation of the *Proposed Design Cycle*.

The NXT-WhIP was first built in SolidWorks®. As described above SolidWorks® has a library of all of the LEGO NXT parts which includes material properties. The SolidWorks® model was then used in the Simulink® Virtual Reality Toolbox for simulation. The Virtual Reality Toolbox allowed for testing the stability of the NXT-WhIP on a multitude of terrain.

The NXT-WhIP was not only simulated using SolidWorks® and Simulink® but it was also modeled using traditional modeling techniques described in the next section.

3.1   Method/Theory

NXT-WhIP Modeling

The NXT-WhIP was modeled using Lagrangian dynamics. The system was analyzed as a simple inverted pendulum with the addition of wheels. This gives the system three degrees of freedom (DOF). *Figure 3* shows the state variables that will be used in the derivations of the model for the system.

The system was modeled using Euler-Lagrange equations[7]. The system was converted to State Space (SS) formation. The state variables are defined in Table 1.

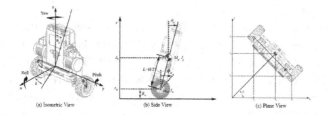

Fig. 3. Views of the NXT-WhIP showing the various angles and lengths. The views shown are the (a) Isometric View, (b) Side View, and (c) Plane View. Note that l and r denote the left and right wheels respectively.

Table 1. States variables used to describe the WhIP system when in state space formation

State Variable	Unit	Discription
$\varphi_{l,r}$	rad	wheel angle
$\varphi'_{l,r}$	rad/sec	wheel angular velocity
θ_p	rad	pitch angle
θ'_p	rad/sec	pitch velocity
δ	rad	yaw angle
δ'	rad/sec	yaw velocity

It is assumed that the system has knife edge constraints when moving in the x,y plane, as seen in *Figure 3* (c), and there is no slip between the wheels and the ground. It was shown from both the simulation and the mathematical model that the uncontrolled WhIP is naturally unstable according to the Routh-Hurwitch Criterion[1]. The systems unstability is shown analytically in Figure 4.

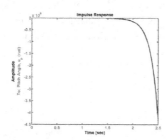

Fig. 4. The plot below displays the impulse response of the WhIP model. The system is shown to be naturally unstable after an impulse input because the system does not approach a steady state any time after the impulse. Please note that the impulse was applied to the wheel angular velocity, φ'.

NXT-WhIP Control

State Variable Feedback (SVF) was used in this case study to illustrate how one would design a functional controller. The block diagram shown in Figure 5 shows the control used to stabilize the pitch angle, θ_p, of the WhIP. The states used for the SVF are θ_p, φ, $\theta_p\grave{}$, $\varphi\grave{}$, δ, and δ'. The reference to the balance controller is $0°$ for θ_p. The gains for the SVF control are found via the use of a Linear Quadratic Regulator (LQR)[1][6]. LQR allows the user to put weights on how important the pitch angle is compared to the other states such as the wheel angular velocity. In the case of the WhIP a higher weight was placed on the pitch angle, θ_p. The yaw velocity is fed forward in to the motor controller to give the WhIP the proper orientation in the x-y plane.

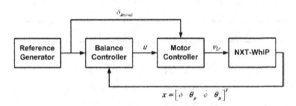

Fig. 5. The block diagram for the control setup for the WhIP is shown below. The states θ_p, φ, $\theta_p\grave{}$, and $\varphi\grave{}$ are fed back to the balance controller to keep the system stable around $\theta_p=0°$. The yaw velocity $\delta\grave{}$ is fed forward in to the motor controller to orient the front of the WhIP to the desired orientation in the x-y plane.

3.2 Experiment Setup

Using the SVF controller with the gains found by using LQR the system was stabilized around the desired pitch angle of $0°$. This control was applied to both the virtual system/dynamic model as well as the physical prototype system. There is a correlation between the simulated results and the prototype system test. The results can be found in Figure 6 below. The simulated results show that the system will reach steady state after approximately 2 seconds. The experimental results show that the real world system took closer to 20 seconds to reach a stable steady state.

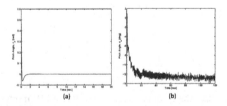

Fig. 6. The plots below show the response of the WhIP to a commanded input of $\theta_p=0°$ at t=0sec with initial conditions of $\theta_p(0)=35°$. Plot (a) shows the simulated response using the Simulink® Virtual Reality Toolbox. Plot (b) shows the response on the physical prototype system in the real world[6].

3.3 Results and Discussion

The results from the case study showed that the experimental and simulated results from the NXT-WhIP tests both reached a stable steady state. Though the steady state value and the settling times were not the same the *Desired Design Cycle* was able to create a stable real world system using simulated results prior to testing. Thus the authors conclude that the *Proposed Design Cycle* does match the *Industry Design Cycle* when replacing the CNC/Rapid Prototype steps in the prototype phase with the LEGO NXT Kit. Thus the NXT-WhIP is a viable alternative to teach students the importance and skills in actually and physically realizing a design.

4 Conclusions

The authors conclude that the NXT WhIP is a viable surrogate for prototype phase of the *Industry Design* Cycle. The NXT WhIP has proven itself to be a Quasner like system which allows the students to "take home" their work because each WhIP cost about as much as a text book. The authors also conclude that this is a viable 10-week course to teach the testing-and-evaluation (T&E) and validation-and-verification (V&V) aspects of mechatronic design: from paper concept to product prototype. By walking the students through the WhIP case study, the final project would be assigned. An example would be to define the technical requirements for a ball-and-beam balance system. Here, students would again prototype in SolidWorks, model in Matlab, design the controller, physically construct in LEGO NXT parts, and test validate their actual design meets simulation.

References

1. Nise, N.S.: Control Systems Engineering, 4th edn. John Wiley and Sons Inc., Chichester (2004)
2. Lofaro, D.M.: Control Design to Reduce the Effects of Torsional Resonance in Coupled Systems. Master's Thesis, Drexel University Department of Electrical and Computer Engineering (May 2008)
3. Kwatny, H.G., Blankenship Gilmer, L.: Nonlinear Control and Analytical Mechanics A Computational Approach. Birkhauser, Boston (2000)
4. Erkkinen, T.: Embedded Coder Robot NXT Demo (updated December 18, 2008), http://www.mathworks.com/matlabcentral/fileexchange/13399
5. Yamamoto, Y.: NXTway-GS Model-Based Design – Control of self-balancing two-wheeled robot built with LEGO Mindstorm NXT. 1st edn., February 29 (2008)
6. Le, T.: NXT-WhIP: NXT Wheeled Inverted Pendulum. Master's Thesis, Drexel University Department of Electrical and Computer Engineering (May 2008)
7. Greenwood, D.T.: Principles of Dynamics, 2nd edn. Prentice Hall, Upper Saddle River (1988)
8. Tony, L., Paul, O.: IDETC Presentation: System Integration Case Study: NXT-WhIP NXT Wheeled Inverted Pendulum. New York City, NY (2008-04-08)

Fostering Development of Students' Collective and Self-efficacy in Robotics Projects

David Ahlgren[1] and Igor Verner[2]

[1] Department of Engineering, Trinity College, Hartford, CT 06106 USA
[2] Department of Education in Technology and Science, Technion—Israel Institute of Technology, Haifa
david.ahlgren@trincoll.edu, ttrigor@technion.ac.il

Abstract. In robot projects student teams develop robots and participate in competitions through collective effort and highly interdependent learning activities. Since it is voluntary, participation in the project highly depends on students' confidence in their individual and team capacity to achieve desired goals and outcomes. In this paper we propose a project guidance approach that aims to achieve high level of both team performance and individual learning outcomes by fostering the development of collective and self-efficacy of team members. The main idea is organizing work in a way that combines collective effort towards performing the team project assignment and individual learning for mastery in desired specific robotics areas. Positive results of implementation of the proposed approach in the projects performed at Trinity College enable us to recommend further development of the proposed approach and its use by other institutions.

Keywords: self-efficacy, collective efficacy, mastery projects, teamwork, robotics.

1 Introduction

The opportunity for teamwork is one of the main strengths of project-based education in robotics. Robot projects promote development of learning situations in which team members, seeking a common goal of designing and building a robot, participate in collective and highly interdependent activities. This engaged learning [1] involves shared cognitive processes in which the students gain compatible and complementary knowledge aimed at solving theoretical and practical engineering problems in the context of need [2]. Robot designs are so complex and the scope of project activities is so wide that students in the team must divide work responsibilities and may acquire expertise in different subject areas [3]. Many robotics educators mention the strong contribution of robot projects on the development of teamwork skills [4-6]. However, only a few studies examine how to cultivate these skills and how to mediate collective goals and individual intentions in the robot project.

In this paper we propose a project guidance approach that aims to achieve high level of both team performance and individual learning outcomes by fostering the development of collective and self-efficacy of team members.

J.-H. Kim et al. (Eds.): FIRA 2009, CCIS 44, pp. 240–247, 2009.

2 Collective and Self-efficacy

Collective efficacy is defined as "a group's shared belief in its conjoint capabilities to organize and execute the course of action required to produce given levels of attainment" [7, p. 477]. The concept of collective efficacy was developed in close connection with the concept of self-efficacy that reflects perceived (i.e. based on real experience) beliefs of the individual in his/her own capabilities to perform the given task self-dependently [7].

Studies of group work in different organizations show that collective efficacy of the whole group and individual self-efficacy of the group members strongly affect the level of performance [8].

These studies also yielded, that collective efficacy of the group is not simply the sum of the individual perceptions of self-efficacy by the group members. The development of collective and self-efficacy can depend on different factors and has to be mediated [9], [10].

In team robot projects, collective efficacy reflects the shared beliefs of the students in their team's capabilities to mobilize the motivation, cognitive resources, and practical activities needed to cope with challenging robotics assignments. The robot team shares knowledge as it designs, builds and programs the robot, and participates in the robot competition. The collective competence acquired by the team in the project is demonstrated through the robot's performance successes at the competition. In the affective domain the project experience leads to the development of collective efficacy of the robot team.

Perceived self-efficacy enables students confidently to explore, solve, and describe their academic projects. Thus, educating students in modes that promote self-efficacy should be a primary consideration in course and curriculum design in engineering. Our belief is that robot design teams are ideal settings for building mastery. The complex and interdisciplinary problems encountered in robot design work argue for solution by teams. The team setting can provide a social environment that promotes peer instruction and offers opportunities to develop mastery of the many subjects (mechanical, electrical, sensing, navigation, for example).

In addition to developing shared, or team, efficacy, it is important to develop the skills and knowledge of each team member. Self-efficacy has been shown to be an important element of student motivation in engineering education [11]. We must solve the problem of how to develop individual skills and knowledge while building the efficacy of the team. Development of individual skills is paramount, as each student must develop a unique skill set that will prepare him or her for a career, and each student must gain perceived self-confidence in the application of his or her skill set to solve engineering problems. Thus our study arises from the need for robot project guidance that directs the team to high-level performance and at the same time effectively facilitates achievement of individual learning outcomes and self-efficacy by all team members.

In this paper we describe a two-year program aimed at promoting self-efficacy among undergraduate engineering students at Trinity College. In the two-year study described here, the independent variable is the teaching method, and the dependent variable is self-efficacy. As mentioned above, team efficacy may be evaluated by the team's performance in competitions. Individual performance is not seen in this way.

As a method to address the self-efficacy issue, we propose individual, or small group, mastery projects that take place within the team framework and are integrated closely with the team's robot projects. Development of self-efficacy is not itself the goal of our work; rather, it is an indicator of development of individuals on the team, who have gained mastery of a subject of high interest to them. We have succeeded when everyone perceives self-efficacy in selected directions by individuals on the team. If the team's overall performance is good, we have achieved team efficacy also.

Our study has two stages—a pilot study that took place during the 2006-2007 academic year and a central study that took place during the 2007-2008 academic year. The goal of this work was to develop and evaluate a new framework for developing mastery, to test and evaluate our framework, and to suggest areas for further investigation.

3 Pilot Study, 2006-2007

This section summarizes the first-year pilot study that was implemented at Trinity College in 2006-2007 [12] as an activity of the Trinity Robot Study Team (RST). The RST comprises 10 – 15 undergraduate engineering students each semester, including students from all four undergraduate years, who design robots to compete in the Trinity College Fire-Fighting Home Robot Contest [13] and the AUVSI Intelligent Ground Vehicle Competition [14]. The pilot study focused on a new program of mastery projects including development of workshops aimed at instructing peers in subjects related to robot design and development. A secondary goal was to evaluate the team learning environment presented by the Robotics Study Team.

The pilot study was spread across two semesters. In the fall, a pre-semester survey asked students to reflect on their backgrounds, interests, and confidence levels in robotics and to state their plans for future studies and careers. In this way the survey aimed to direct students to develop skills and self-beliefs that they would need to realize their plans. They were also asked to describe characteristics of team learning environments that would be most productive and supportive. A second part of the pre-semester survey aimed to help students to identify mastery project topics. In 2006-2007, each RST student was expected to become the team's expert in a mastery topic.

To help students to choose topics, the survey presented twelve topics related to the team's current projects. The topics were related to both contests and included sensors and vision, PCB design, navigation, software development, CAD-based mechanical design, testing and quality control, teamwork/project management, communication systems, energy and power supplies, motors and motor control, electronics and interfacing, and artificial intelligence. Students rated each topic according to four criteria: (1) importance to personal goals; (2) importance to RST projects; (3) level of confidence in the skill area; and (4) the student's priority for this skill area based on his/her individual interest and perceived importance to RST projects. In the second term of the 2006-2007 academic year each student prepared a mastery workshop, aimed at educating other RST members. The primary goal of the workshop was to teach other students about the mastery subject and to assess the mastery level achieved by the presenter. An end-of-year survey provided an overall view of the pilot year experience. The reader is referred to [12] for a full discussion of the pilot project. Important findings were:

- Students took the mastery projects seriously and gained confidence through them.
- Projects were not well integrated with RST design project in fire-fighting robot and autonomous land vehicle design. Lack of integration made mastery project work an extra burden.
- Projects were not required and so participation and attendance were not optimal.
- Each workshop should include a hands-on component.
- Students felt more comfortable preparing mastery projects with a partner.
- The RST offered many opportunities for independent learning as well as peer models and learning scaffolds.

4 Primary Study, 2007-2008

The primary study, which took place during the 2007-2008 academic year, responded to feedback gained through the pilot study survey. Specifically:

- Each project was carried out by two students working closely together. In most cases an experienced student was paired with a beginning student, something that the pilot year survey had suggested.
- Project topic were strongly integrated with RST design projects.
- All students participated.
- Projects were prepared in the fall semester according to a well published schedule of graded milestones including oral and written reports.

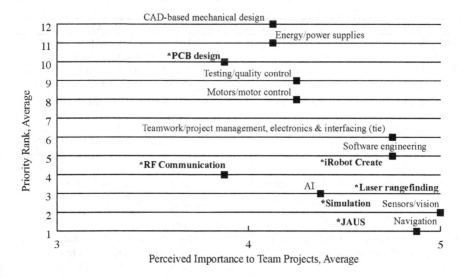

Fig. 1. Average priority ranking by team members vs. average perceived importance to team projects (N = 12 students). Priority Rank axis: 1 = highest priority, 12 = lowest priority. Perceived Importance axis: 3 = medium importance, 4 = very important, 5 = extremely important. Chosen topics are shown by asterisks and boldface type.

- Workshops were scheduled well in advance, and they took place on a regular basis of the week each week in February and March in the spring semester of 2008. All students were expected to attend the workshops.
- Each workshop was expected to include a hands-on exercise to be carried out by the attendees.

As in the first year, a pre-semester survey helped students to choose topics based on perceived importance to his/her design group's needs and to each person's interests or personal priority. In addition, each design group participated in topic development for their team's members. In this way, development of the mastery project and workshop flowed naturally from work the students were carrying out as part of their design projects. Figure 1 plots priority rankings (average taken over all team members) for twelve mastery topics suggested by the pre-semester survey versus average perceived importance to the team's projects. Figure 1 indicates (by asterisks) the topics chosen by the mastery project teams. Two topics were chosen unchanged from the proposed list: RF communication, and PCB design. The Navigation and Sensors and Vision topics were given the highest priority scores and were regarded as the most important to the team. The IGVC design team chose two mastery topics associated with sensors, vision, and navigation: JAUS (Joint Architecture for Unmanned Systems) (JAUS), and Laser Rangefinding. A third related project focused on using the Microsoft Robotics Studio software as a tool to predict performance of both firefighting and IGVC robots being built by the team (Table 1).

Table 1. 2007-2008 Mastery Projects

Project Title	Related Project	Activity
RF Communication	Multi-agent fire-fighting swarm	Design circuitry and protocol for RF communication for swarm.
iRobot Create	Robot able to find lost child	Prepare iRobot Create to compete in firefighting contest.
PCB Design	All RST projects	Introduce PCB design process using Mentor Graphics PADS.
Laser Rangefinding	Intelligent Ground Vehicle	Present Sick laser scanner interface with LabView. Hands-on program development and data analysis.
Robot Simulators	Intelligent Ground Vehicle	Simulate sensors and robot behaviors using Microsoft Robotics Studio. Interface MSR with LabView.
Joint Architecture for Unmanned Systems (JAUS) Software	Intelligent Ground Vehicle	Introduce JAUS coding and philosophy. Show LabView implementation with Q robot.
TReady	Trinity College Fire-Fighting Home Robot Contest	Describe curriculum and outreach program for junior-high students in robotics, culminating in fire-fighting contest participation.

Mastery workshops were held one each week during February and March of 2008. Workshops were one hour long, and each included a demonstration or hands-on exercise for attendees. For example: (1) at the PCB design workshop, each attended developed a printed circuit board design for a simple electronic circuit; and (2) the JAUS workshop allowed attendees to view the JAUS protocol in action via a Lab-View program that students wrote. For a discussion and evaluation of self-efficacy gained through the Primary Study, we refer the reader to [15].

5 Collective Efficacy

As we described in the introduction, a measure of the team's collective efficacy is attained performance in robot competitions. Contest performance demonstrates the team's effectiveness relative to peers from similar backgrounds and institutions. Success in competitions requires technical competence, management of the robot and its support system during a chaotic event, and confidence in preparing a competitive strategy and a facility for managing inevitable errors and unanticipated problems—both technical and operational. The AUVSI Intelligent Ground Vehicle Competition is a challenging, high-level event that attracted more than forty university teams in 2008 to the event site at Oakland University in Michigan [14]. The Trinity College Robot Study Team (RST) has competed in the IGVC since 2001. In 2008 the RST entered the robot "Q" (Figure 2).

Fig. 2. Trinity Intelligent Ground Vehicle "Q" on 2008 IGVC Autonomous Challenge Course

The IGVC consists of four events—the Autonomous Challenge, the Navigation Challenge, and the Design Competition, and the JAUS Challenge. In the Autonomous Challenge robot teams have nine runs in which they attempt to complete a 200 m. outdoor course, navigating between white lines in the presence of barrels, traps, and other obstacles. Most of the IGVC robots use computer vision and laser rangefinders as the main sensors. In the Navigation Challenge robots have three chances to navigate autonomously to within one meter of nine waypoints on an outdoor obstacle course approximately 100m x 100m. The Q robot used an electronic compass, laser rangefinder, and differential GPS (DGPS) to sense position and to detect obstacles. In the JAUS Challenge, teams demonstrate their robot's ability to receive, acknowledge, and act upon JAUS commands sent by judges. Three stages of difficulty in the JAUS Challenge indicate a team's relative level of preparation and knowledge. Finally, the Design Challenge requires teams to prepare a comprehensive design document, submitted to a panel of judges two weeks before the competition, and to give a formal 15-minute design presentation before a panel of judges, each a professional engineer. Thus the IGVC presented a comprehensive, multi-faceted challenge to the more than forty teams that participated in 2008.

The RST demonstrated a high level of success through the 2008 IGVC performance. The Q robot earned seventh place in the Autonomous Challenge, 6th place in the Design Competition, and they won a $500 prize for successful Level 2 performance in the JAUS Challenge. The robot did not complete the Navigation Challenge because the remote electronic stop (E-Stop) was triggered repeatedly by an interfering RF signal, stopping the robot in mid-course. These results suggested that the RST had mastered many of the design problems presented by the IGVC but that there were important areas (navigation algorithms, image processing speed, error analysis) that required further work. As a result, the third-year mastery projects, currently underway in the 2008-2009 academic year, focus on improving mastery in those areas.

6 Conclusion

In this paper we have described a two-year program that encourages development of both self efficacy and collective efficacy among members of the Trinity College Robotics Study Team. This project suggests a general model for developing both self-efficacy and collective, team efficacy (Figure 3).

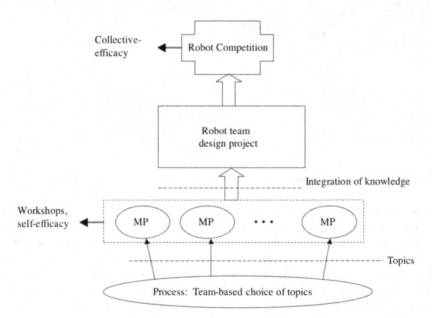

Fig. 3. Model showing development of collective efficacy and self efficacy in a robot team through mastery projects, workshops, robot design projects, and a robot competition

The model relies on a primary robot design project that enables the team to participate in a challenging robot competition. The team's performance in the event indicates collective efficacy gained through integration and application of knowledge gained by individuals through mastery projects. Topics for the mastery projects are chosen by the team to address knowledge gaps indicated, for example, by problems in previous competitions. Mastery project workshops, aimed at educating workshop attendees through hands-on exercises, demonstrate self-efficacy of presenters, broadens the knowledge base of the team, and address the multi-faceted and interdisciplinary nature of the robot design problem. We suggest that this is a powerful model for creating both individual and team efficacy while creating an optimal design given the team's resources. From the positive results of this study we recommend its further development and its use for project guidance in other schools. Wider implementation would pose interesting educational research questions related to the adaptation of mastery projects and their effectiveness in different educational settings.

References

1. Jones, B., Valdez, G., Nowakowski, J., Rasmussen, C.: Designing Learning and Technology for Educational Reform. OakBrook, IL (1994)
2. Cannon-Bowers, J., Salas, E.: Reflections on Shared Cognition. J. Org. Behav. 22, 195–202 (2001)
3. Pack, D., Avanzato, R., Ahlgren, D., Verner, I.: Fire-Fighting Mobile Robotics and Interdisciplinary Design-Comparative Perspectives. IEEE Transactions on Education 47(3), 369–376 (2004)
4. Murphy, R.: Competing for a Robotics Education. IEEE Robotics & Automation Magazine 8(2), 44–55 (2001)
5. Verner, I., Ahlgren, D.: Fire-Fighting Robot Contest: Interdisciplinary Design Curricula in College and High School. ASEE Journal of Engineering Education 91(3), 355–359 (2002)
6. Schneider, D., Leon, M., Van Der Blink, C., et al.: Active Learning and Assessment within the NASA Robotics Alliance Cadets Program. International Journal of Engineering Education 24(6), 1091–1102 (2008)
7. Bandura, A.: Self-Efficacy: The Exercise of Control. Freeman, New York (1997)
8. Whitney, K.: Improving Group Task Performance: The Role of Group Goals and Group Efficacy. Human Performance 7(1), 55–78 (1994)
9. Gibson, C.: The Efficacy Advantage: Factors Related to the Formation of Group Efficacy. Journal of Applied Social Psychology 33, 2153–2186 (2003)
10. Katz-Navon, T., Erez, M.: When Collective-mand Self-Efficacy Affect Team Performance: The Role of Task Interdependence. Small Group Research 36, 437–465 (2005)
11. Ponton, M., Edmister, J., Ukeiley, L., Seiner, J.: Understanding the Role of Self-efficacy in Engineering Education. Journal of Engineering Education 90(2), 247–251 (2001)
12. Ahlgren, D., Verner, I.: Building Self-Efficacy in Robotics Education. In: Proc. of 2007 ASEE Annual Conference, Honolulu (2007)
13. Trinity College Fire-Fighting Home Robot Contest, http://www.trincoll.edu/events/robot/
14. AUVSI Intelligent Ground Vehicle Competition, http://www.igvc.org
15. Ahlgren. D., Verner, I.: Mastery Projects in the Undergraduate Robot Study Team: A Case Study. In: Proc. of ASEE Annual Conference, Austin, TX (2009)

From an Idea to a Working Robot Prototype: Distributing Knowledge of Robotics through Science Museum Workshops

Alexander Polishuk[1,2], Igor Verner[1], and Ronen Mir[1,2]

[1] Department of Education in Technology and Science, Technion
[2] MadaTech, Israel National Museum of Science, Technology and Space

Abstract. This paper presents our experience of teaching robotics to primary and middle school students at the Gelfand Center for Model Building, Robotics & Communication which is part of the Israel National Museum of Science, Technology and Space (MadaTech). The educational study examines the value and characteristics of students' teamwork in the museum robotics workshops.

Keywords: Robotics education, science museum, teamwork, learning behaviors.

1 Introduction

Education for scientific and technological literacy, once considered as luxury, has become an existential necessity for all [1]. Gaining access to scientific innovations, understanding them, and acquiring the ability to function in the rapidly developing high-tech dominated world are matters of individual and communal sustainability [2].

Great emphasis is placed on a combination of formal education, which role is to impart systematic knowledge of basic disciplines, and informal education which is especially effective in broadening horizons, fostering curiosity, and active learning [3]. The central role in informal education is of museums of science and technology. Museum exhibitions and educational programs reinforce, support, and complement the studies of science and technology in schools.

In a museum program, many of the constraints of formal classroom education are left behind, leaving much space for curiosity and creativity. Museums offer interesting, relevant programs that attract an audience and stimulate motivation [4]. Classroom field trips to museums yield cognitive and affective learning outcomes [5].

Teachers believe that museum visits develop scientific and social skills of students [6], as well as provide the pace, general enrichment and fun [7].

2 Robotics Education in Science Museums

In recent years, robotics is widely spread in the informal learning space including dozens of robot competitions around the world. Robotics education is delivered mainly through extra-curricular and outreach programs many of which supported by

J.-H. Kim et al. (Eds.): FIRA 2009, CCIS 44, pp. 248–254, 2009.

universities and colleges [8]. Less experience is gained in robotics education in science museum environments. In literature there are few educational studies on the use of robot exhibits in museum expositions.

Nourbakhsh et al. [9] reported the project in which they developed the Personal Exploration Rover robot, created an interactive learning environment, and implemented it in a number of science museums in the U.S. for education about the NASA's Mars Exploration Program. The project evaluation indicated that expositions of the developed robotics environment effectively served for learning the subject in five national science enters: the Smithsonian National Air and Space Museum (NASM), the Smithsonian Udvar-Hazy Center, the San Francisco Exploratorium, the National Science Center, and the NASA Ames Mars Center. With this, the environment provided only limited interaction of visitors with the exhibit because of physical constraints [10].

A major breakthrough is connected to LEGO Mindstorms kits that allow to study basics of robotics to people of levels of experience, to organize robot competitions and other education programs [11].

This paper presents our experience of teaching robotics to primary and middle school students through extra-curricular programs offered by the Israel Museum of Science, Technology and Space (MadaTech). Our case study focuses on students' teamwork, with an aim to examine characteristics of teamwork in robotics workshops conducted in the museum environment.

In 2008 about 330,000 people visited MadaTech and 170,000 participated in its outreach activities. The Gelfand Center for Model Building, Robotics & Communication is part of MadaTech. It comprises two robotics laboratories, a demonstration hall and other facilities. Each of the robotics laboratories is equipped by a net of 12 computer workstations, robotics software (Robolab, NXT and NQC), and audio visual equipment.

The robotics education activities in the Center include the following: robotics lessons for school classes (K-12); short lab visits (K-12); semester or year-long courses (grade 3 to 9); training courses for kindergarten and elementary school teachers; international summer programs; special lessons for girls (grade 9), immigrants, and families; Science Night program of the European Community; holiday workshops; and Mobile Laboratories.

3 Learning Activities

In this paper we discuss three robotics workshops. In two of them middle school students had once-a-week meetings at the MadaTech, in which they designed, built, and programmed robots using LEGO bricks and RCX microprocessor. Both workshops used the robotics environment and activities in order to introduce students to the issue of road safety, which is of high social concern in Israel. The third robotics workshop was given at one of primary schools in Acre, while the hands-on activities were supported by the MadaTech Mobile Laboratory [12].

In the workshops the robotics studies were connected to the issue of high social concern in Israel - road safety. For most of the students this was the first experience in robotics. The students performed robot projects related to a number of subjects including the following: road intersection traffic light; autonomous vehicles with wireless communication; Mars exploration with the Pathfinder robot; a smart crane; line-follower.

For performing the project each student team (two students) got a work place, a computer and an especially dedicated robot kit that consists of Legos, a RCX microprocessor, sensors, and task-specific components.

The first learning assignment in each of the projects was building a traffic light and programming it to control the flow of traffic in time. The second assignment was performed by groups of four students (two teams together). It required to program a double traffic light intersection (traffic flow in two perpendicular directions), using the traffic lights built in the first assignment. The students optimized the traffic flow control with the use of touch and light sensors. The next assignment given to the teams was to construct an autonomous vehicle which can move, first, in the straight direction, and then along a given curve path. When performing the assignment, the students explored different types of mechanical transmissions, sensor configurations, and control algorithms in order to implement the fastest possible slalom movement, while avoiding obstacles (Figure 1). The final assignment of the course was to provide communication among several different robotic devices (such as traffic light, autonomous vehicle, and automatic barrier) in order to integrate them into an entire system. Students made short videos of the robot performance, that they showed at home and at school.

Fig. 1. Primary school students build a mobile robot

The specific feature of the robotics courses in the museum environment is that hands-on activities were integrated with science and engineering experiments and interactive demonstrations in the museum exhibitions and laboratories.

4   Educational Study

The goal of this study was to identify and analyze typical characteristics of teamwork behaviors of the school students participating in the robotics workshops at the MadaTech Gelfand Center. The study was conducted as qualitative research and focused on students behaviors in three robotics workshops given by one of the authors (Polishuk):

- A full scale workshop for a class from one of Haifa's middle schools (20 meet-ings, 18 students).
- A regular workshop for students from several middle schools in Haifa (12 meetings, 10 students).
- Two brief workshops for students from one of Acre's primary schools (6 meet-ings, 12 students in each workshop).

The follow-up data were collected using observations of learning activities, taped and transcribed semi-structured interviews, diary notes and reflections. The study utilized the grounded theory approach [13].

The characteristics of learning behaviors were crystallized through iterations of data collection and content analysis.

5   Findings

The five elicited characteristics of teamwork behaviors in robotics workshops are the following:

- Commitment to team success in achieving the common goal.
- Collective responsibility for performing the team assignment.
- Inclination to partnership within the same gender and cultural background.
- Pleasable experience in the museum environment.
- Wish to work together and make collective decisions.

The observed indications of these characteristics are summarized below.

5.1   Commitment to Team Success in Achieving the Common Goal

For the students participated in the robotics workshops, it was important to ensure that their team-mates share the common goal and take active part in the collective work. They negatively relate to indifferent partners, but not to lower achieving students. The team was very focused on providing the quality of performance of the robot task, while personal ambitions not relevant to the project were ignored. These findings can be illustrated by the following students' reflections:

Middle school male student: *"Sometimes children are not very much like each other, they do not actually work together, do not cooperate ,one can try to build something but not succeed because the other does not help."*

Middle school female student: *"There are cases when guys don't contribute and aren't serious about the project. ... I expect from my partner real effort and contribution to the project, otherwise working together does not make sense."*

MadaTech teacher (mentored the primary school students from Acre): *"One of the teams consisted of two students did not succeed to perform the task – when the first student worked, the second was passive and vice versa. Eventually, the team collapsed and then the students worked separately. "*

5.2 Collective Responsibility for Performing the Team Assignment

Students tended to see the group success - as their own success. But in case of failure, they explained it by technical faults or partner's faults (in case the teams were formed by the school teacher).

> Middle school female student: *"If you work together with a partner, you get more pleasure from the common success. Through discussion and argumentation, you better understand the subject and get to more effective solutions".*

> MadaTech teacher (mentored the primary school students from Acre): *"After completing the team task, the students strive to share their success with others. They talk about their robot, send by e-mail photos and videos of its operation to the friends."*

5.3 Inclination to Partnership within the Same Gender and Cultural Background

The students formed teams with mates of the same gender and cultural background. Israeli-born students were more open in their emotional reactions on learning situations than immigrant students. With this, different teams effectively collaborated in performing complex tasks.

> MadaTech teacher (mentored middle school students): *"Immigrant students are emotionally restrained. Teamwork helps them to become open, to understand that expressing emotions is not bad. ... I suggested the students to manage occasional teams, but they formed the teams by their selves, based on the same cultural background and gender."*

5.4 Pleasable Experience in the Museum Environment

Students noted that teamwork in the robotics workshop differed from that they experienced in science lessons at school:

- The museum workshops didn't require written reports;
- In the workshop the students back and forth turn from development to robot operation test;
- The workshop assignments were not rigorous and competitive;
- At the workshop meetings the students got and demonstrated tangible results of their work - dynamic models, while at school, the results were presented formally at the end of the project.

> Middle school female student: *"At school we are not asked, if we want to do the task or not. Here it is different: we come because we want learn robotics."*

> MadaTech teacher: *"The collection of instructional robots exhibited in the glass cases attracted the students. They opened the glass cases, practiced with the interactive models and asked to explain the principles of their operation. In case of malfunction they together tried to fix*

it and this way better understood the robot structure. This collective practice created a positive climate and a good break from the everyday routine."

5.5 Wish to Work Together and Make Collective Decisions

Usually, the teams did not divide work between their members, except the case of ninth grade student teams that performed advanced tasks. The team members exchange views and reach compromises at all steps of robot design and programming.

Middle school male student: *"We all do together. I made a greater contribution to programming. My partner checked up the program and detected mistakes. In my turn, I helped to fix his program. In our team there was no team leader."*

Middle school female student: *"I do not think that I or my partner lead the team, because each of us makes a contribution and we share project responsibilities."*

MadaTech teacher: *"Girls worked and made decisions together without a leader. They tried to convince each other and get to compromise."*

6 Conclusions

Our study showed the effectiveness of the Gelfand Center for Model Building, Robotics & Communication. The robotics courses, workshops and visits, within an informal learning environment, produce strong learning effects, advanced project-based learning, enhance creative problem-solving and collaborative teamwork. Robotics activities attract students of all school ages, boys and girls, Jews and Arabs, religious and secular.

Our study shows that teamwork in the museum robotics workshops differs of that traditional for formal and informal science education. We found that students prefer intensive teamwork in small groups which is resulted in making a robot – a real tangible system that operates autonomously and executes certain tasks. The specific aspects of teamwork in the robotics workshops are commitment and contribution to team success, and collective responsibility for performing the assignment. The integration of robotics workshops with relevant exhibitions, demonstrations, and laboratory experiments provided by the MadaTech, facilitates students' intellectual development and forsters motivation to advanced studies of science and technology.

Acknowledgement

The authors are grateful to Mark Gelfand for generous support of robotics activities at Technion and Madatech.

References

1. Standards for Technological Literacy: Content for the Study of Technology. International Technology Education Association (2000)
2. Kohn, J., et al. (eds.): Sustainability in Question: The Search for a Conceptual Framework. Edward Elgar, Cheltenham (1999)
3. Scribner, S., Cole, M.: Cognitive Consequences of Formal and Informal Education. Science 182, 553–559 (1973)
4. Miller, D., Nourbakhsh, I., Siegwart, R.: Springer Handbook of Robotics. Springer, Heidelberg (2008)
5. Riley, D., Kahle, J.: Exploring students' constructed perceptions of science through visiting particular exhibits as a science museum. Presented at National Association for Research in Science Teaching (1995)
6. Michie, M.: Factors influencing secondary science teachers to organise and conduct field trips. Australian Science Teacher's Journal (1998)
7. Kubota, C.A., Olstad, R.G.: Effects of Novelty-Reducing Preparation on Exploratory Behavior and Cognitive Learning in A Science Museum Setting. Journal of Research in Science Teaching (1991)
8. Center for Engineering Education Outreach (2009), http://www.ceeo.tufts.edu/ (retrieved March 8, 2009)
9. Nourbakhsh, I., Hamner, E., Dunlavey, B., Bernstein, D., Crowley, K.: Educa-tional Results of the Personal Exploration Rover Museum Exhibit. In: IEEE International Conference on Robots and Automation. IEEE Press, Piscataway (2005)
10. Nourbakhsh, I.: Robots and Education in the Classroom and in the Museum: On the Study of Robots, and Obots for Study. In: IEEE Int'l Conf. on Robots and Automation. IEEE Press, Piscataway (2000)
11. Lund, H., Pagliarini, L.: RoboCup Jr. with LEGO Mindstorms. In: International Conference on Robotics and Automation, pp. 813–819 (2000)
12. Mobil Lab Activities (2009), http://www.MadaTech.org.il/Pages/MenuItemPage.aspx?ContentItem=1663 (retrieved March 11, 2009)
13. Strauss, A., Corbin, J.: Basics of Qualitative Research Techniques and Procedures for Developing Grounded Theory, 2nd edn. Sage Publications, London (1998)

Teaching Electronics through Constructing Sensors and Operating Robots

Hanoch Taub and Igor Verner

Department of Education in Technology and Science, Technion – Israel
Institute of Technology, Haifa, 32000, Israel
{hanocht,ttrigor}@tx.technion.ac.il

Abstract. This paper proposes an approach to integrating electronics studies in robotics courses for high school students and pre-service teachers of technology mechanics. The studies focus on building electronic sensors, interfacing them to the robot platform, programming and operating robot behaviors. The educational study shows the learners' progress in electronics, thinking and learning skills, and attitudes towards engineering.

Keywords: Robotics education, electronics, learning for understanding.

1 Introduction

Technology education in Israel high schools passes a reform aimed to focus it on subjects relevant for hi-tech industry and modern society, as well as to enlarge the numbers of school graduates who choose to study engineering at universities and colleges. The priority trends of the reform are:

- Development of systems thinking as a general outlook of communication and control processes in natural, technological and social systems.
- Project based learning, as a catalyst of creative thinking, learning motivation, self-learning, communication and practical skills.
- Interdisciplinary connections that provide broad perspective, analogical thinking, and foster development of values and self-identity.

The new technology education curriculum includes a number of tracks such as mechanical engineering, electronics and computer engineering, biotechnology, etc. In each of these tracks the studies consist of three subjects: introductory course, core technology course, and majoring subject.

In the mechanical engineering track physics serves as an introductory course, engineering mechanics and machine control is the core subject [1], and there are several optional majoring subjects one of which is mechatronics. A number of challenges aroused when implementing mechatronics in schools. Two of them are directly related to the electronics studies required by the subject: (1) the traditional teaching methods which are used at the electronics engineering track do not fit the needs of the integrated subject and new teaching methods should be developed. (2) The majority of teachers of mechatronics are mechanical engineers with limited background in electronics.

J.-H. Kim et al. (Eds.): FIRA 2009, CCIS 44, pp. 255–261, 2009.

Our study was conducted in the framework of Master's thesis performed by Taub under the guidance of Verner. It addresses the above mentioned challenges and proposes a possible approach to integrating electronics studies in robotics courses for high school students and pre-service teachers of technology/mechanics. This paper presents results of implementation and evaluation of the proposed approach.

2   Didactical Principles of Teaching Mechatronics

Educational mechatronics relies on the theory of constructionism developed by Seymour Papert [2]. This theory is based on the constructivist approach to learning and considers the situation when the learner is involved in making an artifact which serves as an object to think with and to communicate about.

Learning by making artifacts occurs only in constructivist learning environments (CLEs). The methodology of design of CLEs is based on the principles of the activity theory [3-4]. The activity theory provides a framework for studying learners' behaviors, processes of their mental and social development, and instructional tools given to the learners. This theory is relevant for our study, in which electronics is taught in the mehatronic environment through hands-on experimentation with sensors and designing, building and programming robots.

The central issue in developing a constructivist curriculum is defining instructional objectives, i.e. the capabilities which the learner is expected to demonstrate at the end of the studies. The objectives serve for directing the learning process and assessing its outcomes.

Traditionally, learning objectives related to different levels of performance are defined by Bloom's taxonomy [5]. Recently the Bloom's taxonomy has been updated to fit the needs of modern education, particularly technology education [6]. We found the updated taxonomy relevant to our study and used it for developing learning objectives of the proposed course. The important updates made in the revised taxonomy include the following:

1. "Create" becomes the category which expresses the highest cognitive level of performance. The meaning of create here is " to form a coherent or functional whole; reorganizing elements into a new pattern or structure through generating, planning, or producing" [6]. This fits the constructionist approach which focuses on learning by doing and creating artifacts.
2. The taxonomy categories integrate objectives related to cognitive and psychomotor domains. This is relevant to the interdisciplinary learning activities which occur in the mechatronics course.

The revised taxonomy has a matrix structure with the knowledge dimension as the vertical scale and the cognitive process dimension as the horizontal scale. The four knowledge categories are: factual knowledge, conceptual knowledge, procedural knowledge, and meta-cognitive knowledge. The six cognitive process categories are: remembering, understanding, applying, analyzing, evaluating, and creating.

Examples of learning objectives which are designed by using the revised taxonomy matrix are given in Table 1. We designed educational objectives corresponding to all the knowledge and cognitive process categories because of the diversity of the course content and the learning activities.

Table 1. Examples of learning objectives

Issue: Constructing sensors, integrating them in a mechatronic system and programming a control process.	Cognitive process category under-standing	The pupil will be capable: a. To choose a proper electronic sensor from data-sheet -- Factual Knowledge. b. To interpret the link between the sensor's electronic circuit and its function -- Conceptual Knowledge. c. To foresee the output of the program by reading the sensor -- Procedural Knowledge. d. To give a presentation on electronic sensor and the principles of its operation to peers -- Meta-Cognitive Knowledge.

Mechatronics education utilizes problem and project based learning strategies that emphasize such important factors of engineering studies as challenge, curiosity, imagination, design, construction, and teamwork [7-9]. The projects foster learning motivation and facilitate the development of learning skills. In the project the learner takes responsibility for learning [10]. An example of problem based learning is a study of knowledge construction through the robot technology [11]. The conclusions of the study are that the learners: (1) create knowledge by cooperation; (2) acquire skills across science disciplines; (3) acquire technical skills; (4) develop scientific, mathematical and programming comprehension.

3   Course Syllabus and Activities

The topics covered by the course are as follows:

1. Not Quiet C or Lego MindStorms Language and Programming the robot behaviors (4 hours).
    1.1   Basic commands for operating robot subsystems.
    1.2   Programming and using sensors.
    1.3   Questionnaire of programming problems.
2. Introduction of electricity and electronics (4 hours).
    2.1   Preliminary evaluation exam.
    2.2   Basic concepts.
    2.3   Practicing the basic concepts, evaluation questionnaire.
3. Introducing electronic sensors (8 hours).
    3.1 Sensor of light, temperature ..., explanation of it principle operation and applications.
    3.2 Evaluation questionnaire.
    3.3 Building an electronic sensor circuit and connecting it to a robot's interface.
    3.4 Programming control programs that involve the electronic sensors and manipulate the robot behavior movements.
    4.   Optional projects.
    5.   Evaluation exam.

The study of all the listed topics included experiential activities. When studying the programming languages the learners programmed close-loop control operations of

Lego robots. The students practiced basic concepts of electricity and electronics by building circuits, measuring their physical parameters, and comparing factual data with theoretical solutions. Electronic sensors were learned through the learning by doing approach. The students built different sensor circuits, interfaced them to the input of the robot, and programmed robot operations. The audio sensor circuit built by the students in the course is presented in Figure 1A. After building the sensor the students used it as a clap detector for initiating robot operation.

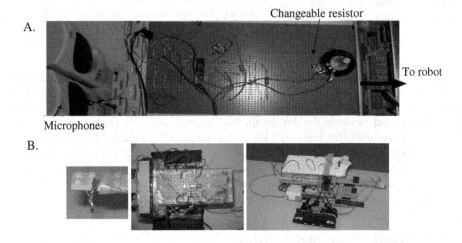

Fig. 1. A. Audio sensor; B. Temperature sensor mounted on the robot

Figure 1B shows the temperature sensor, its electronic circuit and power supply mounted on the mobile robot. With this robot configuration the students measured temperature and programmed the robot to detect heat sources. In addition to the sensors presented in Figure 1, the students built, interfaced and programmed light, IR, LDR, and touch sensors.

4   Educational Study

The goal of our educational research was to evaluate the proposed method of teaching electronics through constructing sensors and operating robots by high school students and pre-service teachers. The research focuses on the following questions:

1. What are characteristics of the learning environment and learning activities that facilitate the acquisition of knowledge and skills in electronics required in high school robotics and mechatronics courses?
2. What effect has the practice in constructing sensors and applying them for operating robot behaviors on understanding concepts in electronics?
3. What are learner's attitudes about the proposed teaching method, outcomes, motivation and learning capability?

The research was conducted as a multi-case study Yin[12], when the cases were follow-ups of the course given in different frameworks and to different categories of learners:

- Eleventh grade pupils studying mechatronics (N=13).
- Scientific extracurricular class of 9th graders (N=12).
- Technion students from the Department of Education in Technology and Science (N=19) participated in the course "Selected problems in design and manufacturing".
- Eleventh grade pupils participated in the Technion International Summer Research Program SciTech (N=4).
- Junior college 14th grade students (N=2) in the framework of graduation projects.

The educational research data were collected by means of knowledge and attitude questionnaires, observations, interviews; products and project reports, and course exams. Quantitative and qualitative methods were used. The quantitative study focused on evaluation of learning outcomes during the course, whereas the qualitative study analyzes the learning process. Results of the case studies were compared in order to increase the reliability of the conclusions about the proposed learning method [13].

In the phase of development of the case studies we based on Kolb's theory [14]; the constructionist approach [2]; methodology of CLE design [3-4], and the revised Bloom's taxonomy [5].

5 Findings

Based on the research data we addressed the first research question and identified the following characteristics of the proposed mechatronics learning environment:

- *Three levels of learning activities*
  While building electronic sensors the learner understands its structure and the function of each of the components. When interfacing the sensor to the robot, the learner understands its functionality in terms of power consumption, communication, and control. Making experiments with the robot involves the learners' peer discourse aimed to understand the physical principles of sensor operation.
- *Linking the levels of learning activities by reflective analysis*
  Through iterations of measuring characteristics of electronic circuits vs. observed robot performance parameters the students practice reflective learning skills and achieve deeper understanding of the electronics concepts.
- *Fostering critical thinking*
  The practice of continual evaluation of measured values of electronics parameters by their comparison with theory-based estimations facilitates development of critical thinking.
- *Fostering development of higher order thinking skills*
  By troubleshooting the robot, the learner develops ability to detect and fix technical problems in integrated systems. By designing the control programs, the learner develops programming skills. By solving the logic problems, the learner develops a logical thinking. By navigating the robot as a challengeable application, the learner develops navigation skills. By learning about the mutual relationships between the

physical concepts, the learner develops general conception about different physical fields, gradient of the field and the common characteristics. By designing and implementing operation of the mechatronics system, the learner develops a systematic view on the system.

With regard to the second research question, the formative and summative assessment showed a progress in learners' understanding the electronics concepts achieved in the courses. In order to evaluate this progress we conducted post-course tape-recorded interviews. In these interviews the learners described the principles of sensor functioning during robot operation and the physical concepts that are behind the measured parameters of the electronic sensor circuit. Data indicating the progress in understanding the electronics concepts were triangulated by means of pre-course and the post-course questionnaires for pre-service teachers, and quizzes for school students. The average grade of the pre-service teachers rose from 59.1% to 74.1%. The school students did not have prerequisite knowledge in electronics. Assessment results indicated their significant progress in electronics studies. The average course grade based on four quizzes was 82.3%.

Based on the data analysis, the progress in understanding electronics concepts can be characterized as follows:

- By "learning by doing" activities, the learner perceived the features of electronic components that can be seen only through practical experience. For example, the learner did not control functioning of the sensor (this was done by the computer) but concentrated on the operation research and troubleshooting. Measuring values of physical characteristics while testing of robot operation helped the learner to understand the links between different electronics concepts. This way, for example, the learners comprehended the link between temperature and voltage by heating the sensor unit (diode) and measuring the output voltage of the sensor circuit, or reading appropriate values from the computer program.
- By integrating sensors to the robot, writing and testing the control program and presenting the project to peers, the learner developed conceptual understanding of the mechatronic system.

Learner's attitudes about the teaching method and learning outcomes, inquired by the third research question, were evaluated using an attitude questionnaire and tape recorded interviews. Our findings:

- Most of the pre-service teachers were very positive about the teaching method and planned to use it in their teaching. As a fact, one of our former students now teaches robotics using the method.
- Most of the learners reported on their significant progress, achieved in the course, not only in electronics, but also in computer programming, control systems and even in mechanics. As the main factors that influenced the progress, the learners mentioned teamwork, learning by doing, the rich technological environment, involvement in the robot design, problem and projects based learning.
- The main motivation factors noted by the learners were: construction of the robot, joyful practice, success in all stages of robot development.

6 Conclusion

Our research shows that electronics studies can be effectively integrated in the high school robotics course and in the teacher training course. The proposed teaching method can facilitate understanding the concepts of electronics, development of higher order thinking and self learning skills, foster learner's motivation and interest in engineering. Based on the experience, we recommend further examination and implementation of the proposed teaching method in robotics education.

References

1. Verner, I.M., Betzer, N.: Machine Control - A Design and Technology Discipline in Israel's Senior High Schools. International Journal of Technology and Design Education 11, 263–272 (2001)
2. Harel, I., Papert, S. (eds.): Constructionism. Ablex Publishing, Norwood (1991)
3. Mursu, A., Lukkonen, I., Toivanen, M., Korpela, M.: Activity Theory in Information Systems Research and Practice: Theoretical Underpinnings for an Information Systems Development Model. Information Research International Electronic Journal 12(3) (2007)
4. Jonassen, D.H., Ronrer-Murphy, L.: Activity Theory as a Framework for Designing Constructivist Learning Environments. Educational Technology Research and Development Journal 47(1), 61–79 (1999)
5. Bloom, B.: Taxonomy of Educational Objectives: The Classification of Educational Goals. Longman, New York (1964)
6. Anderson, L.W., Krathwohl, D.R. (eds.): A Taxonomy for Learning, Teaching and Assessing: A Revision of Bloom's Taxonomy of Educational Objectives. Longman, New York (2001)
7. Waks, S., Sabag, N.: Technology Project Learning Versus Laboratory Experimentation. Journal of Science Education and Technology 13(3), 332–342 (2004)
8. Barak, M.: Learning Good Electronics or Coping with Challenging Tasks: The Priorities of Excellent Students. Journal of Technology Education 14, 20–34 (2002)
9. Doppelt, Y.: Assessment of Project-Based Learning in a Mechatronics Context. Journal of Technology Education 16(2), 7–24 (2005)
10. Frank, M., Barzilai, A.: Project-Based Technology: Instructional Strategy for Developing Technological Literacy. Journal of Technology Education 18(1), 39–53 (2006)
11. Chambers, J.M., Carbonaro, M.: Scaffolding Knowledge Construction through Robotic Technology: A Middle School Case Study. Electronic Journal for the Integration of Technology in Education 6, 55–70 (2007)
12. Yin, R.K.: Case Study Research. Applied Social Research Methods Series, vol. 5. Sage Publications, London (2003)
13. Wiersma, W.: Research Methods in Education: An Introduction. Pearson Education, London (2000)
14. Kolb, D.: Experiential Learning. Prentice-Hall, Englewood Cliffs (1984)

Learning from Analogies between Robotic World and Natural Phenomena

Igor M. Verner and Dan Cuperman

Department of Education in Technology and Science, Technion – Israel
Institute of Technology, Haifa, 32000, Israel
dancup@inter.net.il, ttrigor@technion.ac.il

Abstract. This paper proposes an approach which combines robotics and science education through the development of robotic models and inquiry into natural phenomena. The robotic models are constructed using the PicoCricket kit. The approach is implemented and evaluated in the framework of teacher training courses for Technion students given in connection with outreach courses for middle school and high school students. The educational study indicated that the proposed approach facilitated acquisition of both technology and science concepts and inspired analogical reasoning and crossdisciplinary connections between the two domains.

Keywords: Science-technology education, robotics, natural phenomena, modeling, analogous thinking, PicoCricket.

1 Introduction

Natural Science and Technology are two different domains: the former deals with nature phenomena and the latter deals with man-made creation (Ropohl, 1997). Even though traditionally, these two domains are taught as separate subjects, the progress of science and technology nowadays emphasizes the connection between the domains and modern education is establishing interdisciplinary links between the subjects.

One pathway that connects science and technology is the aspiration of man to create artifacts that imitate or furthermore, attempts to improve solutions existing in nature. As the aspiration to imitate bird fly motivated Leonardo da Vinci's invention of flying machines (Cianchi, 1988), similar aspirations and inventions paved the development of Bionics. Robot design as well, is greatly influenced by the attempt to imitate appearance, functionality and behaviors of nature-made creatures and in particular the human being locomotion and intelligence.

Another pathway that connects Technology and Science is studying natural phenomena by exploring existing, or specially developed technological systems. Neuroscience, for example, uses principals of "information processing theory", which explains mental functions by exploring computer operation (Miller, 2003). These pathways between Science and Technology are both based on analogical thinking.

J.-H. Kim et al. (Eds.): FIRA 2009, CCIS 44, pp. 262–270, 2009.

Analogy can be defined as mapping of structure and attributes from one domain (source) to another (target) which specifies differences as well as similarities between the domains (Hestenes, 2006). Gentner (2008) notes that learning by analogy can occur in three main manners:

1. Learning an unknown domain by inferring from a known domain with similar internal relations.
2. Deducing general categories and rules by comparing internal relations and revealing commonalities between domains.
3. Transforming domain-specific knowledge to create a better match across domains.

Learning by analogy is the focus of modeling activities, where the model is a simplification of a complex system or phenomenon used to aid the formation of knowledge representations (Gilbert, 2005).

Halloun (2004) points out that in science education, modeling natural phenomena is an iterative process, in which any single iteration is a cycle consisted of two main stages: model development and model deployment. One can see that these two stages follow the two abovementioned pathways: model development is based on imitation of the phenomenon, while model deployment extends knowledge about the phenomenon.

Bio-inspired robotics is grounded on the development of robotic models that imitate biological processes. Erlhagen et al (2006) suggests a broader meaning of the imitation, including robot learning as replication of a human learning behavior. Rusk et al. (2007) state that robotics activities can offer rich educational opportunities, but are typically introduced in a narrow way, as building mobile robots. The authors propose four strategies for introducing students to robotics technologies and concepts: (1) Focusing on themes, not just challenges; (2) Combining art and engineering; (3) Encouraging storytelling; (4) Organizing exhibitions, rather than competitions. They suggest and demonstrate implementing these strategies using innovative construction kits, PicoCricket, designed at the MIT Media Lab.

The study presented in this paper develops an integrative approach to science and technology education which combines the new strategies of robotics education (Rusk et al., 2007) and the pathways that connect science and technology through imitational models and analogical reasoning.

2   The Construction Kit

We use the PicoCricket robotics kit which has been designed "for making artistic creations with lights, sound, music, and motion". The kit and its electronic components are presented in Figure 1 (http://www.picocricket.com).

The PicoCricket is a tiny programmable computer that can operate light, sound, and motion actuators, collect data and react on input from a variety of sensors, and provide bi-directional infrared communication with the host computer and other PicoCricket. The PicoBlocks software for programming the PicoCricket is intuitive and uses graphical blocks which are simply snapped together to create a program (similar to LEGO bricks).

(A)

(B)

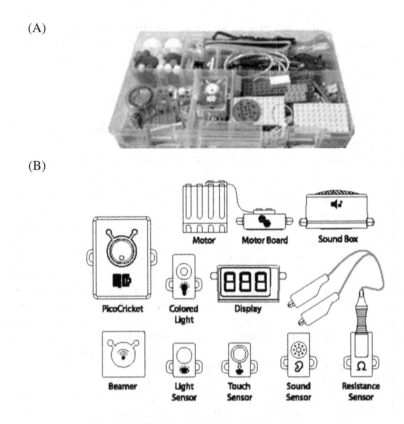

Fig. 1. (A) The PicoCricket kit. (B) Electronic components.

3 Modeling Natural Phenomena with PicoCricket

In this section we will present two of the instructional models that were developed in the framework of teacher training courses and outreach courses for high school students at the Department of Education in Technology and Science, Technion. The first is a sunflower model which demonstrates the heliotropism phenomenon. It represents a series of models developed for demonstrating various types of tropism in plants. The second is the iris model which demonstrates light regulation in the eye. This model is one of our models demonstrating various homeostasis functions in biological organisms.

3.1 Heliotropism: A Sunflower Functional Model

Inquiry into the phenomenon
The nature phenomenon studied and modeled in this project is the solar-tracking movement of plants. This daily movement to follow the sun is called heliotropism (Sherry & Galen, 1998), and is demonstrated most impressively in the sunflowers ("Helianthus annuus"), see Figure 2.

Fig. 2. A sunflower field

The sunflower accomplishes the light sensing by light-sensitive proteins (phototropins) residing not in the flower-head, but in the green leaves, where a plant growth hormone (auxin) is produced. The flower-head movement is caused by differential translocation of auxin to the shaded side of the stem, triggering greater cell elongation in that side (Sherry & Galen, 1998), bending the stem and ending in the flower-head pacing the sunny side. This process takes place in young plants and diminishes as their maturation proceeds.

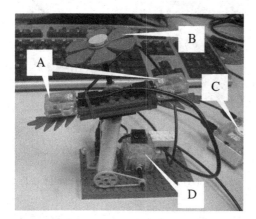

Fig. 3. The sunflower model: A. Light sensors; B. Flower-head; C. PicoCricket; D. dc motor

Building the model
The sunflower model, built using the PicoCricket kit, is shown in Figure 3. It includes two light sensors, a dc motor, driving mechanism, a "flower-head" structure, and some additional Lego blocks. The model functioning is controlled by the PicoCricket which executes the program written by PicoBlocks software .The heliotropism phenomenon is imitated in the following way: the amount of light measured by the sensors is continuously compared, and the "flower-head" structure is turned towards the direction indicated by the higher reading sensor.

3.2 Homeostasis of the Eye: An Iris Functional Model

Inquiry into the phenomenon

The phenomenon studied and modeled in this project is the ability of the eye to regulate the light intensity penetrating through the pupil (presented in Figure 4A), and reaching the retina. The process is controlled by the brain trough a negative feed back loop. The pupil size is determined by the iris, made up of circularly arranged muscles. When light increase, the muscles shrinks the pupil area, and limits the light flux at the retina [11].

(A)

(B)

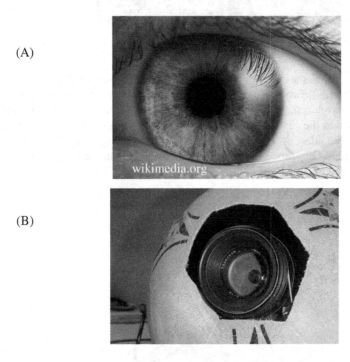

Fig. 4. (A) A human eye; (B) The eye model

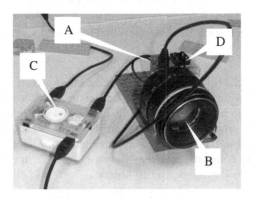

Fig. 5. The mechanism of the eye model: A. Light sensor; B. Iris; C. PicoCricket; D. dc motor

Building the model

The eye model, built using the PicoCricket kit, is shown in Figure 4B. It includes a light sensor, a dc motor, a belt driving mechanism, a lens iris mechanism, a ball shell and some Lego blocks. The model functioning is controlled by the PicoCricket which executes a corresponding program. The homeostasis of the eye phenomenon is imitated in the following way: the amount of light penetrating trough the lens is continuously measured by the sensor and regulated by actuating the lens iris via the DC motor. (presented in Figure 5)

4 Educational Study

The goal of this study is to develop and examine the learning process which combines the design of technological systems and the inquiry into nature phenomena by means of modeling activities and analogical thinking. The research questions derived from this goal are as follows:

1. What are the principles of designing robotic models to be used for studying processes in nature, and what are the characteristics of the learning environment which utilizes these models in the scientific inquiry?
2. What are pre-service teachers' attitudes towards teaching and learning aided by robotic models, and how are these attitudes affected by the teaching practice?
3. What perceptions of the analogies between the natural phenomenon and the technological models are developed in school students following their participation in the modeling activities?

The participants of the study were:

- Technion students majoring in science and technology education (N=11),
- 10th graders who studied advanced high school biology and participated in our outreach course (N=5),
- Middle school students, who were taught by the Technion students in the framework of the teaching practice course (N=10).

We implemented the multi-tiered approach [12] to studying symbiotic educational processes, in which different groups of learners share a common learning environment. The educational objectives differ between the three groups (tiers): the Technion students studied modeling and project based education, the high school students were engaged in modeling biological phenomena using technological systems, and the middle school students were introduced to various nature phenomena and their modeling by means of robotic systems. With these different objectives, all the groups learned through hands on practice with the same PicoCricket kits and in the same departmental laboratory of technology. Data on learners' attitudes and perceptions were collected and analyzed using qualitative methods (questionnaires, interviews, and videotaped observations).

5 Findings

In this section we present findings related only to the third research question about the development of analogical thinking. We will discuss results of the analogous thinking questionnaire conducted among the high school students. The questionnaire refers to different aspects of analogies associated with the modeling process and related to the five types of model structures characterized by Hestenes [4]:

1. Systemic structure - generates analogies related to *composition* of internal parts in the system, *environment* in which the system functions, and *connections* (internal and external).
2. Geometric structure – leads to the analogies that concern a *position* of the system and geometric *configuration* of its parts.
3. Object structure – addresses analogies regarding *intrinsic properties* of the system and the parts.
4. Interaction structure – enables analogies in *functional* links.
5. Temporal structure – introduces analogies reflecting temporal changes in the system.

In the post-course questionnaire the students were asked to scrutinize in the analogy between the studied natural phenomena and the technological models built to represent them. The questionnaire consists of three parts.

The first part investigated the technological and biological concepts acquired by the students through the modeling activities. In this part the students were asked to shortly describe the phenomena and the models. The second part of the questionnaire included open-ended questions. The learners were requested to present their perceptions of the five types of structures regarding the biological systems, in which the phenomena occur, as well as their robotic models. For each structure type, the students were asked, if the model represents an analogy of this type or not. In the final part of the questionnaire the students were asked to give their overall evaluation of the phenomenon-model analogy. This evaluation utilized the following scale: A - full analogy, B - good analogy, C - weak analogy, D - no analogy. The students were also requested to explain their evaluations.

Results of the post-course questionnaire indicated that at the end of the course the students had knowledge of each studied phenomenon and its robotic model. The students neither had prior knowledge of the phenomena, nor studied the phenomena anywhere parallel to the course. Therefore, we claim that this knowledge was acquired due to the course.

The answers to the first part of the questionnaire reflect students' familiarity with the concepts and ability to discuss the phenomena and the models using the right terminology. This can be illustrated by the following citations:

> "When the sunflower plant receives more sunlight in one side then the other, the protein receiving the light passes information to transfer more auxin to one side then the other" (Danny).

> "It enlarges the cells in one side so the plant grows to the direction of the light" (Gabby).

"The light sensors receive light, and the side with more light is detected. The motor moves the model to the side in which the sensor detects more light "(Danny).

The answers to the second part of the questionnaire showed the consistency of students' evaluations of the analogies between the phenomenon and the model, based on the five types of model structures. In particular all the students agreed that the models were analogous to the natural phenomena with respect to the systemic structure (including *composition* and internal *connections*), object structure, and interaction structure.

The students pointed to the lack of analogy between the models and the natural phenomena in the systemic structure (with regard to *environment* and external *connections*), geometric structure, and temporal structure.

Students overall evaluation of the models, given in the third part of the questionnaire, was that there is *good analogy* between the models and the phenomena. This evaluation is explained in the following citations:

"The components are analogous and the response is nearly identical in al aspects. It is very similar" (Igor).

"The model meets all the requirements" (Nathaniel).

6 Conclusion

Modern education calls for introducing robotics activities in a broad way by focusing on thematic inquiry, creativity and self-expression, verbalization and communication of experiential knowledge. Our approach follows this way and directs towards integrating robotics studies with inquiries into natural science phenomena. It stresses the important role of robotics modeling in promoting the study of analogies and differences between the natural and technological systems.

In this approach a natural phenomenon serves a motive for developing a robotic model, while experimentation with the robotic model facilitates deeper understanding of the mechanisms underlying the phenomenon. We observed this effect in the two modeling projects presented in the paper as well as in other projects.

The case studies demonstrated that the proposed approach indeed inspired students' analogical thinking. The students were productive in unveiling both analogies and differences between natural and technological systems. An interesting result is that in spite of finding differences between the model and the phenomenon, the students still accepted the model as having a *good analogy* with the phenomenon. We analyzed the cases of "*good analogy*" and found that these evaluations related to the three types of model structures, namely, internal systemic structure, object structure and interaction structure.

The authors recommend further investigation and implementation of the proposed approach in robotics education.

References

1. Ropohl, G.: Knowledge Types in Technology. International Journal of Technology and Design Education 7, 65–72 (1997)
2. Cianchi, M.: Leonardo da Vinci's Machines, pp. 45–61. Becocci, Florence (1988)

3. Miller, G.A.: The Cognitive Revolution: a Historical Perspective. Trendes in Cognitive Sciences 7(3), 1411–1444 (2003)
4. Hestenes, D.: Notes for a Modeling Theory of Science, Cognition and Instruction. In: Proceedings of the 2006 GIREP conference: Modelling in Physics and Physics Education (2006),
   http://modeling.asu.edu/R&E/Notes_on_Modeling_Theory.pdf
5. Gentner, D., Colhoun, J.: Analogical processes in human thinking and learning. In: Von Müller, A., Pöppel, E. (series eds.) Glatzeder, B., Goel, V., Von Müller, A. (vol. eds.) On Thinking: Towards a Theory of Thinking, vol. 2, Springer, Heidelberg,
   http://www.psych.northwestern.edu/psych/people/faculty/
   gentner/papers/Gentner-Colhoun_in_press.pdf
6. Gilbert, J.K.: Visualization: A metacognitive skill in science and science education. In: Gilbert, J.K. (ed.) Visualization in science education, pp. 9–27. Springer, Dordrecht (2005)
7. Halloun, I.A.: Modeling theory in science education. In: Science & Technology Education Library, p. 186. Kluwer Academic Publishers, London (2004)
8. Erlhagen, W., Mukovskiy, A., Bicho, E., et al.: Goal-directed imitation for robots: A bio-inspired approach to action understanding and skill learning. Robotics and Automation Systems 54, 353–360 (2006)
9. Rusk, N., Resnick, M., Berg, R., Pezalla-Granlund, M.: New pathways into robotics: Strategies for broadening participation. Journal of Science Education and Technology, 59–69 (2008)
10. Sherry, R.A., Galen, C.: The mechanism of floral heliotropism in the snow buttercup, Ranunculus adoneus. Plant Cell and Environment 21, 983–993 (1998)
11. Zeron, E.S.: Positive and Negative Feedback in Engineering. Mathematical Modeling of Natural Phenomena 3(2), 67–84 (2008)
12. Lesh, R., Kelly, A.: Multitiered Teaching Experiments. In: Kelly, A., Lesh, R. (eds.) Handbook of Research Design in Mathematics and Science Education, pp. 197–230. Lawrence Erlbaum, Mahwah (2000)

Integrating Robot Design Competitions into the Curriculum and K-12 Outreach Activities

Robert Avanzato

Penn State Abington
1600 Woodland Road
Abington, PA 19001
RLA5@psu.edu

Abstract. The Penn State Abington campus has integrated several mobile robot design competitions into project-based design activities to provide enhancement for undergraduate engineering and information sciences and technology courses and also to provide outreach to K-12 institutions. The robot competitions, which encourage interdisciplinary design, teamwork, and rapid prototyping, support a wide range of educational outcomes in a variety of courses. A survey of undergraduate students was also implemented to identify the key lessons learned and overall educational quality of the robot competition activities. Overall, the responses on the quality of the robot competition experience were very positive. The strategic selection and implementation of robot design competitions, such as described in this paper, provide a cost-effective approach to enhancing the curriculum, promoting retention, and encouraging interest in science and technology (STEM) careers in K-12 students.

Keywords: robotics, education, robot competitions, design, STEM, outreach.

1 Introduction

Penn State Abington campus (Abington, PA) hosts several autonomous mobile robot design competitions each academic year (since 1995) to support project-based design activities in freshman and sophomore level engineering courses and information sciences and technology courses. The competitions are open to the public and provide outreach to K-12 institutions in the Philadelphia, PA region. The robot design activities are highly interdisciplinary and include topics such as engineering design, mechanical engineering, electrical engineering, computer science, sensors, systems engineering, project management, teamwork, and creative problem solving. As a result of these characteristics, the competitions serve a wide range of educational outcomes and outreach goals.

This paper will first briefly discuss the two indoor mobile robot competitions which have been offered at our campus some 1995. Also a new autonomous outdoor robot challenge, inspired in part by the DARPA Grand Challenge and introduced in 2005, will be outlined. In the second section of the paper, a discussion of the integration of these contests into the curriculum and also into outreach efforts will be presented. Finally, some key results from the student assessment will be provided.

J.-H. Kim et al. (Eds.): FIRA 2009, CCIS 44, pp. 271–278, 2009.

2   Robot Design Competitions

2.1   Robo-Hoops Robot Competition

The Robo-Hoops robot contest challenges student teams to design and implement autonomous, computer-controlled mobile robots which are capable of picking up and shooting or dunking small, 4-inch (10.2 cm) diameter foam balls into a basketball net, which is positioned 12 inches (30.5 cm) above the playing surface. The playing field is 48 inches (121.9 cm) by 80 inches (203.2 cm) in size. The competition is divided into 2 phases. In the first phase, each robot is operated in solo (unopposed) matches, with the goal to score as many points as possible within a 60 second time limit. In the second phase of the contest, the robots compete head-to-head in a double elimination contest, also with 60-second matches. Robots, at the start of the contest, are restricted in size, to 12 inches (30.5 cm) by 12 inches by 18 inches (45.7 cm high), and may expand to a maximum size of 18 inches (45.7cm) by 18 inches by 18 inches after the start of the match. The contest is open to students in K-8, high school, and college and beyond, and robots compete within the same division level. The robots must be fully autonomous for the high school and senior divisions. Middle school students (K-8) may optionally enter robots in an autonomous division or a remote control division. The contest allows participants to choose any hardware or software or combination of technologies for the robot design. The Robo-Hoops contest was first offered in 1995. Information about the specifics of the rules can be found on the Robo-Hoops robot website [1]. Typically, 30 to 40 robot teams across all divisions (K – college) participate each year. Of those, 4 to 6 teams are generally composed of Penn State Abington lower-division undergraduate students. Over 15 middle schools and high schools are typically represented at each event and the competition is held in an auditorium which provides seating for spectators. Figures 1 displays pictures of the Robo-Hoops playing field and event.

Fig. 1. Penn State Robo-Hoops Contest

2.2   Firefighting Robot Competition

The firefighting robot design contest requires computer-controlled, autonomous mobile robots to navigate autonomously though a maze, 8 foot (243.8cm) by 8 foot arena with 13-inch (33cm) high walls, consisting of four rooms connected by a hallway. A lit candle is randomly placed in one of the four rooms, and the goal is to have the robot locate and extinguish the burning candle in the minimum time. Bonuses can be earned by the robots by accomplishing additional tasks such as returning to home base after extinguishing the candle, avoiding furniture (obstacles), starting in response to an audible tone (representing a fire alarm), and others. As with the Robo-Hoops contest, this contest is free and open to the public and robot teams compete within the divisions of K-8, high school, and senior (college and beyond). This contest event is a regional contest for the international Trinity College (Hartford, CT) Firefighting Robot Contest [2, 3, 4] and the official rules are maintained at this site.

There are approximately 40 or more total robots entered annually in the Abington regional firefighting contest, and Penn State Abington generally fields 5 to 10 undergraduate teams in the senior division. (This is similar to participation in the Robo-Hoops contest.) This regional firefighting robot contest was first offered at Penn State Abington in spring of 1995. One of the additional benefits of offering a regional contest is the preparation and encouragement it affords students to move on and participate in the international contest at Trinity College. Figures 2 shows pictures of robots competing in the Abington regional firefighting robot contest.

Fig. 2. Regional Firefighting Contest

Both robot competitions above allow for the choice of any hardware and software solutions, and this enables educators to choose the appropriate technology to achieve desired educational outcomes. For example, the freshman engineering design course or K-12 institution might elect to use Lego Mindstorms™ robot kits and ROBOLAB™ programming (visual, icon-based programming language based on LabView), while a more advanced engineering or robotics course may use more sophisticated hardware (examples: Handyboard, Basic Stamp, PDA, Pontech SV203 board, VEX) and C-based, C#, or Java programming languages. Additionally, contest arena construction and setup for both the Robo-Hoops and Firefighting contests are relatively simple and low-cost, and thereby facilitate integration into classroom and laboratory settings.

2.3 Mini Grand Challenge

An outdoor autonomous robot design competition, the Mini Grand Challenge, was developed at the Penn State Abington campus in 2005 [5, 6]. The contest requires autonomous mobile robots to navigate unmarked, paved pathways, with width of 8 feet (243cm), on a suburban college campus and reach GPS waypoints. Robots must avoid obstacles and robots are also awarded points for interacting and entertaining spectators (see Figure 3). The contest is partly inspired by the DARPA Grand Challenge, but our contest emphasizes accessibility, low-cost hardware and software solutions, spectator interaction, and education opportunities. A successful robot platform constructed for less than $300 and controlled by a laptop running MATLAB software was developed by undergraduate students. The contest, offered annually, is open to students at all levels of education: K-12, college, and beyond. Two of the key features of the challenge are to introduce participants to vision-based mobile robotics and to human-robot interaction.

Fig. 3. Mini Grand Challenge

In 2005, there were 3 robots participating in the outdoor challenge. The interest has been growing each year and in 2008 the Mini Grand Challenge attracted 14 robots. All of the entries in the 2008 contest were at the college and professional level, except for one team at the high school level.

3   Integration into Curriculum and K-12 Outreach Activities

Undergraduate engineering courses which have incorporated the robot design and competition activity include a freshman engineering and graphics design course, an introductory digital design course, circuit analysis course, and a special topics robotics course. These aforementioned courses are at the freshman and sophomore level in the engineering program. More recently, a junior-level information sciences and technology (IST) course in emerging technologies was developed to incorporate hands-on robotics and involvement in the campus robot competitions. Each course has specific educational objectives which can be satisfied by components of the competition event. Most importantly, the design challenge provides a realistic "context" in which to introduce a variety of engineering concepts and techniques.

The freshman design course generally focuses on the engineering design, CAD, general robotics, and project management aspects of the contest design. This course has typically used Lego Mindstorms robotics kits and graphical programming languages in the team designs, and the preparation for the contest is a 4 to 6 week module. The digital design course focuses more on software design, microcontrollers, sensor interfacing, analog-to-digital conversion, serial communication. In a few select cases, laboratory exercises were developed to focus on one or more of these topic areas as student teams (generally 3 to 4 students) developed robots for the main contest event. The hardware for this digital design course has been HandyBoard technology, PDAs, and more recently Vex robotics technology with C programming support. In the electrical circuits and systems course, several laboratory exercises were developed to integrate the robot contest technology into the course material, such as oscilloscope measurement and analysis of servomotor and PWM signals, H-bridge design, actuator driver circuitry, and tone detection circuitry. The benefit of this approach is that the students are being exposed to technology in the context of the robot design activity.

The IST emerging technology course focuses more on systems engineering, robotics applications, software, and project management. These students also used Vex robotics technology and C programming language. For course work that is more focused on software, such as the IST course, it is advisable to provide students with a mechanical base to reduce some of the mechanical efforts that do not support course objectives. The Mini Grand Challenge outdoor robot contest has been the basis for special projects, undergraduate research projects, and honors projects at the freshman and sophomore level. As mentioned, a key technology present in the Mini Grand Challenge robot solutions is computer vision and image analysis. Currently we are using MATLAB to develop all of the control and vision algorithms. One of the unique advantages here is that students can use the MATLAB environment for development, testing, and also deployment. This facilitates rapid prototyping and has proven successful with lower-division engineering students. We expect the Mini Grand Challenge to be integrated more fully as resource and tutorial materials are developed for this contest.

Each of the three contests is open to K-12 participants and each age group of participants competes in a separate division. In the Robo-Hoops and Firefighting contests we have attracted over 20 robot teams at each of the annual events at the high school level. The university also offers a college-level freshman engineering design course for participating high school seniors in the region, and these students (typically 20 to 30 students) participate on a regular basis. Each contest also attracts 3 to 5 middle school teams and approximately the same number of K-5 teams. Registration data is provided on the contest website for examination. The goal of the K-12 outreach program is to encourage interest in STEM careers and promote interest in science, engineering, computer science, artificial intelligence, and robotics. The presence of participants varying in age from grade school to college and professional at the same event greatly improves the networking and sharing of ideas. One of the key features of the contests described above is that these contests support a variety of educational course goals and at the same time support K-12 outreach. This provides a cost-effective tool to enhance educational programs at a variety of levels.

4 Assessment of Robot Contests

A voluntary student survey for the undergraduate students was developed and implemented to investigate the outcomes of the Robo-Hoops and Abington regional firefighting contests offered between 2005 and 2008 (a total of 8 competitions). In all cases, the contest event occurred in the last 2 or 3 weeks of a 15-week semester, and students generally spent between 4 to 6 weeks preparing for the contests. Pictures of undergraduate students working on robot designs in various course environments are shown below in figure 4.

Fig. 4. Student teams in robotics lab

Undergraduate students were asked questions concerning technical challenges, working in a team, time management, key lessons learned, and suggestions for improvements in the robot competition activity. The students participating in the survey were all freshman and sophomore level students. These participants represented many technical majors including EE, mechanical, aerospace, computer science, IST, computer engineering, chemical engineering and civil engineering. A total of 76 students (38%) responded of the survey. A set of key assessment results will be provided below, and a more thorough treatment of the survey results can be found in [7]. Results related to working in teams, technical challenges, and overall educational value will be provided here.

One of the key results of the survey was the response related to questions concerning the value of working on a team. Typical team sizes ranged from 2 to 5 students and 97% (74 out of 76) of the survey participants indicated there were advantages and value in working in a team to develop a robot for the various competitions. Below is a sample of representative comments made by the students.

1. *Yes, working in teams allows for more ideas to be considered before the actual construction begins. Teams also make refining ideas and executing objectives more attainable as well as less stressful.*
2. *No. It is difficult to find people on the same technical level.*
3. *Yes, more minds = more thoughts = better design, concept and implementation.*
4. *Yes because sometimes your partners think of ideas that you would never think of, or they pick up on your mistakes.)*
5. *It is an advantage because you have two or three minds to pull new ideas from or to improve existing ideas. Plus, not every group member is available at all times, but*
6. *Yes because everyone can bring different areas of knowledge and ideas. Your weaknesses may be someone else's strengths.*

The survey also asked students to identify the most important concept or lesson learned from the entire robot design and competition experience. The student responses included comments concerning the importance of testing, simplicity in design, listening to your team members, time management, and others. Overall, as with the teamwork comments above, the major lessons learned were generally consistent across all 8 robot contest events. A collection of representative student responses is presented in the list below:

1. *Keep all designs simple and stay flexible with the predetermined building specifications.*
2. *The difference between theory and reality. "It should work" often doesn't mean it will. There were a multitude of challenges that we faced that were not directly tied to our original goal.*
3. *Listen to your teams' ideas, they might just have a good one that needs work.*
4. *How to develop and test many different ideas and pick the one that works the best*
5. *Time management and organization, expecting the unexpected, adapting to changes.*

Students were also asked to rate the overall educational value of the robot design and competition experience and whether they would recommend the experience to other students in their major. The overall educational value was rated at 4.3 on a scale of 1 (poor) to 5 (excellent). Also, 93% of the students (71 out of 76) indicated that they would recommend the robot design and competition activity to another student. There were no significant differences in the overall ratings among the eight robot design and competition experiences. Students participating in the firefighting robot design and contest in spring 2008 were additionally asked the question, "Do you think hands-on experience with robotics could encourage students to pursue majors and careers in a technical field?" All of the students (N=7; 100%) indicated "yes" to this question. This result indicates the potential of the robot design competition to serve as a retention tool in addition to serving as a course enhancement tool.

5 Summary and Conclusions

The results described above provide successful case studies of robot competitions which serve to provide curriculum enhancement and outreach opportunities in a cost-effective manner. The robot competitions have been successfully integrated into a variety of engineering and information science courses and they provide a context for a wide range of design and experimental activities. The flexibility in hardware and software solutions for the participating robots, and the relative simplicity and low costs involved in hosting the competitions have sustained the Robo-Hoops and firefighting contests over a 10-plus year period. Due to the nature of the design challenges and the divisional structure of the competitions, K-12 students have been able to participate in the same contest and venue as students at the college level and beyond. The goal of this arrangement is to encourage K-12 students to consider careers in STEM areas, and to promote an environment of sharing, mentoring, and networking. Future goals include the assessment and academic tracking of K-12 participants, and also to provide resources for increased participation in the outdoor vision-based robot competition. It is hoped that the results and strategies described here will be of value to other educators who are considering the integration of robot competitions into curricular and outreach enhancements.

References

1. Robo-Hoops Contest,
   http://cede.psu.edu/~avanzato/robots/contests/robo-hoops/
2. Penn State Regional Firefighting Robot Contest,
   http://cede.psu.edu/~avanzato/robots/contests/firefighting/
3. Trinity College Firefighting Robot Contest,
   http://www.trincoll.edu/events/robot/
4. Pack, D.J., Avanzato, R.L., Ahlgren, D.J., Verner, I.M.: Fire-fighting Mobile Robotics and Interdisciplinary Design – Comparative Perspectives. IEEE Transactions on Education 47(3), 369–376 (2004)
5. Penn State Abington Mini Grand Challenge,
   http://www.cede.psu.edu/users/avanzato/robots/contests/outdoor/index.htm
6. Avanzato, R.L.: Autonomous Outdoor Mobile Robot Challenge. To appear in Computer in Education Journal (July- September 2009)
7. Avanzato, R.L.: Assessment and Outcomes of Robot Competitions at Penn State Abington. To appear in Computer in Education Journal (July- September 2009)

Teamwork and Robot Competitions in the Undergraduate Program at the Copenhagen University College of Engineering

Anna Friesel

Copenhagen University College of Engineering,
Lautrupvang 15, DK-2750 Ballerup, Denmark
afr@ihk.dk
www.ihk.dk

Abstract. In today's industry and trade, there is an increasing demand for engineers who don't just have excellent competence in their field of specialization but also a good understanding and practical experience in working with engineering projects, and working as members of a team. These subjects are usually not adequately addressed in engineering degree programs. This paper describes our experience in teaching mathematical modeling, control theory, microprocessors, programming, digital and analogue electronics as part of a robot design project. Robot competition at the end of semester motivates students to study theoretical disciplines. The pass rate compared to the classical courses increased from 60-70% to 85-95%.

Keywords: Mobile Robots, Undergraduate Education, Projects, Teamwork.

1 Introduction

The Electronics and Information Technology Department of Copenhagen University College of Engineering provides the education leading to the Bachelor Degree of Engineering. There is an increasing tendency in engineering to work in teams [1, 2]. Every team-member has to find her/his way how she/he can make a significant contribution to the overall team performance, and therefore has to know her/his specific strengths and weaknesses. A high capacity for team work can have a major impact on high performance output. That is why a specific training concept should be implemented in today's engineering degree programs. Most of the engineering projects especially in automation and robotics are also multi-disciplinary and therefore the students should be trained to tackle such projects. Problem oriented education and teamwork increases the motivation of the engineering students for theoretical subjects like: mathematics, mechanics, physics and control theory [3]. At the same time, almost all engineering jobs require good presentation and communication skills. This includes also the skills to practically apply modern presentation software, like PowerPoint, to present technical aspects in a convincing way to colleagues, to managers, and to customers.

Our students work from the very beginning (first semester) in groups solving practical engineering problems related to the theoretical subjects. The education is

J.-H. Kim et al. (Eds.): FIRA 2009, CCIS 44, pp. 279–286, 2009.
© Springer-Verlag Berlin Heidelberg 2009

described using a study module system. The educational value of a study module is expressed using the European Credit Transfer System (ECTS). The workload of one semester study is equivalent to 30 ECTS credits. In our department - the Department of Electronics and Computer Engineering (EIT) we offer full bachelor programs in Electronics and in Information and Communication Technology, both taught in Danish and in English. Globalization makes it necessary to cooperate on an international platform. A great contributor to globalization is the student mobility program within the EU, like the Socrates-Erasmus program. At the Copenhagen University College of Engineering we have more then 50 active Socrates-Erasmus agreements [4]. The challenge of supervising the international teams is to motivate the students with different prerequisites to study the theory and to work together with other students from very different cultures on a practical engineering project.

2 Objectives of the Robot-Project

The robot-project is an interdisciplinary project at the 4-th semester of 20 ECTS credits, which is 2/3 of the workload of one semester [5].
The theoretical disciplines students learn during this course [6,7,8] are:

- Mathematical modeling and dynamic systems.
- Continuous and digital control theory.
- Applied microprocessors and programming.
- Applied digital- and analogue electronics.

The robot-project challenges the students to find individual solutions to engineering problems and different robot competitions increase the motivation [9]. In principal, this is the integrated project format, where students work in teams. The aim of this project is to design and implement an autonomous mobile robot, executing a compulsory task and an optional task chosen by the students. The students have to deliver the solutions of the mandatory exercises in control theory, related to the robot project then later on include them in the final robot-project report. The students work in teams (the project groups), which consist of 4 to 5 students each. The formation of the groups has to be done during the first week of the semester. The formation of the groups is students' own responsibility, because among the objectives of this course are teamwork and cooperation.

2.1 The Compulsory Task

The compulsory task differs from semester to semester in order to prevent copying. In the following we describe the compulsory task part of the robot-project, carried out by the students in the fall semester 2008 [10]. An example of a track layout that the robot must follow is shown in Figure 1. Figure 2 shows a general overview block diagram of the robot – i.e. which overall modules the robot consists of.

Microcontroller Main Board is the core of the robot. This contains an ATmega32 microcontroller unit (MCU), a programmer for the aforementioned MCU, a LCD display for information output, and several connectors for interfacing the other modules.

Metal Detector Coils are two coils used to detect the position of the robot relative to the tinfoil tape track.

Detector Coils Interface is used to interface the detection coils to the AD converter on the MCU. This module combined with the detector coils is used to obtain a DC voltage difference with a change from when the detector coils are off the tinfoil tape to a different level when the detector coils are on the tinfoil tape. To determine the position of the robot relative to the tinfoil tape students use either detector coils or optical sensors. In order to get the highest resolution of the position of the robot, students choose the metal detector coils together with the ATmega32 on-chip AD converter (ADC) to get a gradual proximity detection of the tinfoil tape.

DC-motors are two armature controlled DC-motors used to run the robot. Each motor has a tachometer for use as means of measuring the number of revolutions of each motor. This is used to control the motors precisely.

DC-Motors Interface is used to interface both motors to the MCU. In the project two given armature controlled DC motors are used to run the robot (type: Faulhaber 2033 012S). The motors' velocity is controlled by applying a voltage in form of a Pulse Width Modulated voltage (PWM). The motors have build-in tachometers to determine the number of revolution performed by each motor. This is used when creating a controller for running the motors precisely. To obtain a simple way to control the motors an H-bridge is used. The motors used in this project require a voltage of 12 V and approximately 200 mA current. The output of the MCU delivers 5 V and a maximum current of 10 mA and by using the H-bridge a larger voltage and current can be controlled from the MCU. To control the velocity of the motors a PWM generated by the MCU is input to the H-bridge which delivers a corresponding PWM of larger voltage

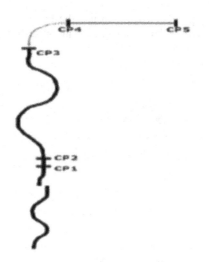

Compulsory Task Track Specifications
The robot must be able to follow the tinfoil tape track.
The robot must stop at checkpoint no. 1 (CP1) and make a 360° turn clockwise.
The robot must stop at checkpoint no. 2 (CP2) and make a 360° turn anticlockwise.
Continue following the track to checkpoint no. 3 (CP3).
Continue to checkpoint no. 4 (CP4) and stop for three seconds (the robot is allowed to be adjusted to straight position).
Continue in a straight path along the blue tape (without sensors).
Stop at check point no. 5. (CP5) The task is completed.

Fig. 1. Example of track layout, compulsory task

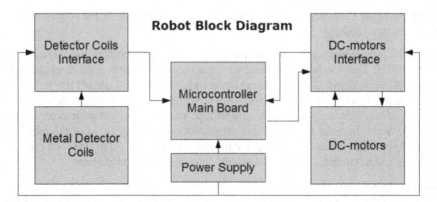

Fig. 2. Block diagram of the robot

and with access to a larger current to the motors. The velocity of the motor is determined by the width of the pulses, i.e. the duty cycle. The H-bridge used in this project is a Texas Instruments L293D Quadruple Half-H Driver which has built-in diodes.

Power Supply is a switch mode boost converter which delivers a constant voltage for the entire robot. It is desirable to have a constant voltage to the motors at all times since changes in voltage would yield changes in the constants for the motor controller which would affect the desired operation of the robot when the voltage of the battery pack falls. In order to maintain a constant voltage level for the motors a switch mode power supply is used.

2.2   Some Design Details

Mechanical design of the Robot. Some groups decided not to use the handed-out robot chassis and wheels, and in such cases they have designed and produced a

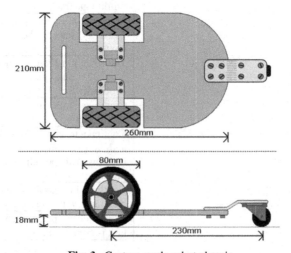

Fig. 3. Custom made robot chassis

custom chassis of an acrylic glass plate and used another set of wheels with a greater diameter. An example is shown in Figure 3.

Microcontroller Main Board Design. The Microcontroller Main Board contains an ATmega32 MCU as the central component. This board contains connectors for interfacing the detector coils, DC-motors, and other peripheral components (in this case only LEDs and a buzzer). The board contains a DEM20485 LCD display connected with the ATmega32. This LCD is used for writing out information to the user of the robot when this is operating. Examples of information available when the robot performs the compulsory task are:

- The current battery voltage level. The robot goes to idle mode when the voltage level of the battery pack drops beneath a certain level in order to preserve the lifetime of the rechargeable elements.
- The current speed of the robot in m/s. It can at all times be seen if the robot meets the requirement of maintaining a minimum speed of 0.5 m/s on the tinfoil tape during the compulsory task.
- The number of check points passed during the execution of the compulsory task program.
- The state of the state machine used to execute the compulsory task (most interesting for debugging).

The display will also write out if an error occurs, when the compulsory task is completed, or if the battery level drops below the set limit. Furthermore the LCD is a quick way to write out debug information when designing/altering the C-program. A general power plug is available to which the output from the switch mode boost converter is connected. This delivers power to all the other modules. Furthermore a charging plug is available from where the batteries can be charged and/or the robot can be run from. The main board contains an onboard programmer for the ATmega32 such that this can be programmed directly from a PC with a USB cable, i.e. no external programmer is needed.

Compulsory Task State Machine Design. A state machine can be viewed as a form of Artificial Intelligence (AI) where each state contains information about a specific action to be carried out.

Design of Controllers and Simulations. The chosen controllers for this project are usually two P-controllers, one for each motor, and a common I-controller. The I-controller synchronizes the velocities of the right- and left wheels of the robot. The control law calculations are included in the mandatory exercises 1 and 2 (ME1 & ME2), given to the students during the lectures in Control Theory. Students use linear control methods for control law, and make simulations in MATLAB and SIMULINK [11]. SIMULINK allows them to include nonlinearities which are present in the system and adjust the values of the controllers. In general, before implementing the controllers, in order to learn different controller's advantages and disadvantages, students make a lot of simulations both in MATLAB and in SIMULINK to simulate different kinds of controllers and values.

2.3 An Example of the Free Task Design

The free task as defined by Group 2, Spring 2008 [10]. This free task is based primarily on implementation of a bluetooth module on the robot making it possible to remote control the robot from another bluetooth enabled device and to receive information from the robot. The devices used to communicate with the robot are:

1   A bluetooth enabled PC running the custom Java program called »Robot Control Center«.
2   A bluetooth enabled mobile phone running another custom Java program called »Robot Control Center Phone«.

The communication protocol created to connect another device with the robot is able to send commands from the connected device to the robot for remote control – the commands determine the speed, direction, etc. on the robot. The protocol is also able to send commands from the robot to the connected device containing information on the battery level, actual speed, motor controller values, etc.

Requirement Specifications of the Free Task. As listed in the following table the group has set a number of requirements to the free task:

Requirement Number	Specification
R1	The robot must implement a bluetooth module and be able to communicate with another bluetooth enabled device such as a PC or mobile phone with a software program able to send and receive commands.
R2	The robot must be able to be remote controlled from the connected device.
R3	The connected device must be able to receive data from the robot with information about actual speed, battery level, etc.
R4	The robot must implement distance sensors to prevent it from crashing into a wall or other objects.
R5	Timing issues and improvements of the already implemented parts must be optimized.
R6	The robot must be able to switch between the task required for part one and the task required for part two.

2.4 The Evaluations and the Competition

The progress in the project is evaluated during the semester, in accordance to the project scheme (milestones) and in accordance with the plan made by the group and approved by supervisor at the beginning of the semester. The supervisor has the right to refuse admission to the examination for her/his group, if the agreements regarding the progress of the project are not kept. The mandatory course assignments must be approved in order to enter the examination. The examination is an oral examination, the external examiner (from other technical university) and all supervisors are present. At the examination each student is allocated 30 minutes. The evaluation is based on a general impression of the level achieved by the student relative to the objective of the course. The evaluation is based on the report, the oral performance and the functionality of the project.

We also perform students' evaluations both in the middle and at the end of the semester. The results of these evaluations show the following:

1. High score:
   - teamwork
   - applying the theoretical knowledge from previous semesters on a practical project from day one,
   - last day event and 4STARS competition,
2. Low score:
   - workload during semester (very high),
   - problems with some components (lead times if components break etc).

Every semester we have some exchange students taking this course and many of them have not tried working in teams the way we do it. All the evaluations we have made after they finished the exchange semester at our university show, that they are very happy about combining theory with practical projects and the possibility to use the laboratories 24 hours a day.

The department of Electronics and Information Technology has a tradition for inviting the students' family and friends for the last day event, which takes part on the last day of the tuition period, before the examination period begins. Students from all semesters present and demonstrate their projects for the guests and all the other students. A part of this tradition is the robot competition. The 4 STARS robot competition is held for all the robot project teams, where the best free task, the fastest, most precise and most elegant robots win prizes. Students work very hard in advance to optimize their robots, but the last day event is an extraordinary motivation to optimize their robots to win the prizes. Some of the teams actually continue with new mechanical constructions and optimize control algorithms in order to make the robot run faster.

3 Conclusions

Combining different engineering disciplines with teamwork during the robot-project improve the students' learning process. The motivation of the engineering students increases as they get the possibility to work with engineering design problems [12]. Another motivation factor is the freedom to choose their own solutions, freedom to choose the components and the combination of different theoretical disciplines in a project. The international exchange students participating in this project adapt very well to this form of studying. It takes from 3-5 weeks on average for exchange students to adapt to "teamwork" and to study independently with the project. Exchange students are usually good in theoretical skills, but too often look for "the right solutions". About 60-65 % of our students achieve grades above the average, their motivation to learn is very high and the teams work very hard to make their robots capable of winning the robot competition at the end of the semester. This is an additional motivation factor for the teams. Compared with other engineering departments, where the students follow traditional control theory courses, the pass rate for robot-project is increased to 85-95% from 60-70%.

Acknowledgments

I would like to thank my colleagues, the professors Henrik K.Palle and Ole Shultz at the Copenhagen University College of Engineering, for our good discussions and close cooperation during this course. Without their support the project would not been able to run in its current form. Thanks to all the students for valuable discussions on future development of this course, and special thanks to Group 2 (Ch. Dunweber, E. Hansen and N. Svendsen). Thanks to I.Stauning and J.Greve for technical support in connection to the practical part of the project.

References

1. Andersen, A.: Implementation of engineering product design using international student teamwork – to comply with future needs. European Journal of Engineering Education 26(2), 179–186 (2001)
2. Denton, A.A.: The role of technical education, training and the engineering profession in the wealth-creating process. In: Proceedings of the Institution of Mechanical Engineers, Part B, vol. 212, pp. 337–340 (1998)
3. Fink, F.K.: Integration of Work Based Learning in Engineering Education. In: 31st ASEE/IEEE Frontiers in Education Conference (October 2001)
4. http://www.ihk.dk/international/exchange-students
5. Friesel, A.: Learning Robotics By Combining The Theory With Practical Design And Competitio. Undergraduate Engineering Education. AutoSoft Journal, International Journal on Intelligent Automation and Soft Computing; Special Issue on Robotics Education
6. Jones, J.L., Flynn, A.M., Seiger, B.A.: Mobile Robots, 2nd edn. A.K.Peters (1999)
7. Ulrich, N.: Mobile Robotics, A Practical Introduction. Springer, Heidelberg (2000)
8. Nise Norman, S.: Control Systems Engineering, 4th edn. Wiley, Chichester (2004)
9. Ahlgren, D.J., Verner, I.M.: Robot Projects as Education Design Experiments. In: Proceedings of International Conference on Engineering Education, vol. 2, pp. 524–529 (2005)
10. Dunweber, C., Hansen, E., Svendsen, N.: TVP4E Robot Project – Compulsory Task, The Bug, Copenhagen University College of Engineering, 4th-semester Report (2008)
11. Tewari, A.: Modern Control Design with MATLAB and SIMULINK. Wiley, Chichester (2002)
12. Larson, E.C., La Fasto, F.: Team Work. McGraw-Hill, New York (1989)

Multiagents System with Dynamic Box Change for MiroSot

Mikulas Hajduk and Marek Sukop

Technical University of Kosice
Faculty of Mechanical Engineering, Department of manufacturing system and robotic,
B. Nemcovej 32, 042 00 Kosice, Slovakia
mikulas.hajduk@tuke.sk, marek.sukop@tuke.sk

Abstract. In multiple robotic system, from the point of view of control, there are crucial knowledge of the vision of the states on the playground, communication and cooperative. Our article describes a further significant factor of multirobotic system, and namely a dynamic exchange of strategic positions and actions of robots. In real football, if a player whose dominant role in a team is defence, but if there arises a suitable situation for the attack, he carries out this action but in the same moment another his team-mate takes up the function of a defender in case he receives a ball and an opponent starts to attack. The article defines a set of actions for the robot within the strategy. In the article, dynamic change of agents means that the software agents exchange control of robots, what means that there occurs a commutative substitution of identity. This approach is described in two basic actions of SjF TUKE Robotics team.

1 Introduction

Robosoccer is one of the best games for testing new approaches of development of multirobotic systems. Category Mirosot belongs to very interesting category of robosoccer games, mainly for its high dynamics of the game. Mobile robots in this category reach speed of movement up to 3.5-4 m/s and another of its characteristics is also number of players, namely 11. While further development in this category refers to wide application of methods of computer intelligence, implementation of sensors for actual detecting of surrounding, new methods of direct communication among robots, a higher level of visual systems, respectively equipment of a robot by its own visual system, and of course the development of new sophisticated software. Development may be expected in structural design of robots, particularly use of new materials and robot components manufacturing technology, smaller drives but more powerful but also intelligent ways of their controlling, providing higher manoeuvrability of robots, for example by using four-wheel robots, using smaller but more powerful batteries to obtain the space to be able to equip the robot by other mechanical and electronic hardware, such as a miniature camera or a mechanism to kick-off a ball.

In our development, SjF TUKE Robosoccer Robotics team have devoted great attention to construction of robots, we again came as first with the concept of 4 wheels but also to development of control and coordination of robots.

J.-H. Kim et al. (Eds.): FIRA 2009, CCIS 44, pp. 287–292, 2009.

In general, current approaches of control of robots are based on the central MAS structure where each agent is statically assigned an identity of a robot, which still does not mean that there is no confusion of position among robots in the course of a game but individual agents have relatively complex internal structure. Our paper describe dynamic box change of agents – robots on the base actual situation on the playground.

2   The Structure of the MAS for MiroSot

Fig. 1 depicts the hierarchical structure of a multiagent system for MiroSot with five robots proposed and used by us. At the highest level in this scheme, there is an agent "Master", whose main task is to decide on the choice of action on the basis of which the decision on the ongoing strategic position of individual robots "players" on the field will be made. Evaluation of the action on chosen strategy takes place according to the actual positions of all players and a ball on the field, as well as assumed future positions of these objects located on the playground.

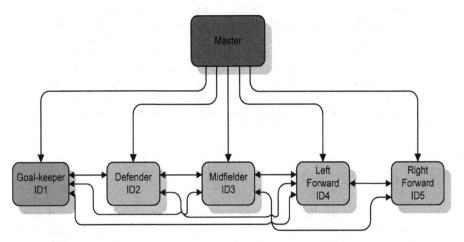

Fig. 1. Applied hierarchical structure of MAS with communication flows among agents

3   Dynamic Box Change of Identities of Agents and Their Existence

Depending on the type of selected action the master schedules areas in the playground in which the players should operate. On this basis, there is a decision on selecting the most appropriate agents "players" to accomplish this task. The selection is done from a set of 10 elementary agents Fig.2. at a given moment and for any action there are always 5 agents selected from the set. In the subsequent one there follows assignment of such agents to the individual mobile robots in the real field, i.e. identities are assigned to individual players. During this deciding, last assigned identities are taken into account. By the described algorithm the master performs confusion of individual

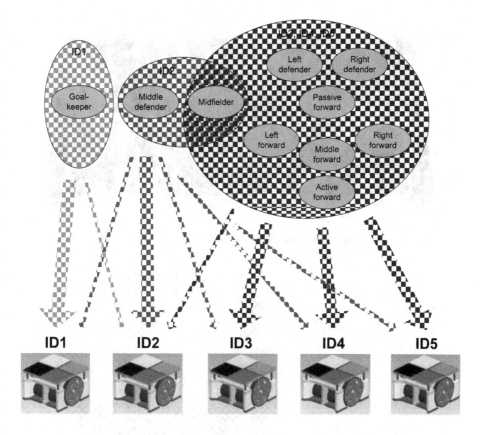

Fig. 2. Agents "players" and their possible assignment of identities

players among one other according to the suitability of their position on the field. The individual elementary agents "players", in terms of their internal functional structure, are becoming simpler and more transparent. It is these features that help developers in the more simple orientation in the whole system.

In continuation of the article there are introduced, as an example, 2 states in the playground, according to which the master makes decision on selection of an action : Total defensive and Attack.

4 Action - "Total Defensive"

Total defensive is occur, if ball is near our goal area and competitor attack. Figure 3 depicts the unambiguous situation when the master makes decision on the most defensive action. The selection of agents for this case is in the Fig. 4 marked as yellow. Selected agents ensure the most defensive strategic arrangement of players on the field. In this layout of players on the field players must pay attention to the violation of rules. If in a smaller goal area there is more than one player, the referee blows the whistle for penalty. Also, in a larger goal area there can be located up to two players.

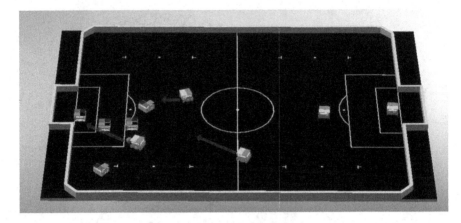

Fig. 3. The situation on the field "Total defensive"

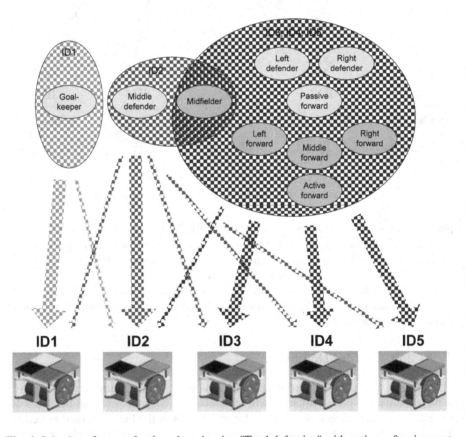

Fig. 4. Selection of agents for the selected action "Total defensive" with options of assignment of identities

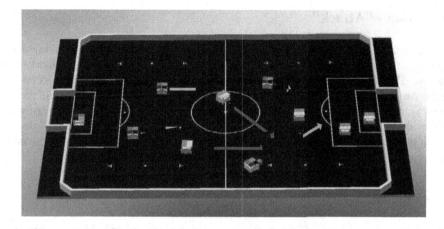

Fig. 5. The situation in the playground "Attack"

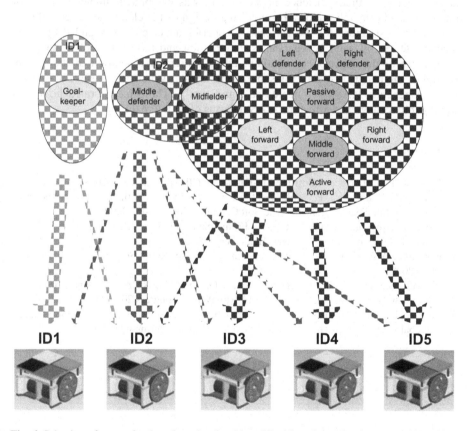

Fig. 6. Selection of agents for the selected action "Attack" with options of assignment of identities

5   Action - "Attack"

In Fig. 5 there is shown an unambiguous situation where the master is making decision on offensive action. The selection of agents for this case is in Fig.6 marked in yellow. The selected agents ensure offensive strategic arrangement of players on the field. An active attacker is a player who has the ball under control and whose movement is in the direction towards the gate. This player, when in close enough distance, makes decision on endings by a "kick". Execution of a kick is on the agent "player" and it is only him where he directs the ball in the direction towards the gate (countermovement of a goalkeeper or a freer side of the goalmouth).

6   Conclusion

The designed and revised diversification of functions of individual agents had been applied to our team. The algorithm of dynamic changes of identities of subordinated agents and their existence in the system was elaborated. Implementation of the introduced principles allowed simplifying design of the elementary agents due to their internal structure for the actions. A superior agent considers and decides on their existence, by what he guarantees the creation and termination of agents. The complexity of the superior agent compared to conservative solution of the system is slightly higher. Complexity of subordinate agents, which are of several species in heterogeneous MAS, was decreased significantly. These incorporated principles significantly change and affect the MAS in terms of its internal life.

References

1. Kopáček, P.: Robotsoccer: Post-Present-Future. In: CIRAS 2008, Linz (2008)
2. Kim, J.-H., Vadakkepat, P.: Multi-agent systems: A survey from the robot soccer perspective. Int. J. Intelligent Automation and Soft Computing 6(1), 3–17 (2000)
3. Jesse, N.: Autonomous Mobile Robots – From Science Fiction to Reality. Studies in Fuzziness and SOFT computing. Springer, Heidelberg (2005)
4. Hofer, G.: A Agent Based software concept for Mirosot Robot Soccer. In: CURAS 2008, Linz (2008)
5. Krywult, S., Deutsch, T., Bader, M., Novak, G., Onrubia, A.: Autonomons Mirosot the autonomons way of playing Mirosot. FIRA RoboWorld Congress 2006, Dortmund (2006)
6. Choi, S.H., Park, I., Cho, S., Jeong, I., Kim, J.H.: The revision method for dostorted image in global vission system. In: FIRA RobotWorld Congress, Dortmund (2006)

Multi Block Localization of Multiple Robots

TaeKyung Yang, JaeHyun Park, and JangMyung Lee

Dept. of Electrical Engineering
Pusan National University
Busan, Geumjeong Gu, 609-735, Republic of Korea
{yangpa,jae-hyun,jmlee}@pusan.ac.kr

Abstract. The multi block localization method for multiple robots using ultrasonic beacons provides a high accuracy solution using only low price sensors. To measure the distance of a mobile robot from a beacon, the mobile robot wakes up one beacon to send out the ultrasonic signal to measure the traveling time from the beacon to the mobile robot. When multiple robots are moving in the same block, it needs a scheduling to choose measuring-sequence in order to overcome ultrasonic signal interferences among robots at every time. But the increased time delay to estimate the positions for the multiple robots degrades the localization accuracy. This paper proposes an efficient localization algorithm for the multiple robots, where the robots are grouped into one master robot and the other slave robots. In this method, when a master robot calls a beacon, all the robots simultaneously receive the identical ultrasonic signal to estimate their positions.

Keywords: localization, multi block algorithm, group scheduling.

1 Introduction

For the localization of the mobile robot in the indoor environment, there are several schemes using IR [1], laser [2], vision [3], Ultrasonic and etc. The ultrasonic sensors widely used indoor localization system, since they are cheap and easy to be controlled with high accuracy [4]. Even so, they are susceptible to environmental noise [5] and reflection from their propagation characteristics, and they are hard to transmit to long distance because of their decay phenomena.

The typical Systems using the ultrasonic sensors are Active Bat [6] and iGS [7]. The abovementioned systems have a high accuracy of localization about a single mobile robot. However, if the number of robots increases in the same area, they produce a signal collision for increasing signals. That is, they will cause problems for localization. If they do not know concurrently positions of multiple robots, it is not useful because of only applying a single robot.

In this paper, the localization of multiple robots using ultrasonic sensors has been introduced to analyze problems. And efficient method to overcome problems has been proposed.

This paper consists of five sections including this introduction section. In section 2, the indoor localization system, iGS, has been introduced in detail and section 3

J.-H. Kim et al. (Eds.): FIRA 2009, CCIS 44, pp. 293–299, 2009.
© Springer-Verlag Berlin Heidelberg 2009

describes the concurrent localization of multiple robots that is the major contribution of this paper. In the section 4, the effectiveness and usefulness of the master & slave algorithm have been verified by the simulation. Finally, in the section 5 concludes this research, work and mentions future studies related to this research.

2 Indoor Global-Localizaiton(IGS)

2.1 iGS Basic Principle

The iGS is composed of a localizer, beacons, and a PC for the user. At first, the active beacon sensor consists of a radio frequency(RF) receiver and an ultrasonic transmitter. A mobile robot can select a specific beacon that has its own ID and position information during the navigation by sending a desired beacon code via RF. When a beacon receives its own ID from the localizer, it sends back an ultrasonic signal to measure the distance from the beacon to the localizer using the time of flight (TOF). Using the distances and the relative beacon position information, the robot position can be computed using the trilateration method. Figure 1 illustrates the basic configuration of iGS.

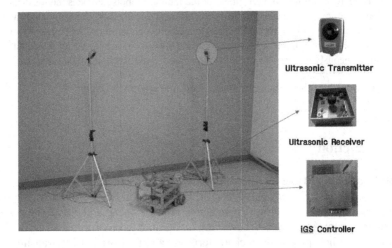

Fig. 1. Configuration of iGS

2.2 Position Measurement

Localization of mobile robot is measured by distances from the beacon to the localizer using TOF of ultrasonic. The distance between beacon and localizer, r, can be obtained as the multiplication of v and TOF.

$$r[m] = v[m / \sec] \cdot TOF[\sec] \ . \tag{1}$$

The speed of the ultrasonic signal is function of environment temperature, T, and it is represented as

$$v[m/\sec] = 331.5 + 0.6 \times T[°C] \ . \qquad (2)$$

In the Equation (2), the transmission velocity of the ultrasonic signal is changing according to the environment temperature. The temperature is assumed to be 20°C in this research for simplicity in these experiments. The TOF of ultrasonic signal is represented as

$$TOF = n \times T_c - T_d \ . \qquad (3)$$

Where n is time counter, Tc is counter clock, Td is the delay of the ultrasonic signal in the circuit.

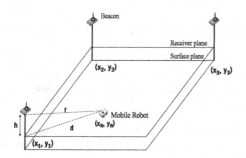

Fig. 2. Mesurement of distance between the beacon and the receiver for trilateration

Figure 2 is concept of the distance measurement of iGS. Since the height, h, from the localizer to the beacons is known as a constant, the distance from a beacon, r, is mapped to d, which is the horizontal distance. With the three distance d1, d2, d3, from the three beacons and the pre-specified beacon locations, (x1, y1), (x2, y2), (x3, y3), the location of the mobile robot, (xR, yR), can be obtained as the intersection of the two circles, and they are represented as

$$d = \sqrt{(r^2 - h^2)} \ . \qquad (3)$$

$$\begin{bmatrix} (x_R - x_1)^2 + (y_R - y_1)^2 \\ (x_R - x_2)^2 + (y_R - y_2)^2 \\ (x_R - x_3)^2 + (y_R - y_3)^2 \end{bmatrix} = \begin{bmatrix} d_1^2 \\ d_2^2 \\ d_3^2 \end{bmatrix} \ . \qquad (4)$$

3   Localization of Multiple Robots

3.1   Localization of Multiple Robots

Localization of multiple robots currently having used iGS has some problems. If mobile robots are more than two in the same area, where consist of three beacons, they separately select each beacon for localization. And then a selected beacon and other beacons send ultrasonic signals at once. At this time, mobile robot cannot recognize which beacons send out a useful signal, since signals are concurrently submitted within the transmission time of the ultrasonic signal.

In order to overcome abovementioned problems, each robot and beacons need synchronization for localization. There are two ways of beacons synchronization and mobile robots synchronization. In case of the beacons synchronization, all of the beacons are synchronized since a beacon sends out ultrasonic signal once. And at the same time, mobile robots are synchronized in order to recognize a specified beacon that sent out ultrasonic signal. It has a disadvantage to add more hardware since the beacons consist of only RF transmitter not receiver. The other way of mobile robots synchronization measure a robot position at once or synchronize a call time of the beacons. It has advantage to have a high accuracy and measuring time in a small number case of robots. But if a number of mobile robots increase, time of measuring the robot position will also increase.

Table 1. The time of measuring position of a robot

d_1	d_2	d_3	Trilateration
40ms	40ms	40ms	80ms

Table 1 shows the sampling time of measuring position of a robot. $dn(n = 1,2,3)$ is the sampling time measuring the distance from the beacons to the mobile robot. All amount of the sampling time measuring the position increase as n times because of increasing in a number of robots. And mentioned two ways have a disadvantage to add controller for synchronization.

3.2 Master and Slave Method

The method proposed in this paper is kind of ways synchronizing mobile robots only. One robot is designated as the master and the other robots are designated as the slaves. Figure 3 shows the basic structure of the master& slave method.

This method is that only the master robot calls the beacons. Slave robots synchronized with the master receive the ultrasonic signal from a same beacon calling master robot concurrently. The slave is always watching and waiting master's synchronization signal. If the master robot sends out a synchronization signal, slave robots wait ultrasonic signal. And then the master and the slave calculate distances from a beacon

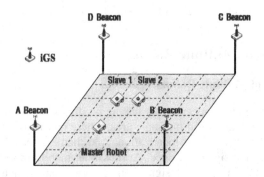

Fig. 3. Basic structure using Master & Slave method

to each mobile robot. Finally they measure each position themselves after calculating distances from the next beacons in serial. The master robot considers max transmission time that arrive at each robot because the arrival times of ultrasonic signal are difference. Figure 4 summarize a flowchart of master& slave method.

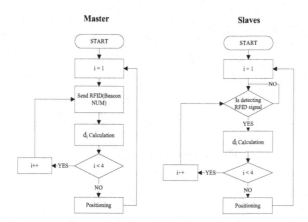

Fig. 4. Flowchart of Master& Slave algorithm

3.3 Master and Slave Method

This paper also proposes the multi zone algorithm. Beacons are arranged at the apex of squared plane. In the iGS system, only one selected beacon can send ultrasonic signals at a time. However the distance that ultrasonic signals can reach is limited. When the distance between beacons is far enough, the beacons can be grouped into several groups. The beacons in the same group can send a signal simultaneously. Ultrasonic signal can be detected up to about 7meters and the signal's distance depends on the height of beacon.

Figure 5 shows 6 blocks with 12 beacons and 9 tags. Creating each group makes it possible to send more than 2 signals at the same time as well as to prevent signal's interference. Group scheduling reduces sampling time and is suitable for using numerous beacons.

We are assuming that the tags' initial location is already known. Tags are synchronized with each other and T1 is given the highest priority so that it becomes a master robots. The master robot, T1, calls beacons from the group A to the group F in order. B1 and B7 in the group A, for example, can send a signal at the same time because there is no ultrasonic signal interference between two beacons. Group scheduling can reduce sampling time so as to increase the blocks. When T1 goes out of blocks or disappears, T2 becomes the master robot by the priority. Each robot can recognize which signal they will use because initial location is known.

From group A to group F, the ultrasonic signals are sent in regular sequence by the master robot's signal. In each turn, master robot calls the specific beacon and the other synchronized slave robots wait ultrasonic signal.

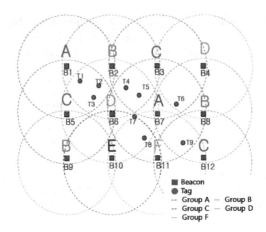

Fig. 5. Multi block & Multi tag

4 Simulation

As you can see in figure 6, simulation is based on the localizer and the beacon below. The localizer that was attached to the robot to receive the RF and ultrasonic signals and to measure the distance from the beacons was designed in the Intelligent Robot Laboratory in Pusan National University and Ninety Systems Corp. using DSP TMS320C2406, and the ultrasonic transmitters were designed using MSP430 [Fig. 5].

In case of measuring position of one robot, the sampling time is about 200ms. As the number of robots is increased, the sampling time grows in arithmetical progression. In the 6 blocks with 12 beacons and 9 tags, it takes 1800ms to measure all robots' location. With the proposed algorithm, the sampling time can be reduced remarkably. All the beacons send the signal within 100ms.

For example, in order to measure T7's location, it takes about 1400ms because T7 should wait until the other robots finish the localization. However, the localization with group A, D, E and F can be completed within 240ms[Fig.7]. We assume that the position of each robot will not move at fixed position.

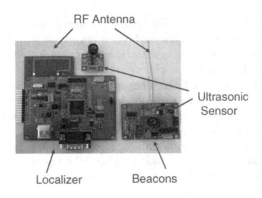

Fig. 6. The beacons and localizer

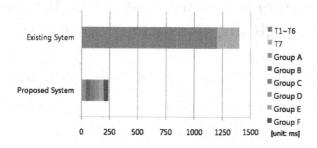

Fig. 7. Sampling time of T7's localization

5 Conclusion

In this paper, the multi block localization of multiple robots using iGS based on the ultrasonic sensors has been proposed. For the localization, master robot is assigned for each block which wakes up the iGS to send out the ultrasonic signal to all the robots in the block. With the communication between the master and the slave, ultrasonic signal can be received without interference. Through the expansion of blocks and group scheduling algorithm, it can be applied to wide area localization with the minimum sampling time.

Acknowledgments. This work was supported by the IT R&D program of MKE/IITA [2008-S033-01, Development of Global Seamless Localization Sensor].

References

1. Wang, J., Cipolla, R., Zha, H.: Vision-based Global Localization Using a Visual Vocabulary. In: IEEE int. Conf. on Robotics and Automation, April 2005, pp. 4230–4235 (2005)
2. Zhou, Y., Liu, W., Huang, P.: Laser-activate RFID-based Indoor Localization System for Mobile-Robots. In: IEEE int. Conf. on Robotics and Automation, April 2007, pp. 4600–4605 (2007)
3. Lin, W., Jia, S., Abe, T., Takase, K.: Localization of mobile robot based on ID tag and WEB camera. In: IEEE int. Conf. on Robotics, Automation and Mechatronics, December 2004, vol. 2, pp. 851–856 (2004)
4. Tsai, C.C.: A Localization system of a mobile robot by fusing dead-reckoning and ultrasonic. IEEE Trans. on Instrumentation and Measurement 47, 1399–1404 (1998)
5. Premans, H., Campenhout, V.: High resolution sensor based on tri-aural perception. IEEE Trans. on Robotics and Automation 9(1), 36–48 (1993)
6. Harter, A., Hopper, A., Steggles, P., Ward, A., Webster, P.: The Anatomy of a Context-Aware Application. In: 5th ACM International Conference on Mobile Computing and Networking (Mobicom 1999), August 1999, pp. 59–68 (1999)
7. Seo, D.G., Lee, J.M.: Localization Algorithm for a mobile robot using iGS. Jour. of inst. of Control, Robotics and system 14(3) (March 2008)

Soty-Segment: Robust Color Patch Design to Lighting Condition Variation

Seung-Hwan Choi, Seungbeom Han, and Jong-Hwan Kim

Department of Electrical Engineering and Computer Science,
KAIST, Daejeon, Republic of Korea
{shchoi,sbhan,johkim}@rit.kaist.ac.kr
http://rit.kaist.ac.kr

Abstract. Significant developments have been seen over the last few years in the robot soccer domain. Considerable improvements have been made in the areas of robot control, strategy generation, vision processing, etc. In particular, in the vision area, there have been many researches aimed at decreasing processing time. However, many other issues remain in the development of ideal vision system. Despite MiroSot being a well established competition, development of a vision system is still considered as a difficult task. In global vision systems, the development of insensitive system to lighting condition variation is an important issue. This paper proposes a novel color patch design which possesses some advantages from the vision processing viewpoint. It is insensitive to lighting variation, greatly reducing vision setting time as well as providing improved recognition. The effectiveness of the proposed Soty-Segment color patch is demonstrated through real experiments.

Keywords: Color patch design, localization, MiroSot, robot soccer.

1 Introduction

Micro-Robot World Cup Soccer Tournament (MiroSot) has witnessed considerable technical improvement as well as continuous development of robust system over the last decade. MiroSot development is based on the standard architecture for robot soccer, which is composed of cameras, robots, color patches and RF transmission modules [1]-[4]. MiroSot has expanded to the Middle League MiroSot (5 a-side) and Large League MiroSot (11 a-side) with the progress in the related technologies [5]-[9]. The most important issue associated with these expansions is the processing time for the vision system. It is not easy to obtain optimal digital camera performance completely for larger playground. Most previous researches focused on this problem and a wide variety of solutions was developed.

The vision system of Large League MiroSot shares many common elements with that of the Middle League MiroSot. The vision system recognizes each robot with color patch through the camera located on top of the field. Field lighting condition is thus an important factor, as the lighting condition is not uniform continuously. If lighting condition varies during the game, the vision

J.-H. Kim et al. (Eds.): FIRA 2009, CCIS 44, pp. 300–309, 2009.

system confuses colors such that it may not be able to recognize the robots. Therefore, during vision setting time, all colors are sampled considering various lighting conditions on the field. It is obviously a tedious and time consuming task.

These days color patch design usually comprises of a team color and a combination of ID colors for distinguishing each robot. Accordingly, with an increase in the number of robots, more independent ID colors are needed and consequently this makes vision setting time longer. Many teams implement a wide variety of strategies, depending on their experience and know-how, for reducing this time-consuming effort. For example, a simpler and more comfortable interface is programmed and utilized or alternately a large number of color patches are placed on field during vision setting time. Nevertheless, these have been no ideal solutions.

The proposed novel color patch design on top of robot uses the state of each segment representing an identification (ID) number of each robot to distinguish each other, where the ID number is placed on team color. Lens distortion revision, perspective distortion revision and image conversion algorithm in image preprocessing are employed to calculate the correct position of each segment. The heading direction of the robot is calculated by simply detecting the orientation of the direction color region.

Proposed design reduces the number of patch colors and saves time for vision setting and processing time. Also, the vision system becomes more stable than the existing ones and is independent of lighting condition variations due to the increased number of margins that are allowed in RGB color space with less number of colors. Moreover, its human friendly design leads to much easier operation of human operators as it can be easily identified by the number on top of robot. Additionally, as the robot soccer system can be utilized as a test bed for experiments in mobile robot navigation, multi-agent cooperation, and so on, the proposed design will be useful in the research of mobile robot.

This paper is organized as follows: Section 2 explains the previous and recent color patch designs. Section 3 proposes a novel color patch design and its recognition algorithm. Section 4 presents experimental results and concluding remarks follow in Section 5.

2   Previous Color Patch Designs

Fig. 1(a) shows two representative types of earlier color patches for soccer robot. These color patch designs used ID colors to identify each robot. Fig. 1(b) shows the modified color patch design which uses just two ID colors. To the best of the authors knowledge, this is the first attempt at reducing the number of ID colors. This color composition became a basic idea for later color patch designs. Fig. 1(c) is a recent color patch design used in Middle League MiroSot and Large League MiroSot. The characteristic feature is the usage of more than two colors with increase in the number of robots. Such color patch designs are considered as a relatively effective method. However, in Large League MiroSot, the number of robots is more than twice of Middle League MiroSot, and two cameras

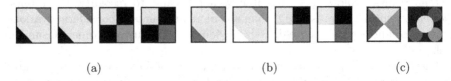

(a)   (b)   (c)

Fig. 1. Various color patch designs. (a) conventional ones, (b) improved ones and (c) the most recent ones.

should be set up. Therefore the conventional color patch design proves inefficient in Large League MiroSot. Setting up one more ID color is not a trivial issue. As the size of a ground increases, the lighting conditions become more variable. Consequently, the possibility of failure to recognize exact color is greater. This is the principal reason necessitating much time consumption for the preparation of Large League MiroSot. The problem is further compounded due to the usage of two camera systems. Furthermore, the conventional color patch design presents another problem. It is difficult for the human operators to identify the robots quickly and intuitively. Distinguishing robots by various compositions of colors is an inconvenient task for human operators. This further complicates the communication among human operators.

3   Soty-Segment Color Patch Design

As described above, a fundamental problem of conventional color patch designs is the usage of too many ID colors. To solve this, the color patch design for distinguishing the robots should be based on a specific pattern. At the same time, the algorithm must be simple to avoid impeding the performance of the system.

Fig. 2 depicts the proposed color patch design, the Soty-Segment color patch. Each robot can be distinguished by the segments on team color instinctively. The heading direction of the robot is calculated by simply detecting the orientation of the direction color region (white), which thus makes the algorithm faster and simpler.

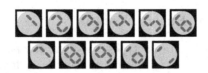

Fig. 2. The Soty-Segment color patch design

3.1   Preprocessing

The image through a camera is distorted image due to lens. Image distortion also occurs in the case that a camera is not installed horizontally. This distortion has to be revised to suitably recognize small patterns like the color patch of soccer robot.

1) Revision of Lens Distortion: A camera, which can cover wider range, has larger distortion near the edge region. For this reason, a pixel in an image is not exactly located at a real position in the image. It is known as radial distortion. To recover the distortion, One of well-known method is mapping the camera image position r to the perspective image position r' through 2nd-order or 3rd-order polynomial in polar coordinates (Fig. 3) [10] as follows:

$$r' = R_{lens}(r) \tag{1}$$

To use this method, lens parameters should be examined. However, in the absence of this knowledge, the parameters should be found experimentally.

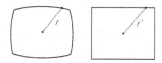

Fig. 3. Relationship between camera image position and perspective image position

2) Revision of Perspective Distortion: Because of imperfect alignment of a camera, an image from a camera suffers perspective distortion. This paper proposes its restoration using 2D homography. 2D homography is a transformation between 2D planes in images. It includes translation, scaling, rotation, perspective transformation, and also general 2D transformation. The transformation can be represented as a 3×3 matrix whose degrees of freedom (DOFs), is eight [10]. To use the matrix, position of pixels should be represented by a homogenous coordinate (x,y,w), where w is a scale factor, which is generally 1 in an image. One missing DOF results from its scale-less property, that is, scalar multiplication of the matrix makes the same transformation. To transform an arbitrary plane to a desired rectangular plane, the homography, that is, the 3x3 matrix should be evaluated. Generally the (normalized) four-point algorithm can be applied, and it needs four pairs of corresponding points. Each pair of corresponding points has a constraint as follows:

$$\begin{pmatrix} x' \\ y' \\ 1 \end{pmatrix} = H \begin{pmatrix} x \\ y \\ 1 \end{pmatrix} \tag{2}$$

Therefore four pairs are enough to estimate the matrix H. Using DLT (direct linear transformation), the matrix H can be calculated as a solution of homogenous linear system of equations [10]. Fig. 4 shows the result of revising perspective distortion.

3) Image Conversion Algorithm: Revising distortion of all pixels in the image consumes too much operation time. Therefore applying image conversion algorithms to the necessary pixels only is effective for robot soccer system which needs fast operation (Fig. 5). It will be demonstrated in following section.

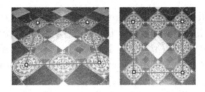

Fig. 4. Relationship between perspective image position and real location

(a)   (b)

Fig. 5. Revision process for lens and perspective distortion. (a) revise entire image, (b) revise just only reference points.

3.2   Algorithm of Color Patch Recognition

Table 1 describes the colors used in the recognition algorithm of the Soty-Segment color patch, composed of the Position-color, Direction-color and Team-color regions.

1) Pixel Sampling: Two-dimensional image information is recorded in a one-dimension array. To reduce time required for scanning of the entire image, color information is only extracted from every kth pixel. Fig. 6 shows extracted points when k is 180.

2) Locate Position-color Region: If the extracted color is Position-color, begin locating the Position-color region by flood-fill algorithm which searches the circumference color region in order.

3) Locate Direction-color Region: During locating Position-color region, if the extracted color, which is the detected contiguous position of the Position-color, is Direction-color, then Direction-color region is located with the same method as in locating the Position-color region. This method prevents the spurious detection of Direction-color, extracted from opponent team color patch or noises on field.

4) Calculate Position and Orientation of Robot: The position and orientation of the robot are calculated by using center positions of the Position-color and Direction-color regions. However, these two center positions are from distorted

Table 1. Types of colors used in Soty-Segment color patch design

Type	Color
Position-color	Team-color + (Segment-color)
Direction-color	WHITE or ALL–(BLACK+Position-color)
Team-color	Yellow or Blue

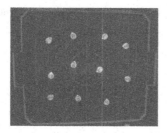

Fig. 6. Image scan with k-pixels jumping (k=180)

image, due to which the actual center positions must be calculated from a revised image. Each position should be revised by an image conversion function. Direction-color region of the proposed Soty-Segment color patch is placed 45 degrees counter clockwise. Therefore the positions and the orientations are calculated as follows:

$$
\begin{aligned}
Q_{pos} &= H(R_{lens}(P_{pos})) \\
Q_{dir} &= H(R_{lens}(P_{dir})) \\
V_{dir} &= Q_{dir} - Q_{pos} \\
Orientation &= Direction\ Of\ Vector(V_{dir}) - 45°
\end{aligned}
\tag{3}
$$

5) Identify the Segments: When the position and the direction are calculated, the exact locations of each segment are determined. Circumference colors are extracted from each segment to increase accuracy of detection, as shown in Fig. 7. ON/OFF condition of segment is decided as follows:

```
NumberOfSegmentPixels = 0
For Each pixel of 9 pixels
    IF pixel does not belong Team-color
    THEN NumberOfSegmentPixels++

IF NumberOfSegmentPixels >= Threshold
    THEN Segment is ON
    ELSE Segment is OFF
```

Threshold value is appropriately controlled according to the environment. When the states of segments are decided, the identification number of the robot is also decided by the predefined table as shown in Table 2. The decision accuracy

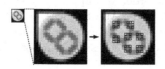

Fig. 7. Segment state decision with nine circumference points

Table 2. States of seven segments for each robot

State of 7 Segments	Robot ID										
(ON:○, OFF:×)	1	2	3	4	5	6	7	8	9	10	11
1	×	○	○	×	○	○	○	○	○	○	○
2	×	×	×	○	○	○	×	○	○	×	×
3	○	○	○	○	×	×	○	○	○	×	×
4	×	○	○	○	○	○	×	○	○	○	×
5	×	○	×	×	×	○	×	○	×	○	×
6	○	×	○	○	○	○	○	○	○	○	×
7	×	○	○	×	○	○	×	○	○	○	○

can be also increased by using the whole shape of Soty-Segment color patch. However, referring to the corresponding color in each segment is more efficient from the viewpoint of processing time.

3.3   Color Demarcation

Demarcation methods of Position-color, Direction-color and Team-color differ depending on their usage in the algorithm. Position-color is assigned as the summation of team color and segment color. It has the widest possible region in RGB space because it is the most fundamental color for identification of robot position. Direction-color can be assigned in two ways. One is to simply extract the white color, while the other is to extract the rest part excluding black color, boundary color of color patch, and Position-color in RGB space. A characteristic of Direction-color is applied, which is only dependent on the circumference of Position-color region and independent of its actual color itself. Since Team-color is only used for identifying the state of the segment, it is arbitrarily assigned in region excluding the segment color. The black and white regions are the easiest colors to be demarcated. Thus the only critical factor in color demarcation is to identify the Position-color region in RGB space.

4   Experiments

To test the proposed methods, many images were captured at various heights and angles. Images were captured in the environment of Small League MiroSot game field using a UNIQ vision UC-685CL digital camera which further utilized a PANTAX 6.5mm C-mount lens. The camera was installed at a height of 2.5 meters above the field. In order to capture whole rectangular game field at once, the field for Small League MiroSot was used for testing. The proposed Soty-Segment color patch design proved its applicability and effectiveness in Middle League MiroSot and Large League MiroSot at FIRA Robot World Cup 2008.

4.1   Revision of Distortion

Fig. 8 shows revised lens and perspective distortions. For convenience of analysis of the results, all pixels in the image were revised. In practice, however only a much lower number of pixels need to be revised depending upon the color patch recognition algorithm. Fig. 9 shows the located regions of Position-color and Direction-color. The numbers on image mean the number of pixels in each region.

4.2   Color Patch Recognition

Fig. 10 shows segments recognition processing. Fig. 10(a) shows the original patch image. Fig. 10(b) illustrates that the positions of segments are fixed by position and direction of the robot. Only Team-color regions are shown Fig. 10(c). The numbers on image indicate the number of segment pixels, which was described in Identify the segments in Section III. If the threshold is set to seven, seven segments have all ON state. Then, the robot ID number is decided to

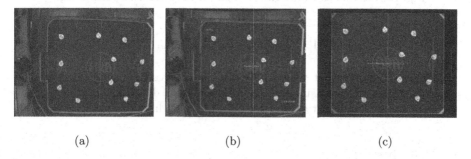

(a)   (b)   (c)

Fig. 8. Revising lens and perspective distortion. (a) original image, (b) revising lens distortion and (c) revising perspective distortion.

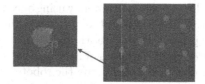

Fig. 9. Position-color and Direction-color region location

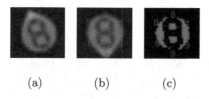

(a)   (b)   (c)

Fig. 10. Segment recognition

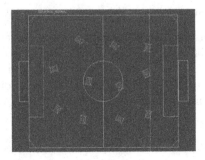

Fig. 11. Recognition result with 11 robots

Fig. 12. Possible patch design based on the Soty-Segment patch design

be 8 from Table 2. Fig. 11 is the final recognized image with 11 robots on the playground.

It should be noted that various the color patch design is possible using the same algorithm as shown in Fig. 12. This different design is suitable for the opponent team to differentiate between the two matching teams.

5   Conclusion

This paper proposed a novel advanced Soty-Segment and demonstrated its effectiveness and efficiency through experiment. The Soty-Segment color patch design was able to identify and distinguish each robot by using only Team-color, Direction-color and Position-color. As the number of colors needed is lower, the effort needed in the color recognition process is significantly reduced. Furthermore, this design stabilizes the vision system from variation of lighting conditions. This color patch design is easy for human operators to identify the robots, and thus posses an additional advantage in system management. The Soty-Segment color patch design thus makes the robot soccer game more realistic as the playing robots are easily identified by the ID number like the back number of human soccer players.

References

1. Kim, J.-H., Shim, H.-S., Kim, H.-S., Jung, M.-J., Choi, I.-H., Kim, K.-O.: A Cooperative Multi-Agent System and Its Real Time Application To Robot Soccer. In: Proceedings of the IEEE International Conference on Robotics and Automation, Albuquerque, Albuquerque, New Mexico (1997)

2. Shim, H.-S., Kim, H.-S., Jung, M.-J., Choi, I.-H., Kim, J.-H., Kim, J.-O.: Designing Distributed Control Architecture for Cooperative Multi-agent System and Its Real-Time Application to Soccer Robot. Journal of Robotics and Autonomous System 21(2), 149–165 (1997)
3. Kim, J.-H. (ed.): Special Issue: First Micro-Robot World Cup Soccer Tournament, MiroSot. Robotics and Autonomous System 21(2) (September 1997)
4. Kim, D.-H., Kim, Y.-J., Kim, K.-C., Kim, J.-H., Vadakkepat, P.: Vector Field Based Path Planning and Petri-net Based Role Selection Mechanism with Q-learning for the Soccer Robot System. Intelligent Automation and Soft Computing 6(1), 75–88 (2000)
5. Shim, H.-S., Jung, M.-J., Kim, H.-S., Kim, J.-H., Vadakkepat, P.: A Hybrid Control Structure for Vision Based Soccer Robot System. Intelligent Automation and Soft Computing 6(1), 89–101 (2000)
6. Kim, J.-H., Kim, D.-H., Kim, Y.-J., Seow, K.-T.: Soccer Robotics. Springer Tracts in Advanced Robotics. Springer, Heidelberg (2004)
7. Kim, J.-H.: Soccer Robotics Micro-Robot World Cup Soccer Tournament. Intelligent Automation and Soft Computing 6(1), 1–2 (2000)
8. Han, K.-H., Lee, K.-H., Moon, C.-K., Lee, H.-B., Kim, J.-H.: Robot Soccer System of SOTY 5 for Middle League MiroSot. In: Proceedings of The 2002 FIRA Robot World Congress, May 2002, pp. 632–635 (2002)
9. Koo, M.-H., Lee, Y.-K., Lee, K.-H., Kim, T.-H., Lee, J.-K., Kim, J.-H.: Development of Robot Soccer System for 11-A-Side MiroSot. In: Proceedings of 2004 FIRA Robot World Congress (October 2004)
10. Zhang, Z.-Y.: A Flexible New Technique for Camera Calibration. IEEE Transactions on Pattern Analysis and Machine Intelligence 22(11) (2000)

Task-Based Flocking Algorithm for Mobile Robot Cooperation

Hongsheng He[1], Shuzhi Sam Ge[1], and Guofeng Tong[2]

[1] National University of Singapore, Singapore 117576, Singapore
elegesz@nus.edu.sg
http://robotics.nus.edu.sg/sge/
[2] Northeastern University, Shenyang 110004, China

Abstract. In this paper, one task-based flocking algorithm that coordinates a swarm of robots is presented and evaluated based on the standard simulation platform. Task-based flocking algorithm(TFA) is an effective framework for mobile robots cooperation. Flocking behaviors are integrated into the cooperation of the multi-robot system to organize a robot team to achieve a common goal. The goal of the whole team is obtained through the collaboration of the individual robot's task. The flocking model is presented, and the flocking energy function is defined based on that model to analyze the stability of the flocking and the task switching criterion. The simulation study is conducted in a five-versus-five soccer game, where the each robot dynamically selects its task in accordance with status and the whole robot team behaves as a flocking. Through simulation results and experiments, it is proved that the task-based flocking algorithm can effectively coordinate and control the robot flock to achieve the goal.

1 Introduction

The cooperation and control of a group of mobile robots effectively for a common goal is a widely studied topic in robotics and autonomous vehicles [1]. Mobile robot flock can finish the task that single robot can not achieve. In the cooperation, each robot takes its responsibility and performs certain role to contribute the global goal. Currently, mobile robot team is employed in various applications such as rescue, exploration, communication and surveillance missions and so on [2].

However, effective coordination of a robot team for a common goal is a difficult task [3]. In the real world applications, the cooperation efficiency of the mobile robots team is important for collaboration work especially for competition. Generally, we catalog the cooperation techniques into two classes: the top-down approach and bottom-up approach. In the former approach, the cooperation criterion emphasizes the formation. The work is first interpreted into certain formation and the robots must follow this formation law to finish the job. There are mainly three techniques to control the formation of mobile robot team: behavioral, leader-following and graph-based [4]. [5] proposed a new reactive behaviors that implement formation in mobile robot teams and [6] presented an

J.-H. Kim et al. (Eds.): FIRA 2009, CCIS 44, pp. 310–321, 2009.

approach for representing formation structures with regard to queues and formation vertices. Both of them reported good simulation results of formation control but no consideration of the cooperation performance. In the latter approach, the cooperation criterion highlights the individual behaviors. There is no strict constraint or predefined rules for the formation maintenance. Each robot chooses the best-suited role according to its local status and area information. Flocking algorithm is one of the best approaches to maintain multiple mobile robots. There are roughly two kinds of flocking model: with-leader and without-leader [7]. In the first model, the robots are classified into leader and lead-followers, and the followers follow the behavior of leader to maintain the formation of the team. As for the without-leader model, all the robots in the the team are equal and all the robots know the destination. Recently, flocking algorithm has also been successfully applied in other research fields. [8] applied the flocking intelligence to govern migration in the distributed genetic programming algorithm, and the results confirm the performance of the modified algorithm.

In this paper, a task-based flocking algorithm(TFA) is presented to address the problem of multiple robot cooperation. TFA provides a effective framework for coordinating a team of mobile robots. This algorithm adopts the basic principle of the flocking algorithm for formation maintenance but emphasizes the common goal of the team through task self-allocation. TFA can be used in the case that all the boids have similar or same functions and properties. If there is no obvious difference of abilities between the boids in the team, flocking algorithm is the best consideration since it emphasizes the balance of the chance for all the team members. In TFA, the destination of the task is highlighted and original flocking algorithm is applied to control the formation that is of high flexibility. The contributions of this paper are highlighted as follows:

- proposing one task-based flocking algorithm framework for effective cooperation in multi-robot system and the task-design and performance feedback criteria;
- introducing flocking energy function to define and analyze the stability of flocking, and proving the relation of local task stability with flocking global stability;
- evaluating and comparing the algorithm through simulation on the standard platform. Through the experiment result, the task-based flocking algorithm is efficient and be used in real world applications.

2   Task-Based Flocking Algorithm

2.1   Task-Based Flocking Algorithm

Flocking is a collective behavior of a flock of individuals acting for certain purpose [9]. In 1987, Reynolds first presented the algorithm for simulating flocking behaviors of a flock of fish, birds and animals [10]. The simulated creatures can keep the form and avoid obstacles like terrains, airplanes or other moving objects with run-time performance. The three rules he defined for flocking behaviors are:

- Cohesion: attempt to move towards the average position of flock mates;
- Separation: avoid the collision with nearby flock mates;
- Alignment: try to keep consistent with the average heading of flock mates.

These rules can be interpreted as flocking centering, collision avoidance and velocity matching [11]. The cohesion behavior endows the boid with the ability of cohering with the global action of boids and the related cohesion steering is determined by the average position of neighbors. This long range attraction of steering drives the boid towards the direction of average position. The separation is short range repulsion of steering which maintains a certain distance between boids, avoids their colliding with each other and spans a wide working sphere. Alignment steers the boids towards average heading of neighbors. This simple alignment direction can be computed by averaging the alignment direction of the local neighbors or all the boids in the flocking. In flocking algorithm, there is no global control or formation control. The boid acts autonomously as an individual. The behaviors of all the individuals in the flock construct the global behavior of the flocking. Each boid can read the whole scene information but only interact with the flock mates in the small sphere. Flocking algorithm presents a cooperation model for multi-robot system without central formation control.

However, the original flocking algorithm only considers these three rules without a common objectiveness of the flocking. Task-based flocking algorithm(TFA) extends the basic flocking algorithm for mobile robot cooperation by allocating tasks to each robot for a common job. As for the basic flocking algorithm, it is behavior-based and all the robots in the flocking comply with the same behavior rule which maintains the flocking. There is no strict task control in the standard flocking algorithm. TFA incorporates task into the flocking algorithm and is more object-oriented than the original flocking algorithm. The task of TFA for the robots is scheduled for efficient cooperation and each robot has the same priority to choose the task just like the standard flocking algorithm(this rule can be adjusted in the algorithm implementation). There are more or less tasks comparing with the numbers of the robots in the cooperation which depends on the complexity of the work. For algorithm performance and simplicity, each robot in TFA only considers the robots in the neighbor sphere for flocking criterion, where there should be at least one robot in the neighbor sphere. In the stability analysis, we prove that the local sphere flocking guarantees the global stability of the flocking algorithm.

2.2   Flocking Model

Consider a flock of m autonomous mobile robots, where individual robot is denoted as r_i, $1 \leq i \leq m$. We model each robot as dynamic solid point that can freely move on a two-dimensional plane in all direction. Then the dynamic the dynamic model of single robot is described as $F_i(P_i, V_i, \psi_i)$ $1 \leq i \leq m$, where $P = P(x, y)$ is the position vector, $V = [v_x, v_y]$ is the velocity vector and ψ_i denotes robot heading angleFrom the figure, it is straightforward that

$$V_x = \dot{P}\cos\psi$$
$$V_y = \dot{P}\sin\psi$$
(1)

The gesture of the single robot can be uniquely determined by position and heading angle. Thus the flocking model is described as $F = [F_1 \cdots F_m]^T$. The assumption of this flocking model is that initial state and distribution of the robots in the flock are arbitrary and unknown. There is no leader and coordinator in the flock [2]. In addition, $\overline{V_i}$ and $\overline{\psi_i}$ denote average position of the flock mates and the average angle of the flock mates in interested area respectively. In origin flocking algorithm, the alignment angle for a flock mate is chose as the mean heading angle of the flock mates in the interested region. The fundamental describers for the flocking are the magnitude and velocity of the flocking. The characteristics of flocking behavior can be captured into two aspects [12]: size describer of the flocking based on the average distance between flock-mates and entropy describer of the flocking based on the degree of the alignment of flock mates. In the study, $\sigma_p = \frac{1}{R_0}\left(\frac{1}{m-1}\sum_{i=1}^{m}(P_i - \overline{P})^2\right)^{\frac{1}{2}}$ and $\sigma_v = \frac{1}{v_\infty}\left(\frac{1}{m-1}\sum_{i=1}^{m}(V_i - \overline{V})^2\right)^{\frac{1}{2}}$ define the standard deviation the flock mate position and the velocity respectively. As a system with multiple particles, the global and effective evaluation of the flocking state is the system energy. Consider flocking describers, we obtain the energy of flocking F,

$$E_F = \sum_i E_{F_i} = \frac{1}{2}\sum_i K_k \|V_i\|^2 + \frac{1}{2}\sum_i \sum_j K_{p_{ij}}(\|P_i - P_j\| - Eqi_{ij})^2 \quad (2)$$

where K_k and K_p are the kinetic energy coefficient and potential energy coefficient respectively, and Eqi_{ij} represents the equilibrium distance between F_i and F_j. K_p is defined as

$$K_{p_{ij}} = \begin{cases} K & P_i \neq P_j \\ -\infty & P_i = P_j \end{cases} \quad (3)$$

Eqi_{ij} is one of the most important parameters of the flocking which indicates the relationship of flock mates and influences the task allocation and cooperation. In a balance state, that is all the mobile robots in the team have the same priority and relationship, Eqi_{ij} is constant. Flocking energy performs as the kernel of mobile robot flocking that determine or influences two key performance of the flocking: stability and task-selection criterion. A general assumption we make in the algorithm is that the energy function E_{F_i} reflects the activity of the robots in R_{Ni} around F_i. If the robots in a region is more active, then the robots is usually running faster for the task destination. In addition, the distance between robots should also severely be apart from the equilibrium distance. Those two factors will directly influence the the energy function. Thus, in real applications, this assumption generally holds. In the implementation of the algorithm, we use E_{F_i} to identify the importance of the robot in the flocking i.e. the most active robots deserve the highest priority. We will discuss these two aspects in detail in following sections.

2.3   Task and Switching Criteria

The efficiency of the mobile robots cooperation is determined by task design and task switching criterion. Task-design depends on the functions and complexity of the specific cooperation work. The typical task design for mobile robot is listed in table 1. Properties of each task are evaluated before being registered into the task pool in each cycle with updated information. The updating rule depends on the specific mission of the task. For instance, following task is initialized with the information of the tracked object.

Table 1. Task Descriptor

Task property	Description
ID	the global unified identifier
Destination	the point the robot runs for
Speed	the speed the robot should maintain when reaches the destination
Angle	the angle the robot should maintain when reach the destination
Robot	queue Q_R stores the robot that chooses this task
Ticks	Record the maximum cycles left for the current robot to finish this task; this value is initialize with Pri_{ij} when the task is assigned, and decreases each cycle
Specification	the robot number which this task can only be or not be assigned
Mutually exclusive	index whether this task can be assigned to multiple robots

In TFA, the mobile robot in flocking autonomously select the task according to the status of itself and neighbors within its interest region. The task switching criterion should sort the suitability of each task using a universal rule since the different tasks may drive the robot to the same state. For autonomous mobile robots on a 2D plane, the basic movements are *running to a point, turning* to a angle and *avoiding an obstacle*. For simplicity, we define task switching criterion through priority of the task for robot F_i. The robot should always select the available task with highest priority. One simple priority definition of robot F_i chooses task T_k is

$$\frac{1}{Pri_{ik}} = \alpha \frac{\|\Delta dist\|}{\|V_i\|} + \beta \|\Delta angle\| \|V_i\| + \gamma \frac{1}{\|\Delta obs_i\|} + Bias_{ij} \qquad (4)$$

where $\Delta dist$ is the error distance to go, $\Delta angle$ is the error angle to turn, Δobs_i is the distance between the robot and its nearest obstacle in V_i direction and α, β, γ are weighing coefficients. $Bias_{ij}$ is the bias of the priority which increases the flexibility of adjusting priority, which also play an important role in keeping the continuity of task and task performance feedback. The influence of velocity is covered in the basic movement. The physical meaning of this priority definition is straightforward. The higher the priority is, the robot is more suitable for this task. In order to improve the real time performance, $\frac{1}{Pri_{ik}}$ is used instead since $priority > 0$. In realization of the algorithm, the main computation burden comes from the priority since this function is computed for every robot in each cycle. The effectiveness of task switching criteria can be greatly improved if $Bias_{ij}$ is well-tuned although $Bias_{ij} = 0$ most of the time. We mainly adjust $Bias_{ij}$ in four typical cases: (i) For keeping the continuity of the task and avoiding the frequently useless task switching. In this case, $Bias_{ij} > 0$ and is slightly increased if the task is always chose as the best suit task in a few continuous cycles. (ii) For adding artificial influence into the task switching. When specific events takes place, it is common that the mobile robot must or must not choose certain tasks. At this point, we can simply increase or decrease the $Bias_{ij}$ to the peak values. (iii) For task performance and feedback. Although the tasks and task-switching criterion is well designed, there are also many factors we cannot cover and well represent. Thus, we adopt learning mechanisms into the algorithm. The selection of the task is praised or punished according to the feedback of the task performance. The detailed feedback technique is illustrated in the next section. (iv) For flocking stability. The TFA regularly checks the stability of the flocking. When certain task tends to render the flocking unstable, the checking routine sets the $Bias_{ij}$ accordingly.

2.4   Performance Feedback Criterion

After a mobile robot in flocking takes one specific task, the system needs to evaluate how well the robot has performed or finished the task in order to judge whether the robot is suitable to continue doing this task. Although this weight is relatively mute compared with other parameters, it influences the task-switch especially when the robot has been performing the same task for a sequence of cycles. The judgment criterion considers the predicted time for the robot to finish the task T_k.

$$Perf_{ik} = \eta \frac{\|\Delta dist\|}{\|V_i\|} + \frac{\|\Delta angle\|}{\left\|\dot{\psi_i}\right\|} \tag{5}$$

where η are the weights for distance and angle errors. By comparing the $Perf_i$ and the ticks stored in the task descriptor, if $Perf_{ik} \leq ticks$, this priority of this task is enhanced; otherwise, suppressed. An alternative for the task performance feedback is modified the weight in $Perf_{ik}$. However, the method renders the algorithm hard to maintain, so it is not applied in the implementation of the algorithm.

3   Algorithm Analysis and Implement

3.1   Stability Analysis

The basic problem for mobile robot cooperation is the stability analysis of the team, which guarantees the existence of the robot group. In this section, we mainly analysis the constraint for the task selection to guarantee certain sense of stability of the team behaviors. Since in TFA, formation is not treated as the important factor for effectiveness of robot cooperation, so we cannot adopt the stability definition in graph [13]. We propose the global definition of stability for robot flocking based on flocking energy E_{F_i}.

Definition 1 (Global Stability). *The global stability of the flocking is defined as the flockmates neither converge to a point nor break up into separate parts. In sense of flocking energy, the flocking is globally stable if $E_{min} \leq E_F \leq E_{max}$.*

Definition 2 (Internal Stability). *The internal stability of the flocking refers to that the flocking is globally stable and the relations of boids in the flocking do not exchange frequently and repeatedly.*

In other words, no frequent status-switch between robots happens in the same scenario even though the solution for this scenario is not unique. Some of the flock mates that are shuttling and chasing for each other's state will cause oscillating of the E_F and destroy the internal stability.

3.2   Rule of Task

Tasks are the fundamental element for scheduling in TFA. In section 2.3, we have demonstrated the framework of the basic task for mobile robots. Here we discuss the rule of task design and selection while ensuring the stability of the flocking. In the task selection, there is no need considering all the flock mates which increases the futile computation burden since robot F_i mainly interacts with the robots in R_{Ni}. The rule of task design based on local information of interested region R_{Ni} for robot F_i is proposed as,

Theorem 1. *For task assigned to robot F_i , it maintains the stability of the flocking if*

$$\begin{cases} \|V_i\| \leq V_{max} \\ 0 < \|P_i - P_j\| \leq P_{max} \quad F_j \in R_{Ni} \\ \sum j > 0 \qquad\qquad F_j \in R_{Ni} \end{cases} \tag{6}$$

Proof: if $\forall F_i \in F$, $\|V_i\| \leq V_{max}$ and $\|P_i - P_j\| \leq P_{max}$ holds, then from Eq.2

$$E_F \leq \sum_i \frac{1}{2} K_k V_{max}^2 + \sum_{P_{max} \leq 2Eqi_{ij}} \frac{1}{2} K_{p_{ij}} Eqi_{ij}^2$$

$$+ \sum_{P_{max} > 2Eqi_{ij}} \frac{1}{2} K_{p_{ij}} (P_{max} - Eqi_{ij})^2 \triangleq E_{max} \tag{7}$$

for $0 < \|P_i - P_j\|$ and $\sum j > 0$, we obtains

$$E_F \geq \sum_{Pmax \leq Eqi_{ij}} \frac{1}{2} K_{p_{ij}} (P_{max} - Eqi_{ij})^2 \triangleq E_{min} \tag{8}$$

from Eq. 7 and Eq.8, it is obvious that

$$E_{min} \leq E_F \leq E_{max} \tag{9}$$

Therefore the flocking is globally stable and this task rule maintains the stability of the flocking. This rule interprets that if the local stability is satisfied based on local information in R_{Ni} then the global stability can be achieved. This theorem gives a sufficient condition in task stability: the robot F_i with a limited velocity and there is at least one neighbor in R_{Ni} can guarantee the global flocking stability. For internal stability, the rule is that the destination of one task cannot be the position of another task. By following this, the phenomenon that the robots repeatedly pursue each other's position can be effectively avoided. Thus, the internal stability of the flocking is fulfilled in certain sense.

3.3   Algorithm Implement

The basic framework of TFA and original flocking algorithm are shown in Alg.1 and Alg.2. For robots that cause the instability of the flocking and those do not take any tasks, the original flocking algorithm is called to maintain the stability and formation of the flocking. There are many more improvements can be merged into the algorithm such as the modification of $Bias_{ij}$ as discussed in section 2.3.

Algorithm 1. TFA()

---

update the status of flock mate $F_i(P_i, V_i, \psi_i)$;
initialize task with the updated information in the task pool;
for *each F_i in F* **do**
    compute flocking energy E_{F_i};
if *E_{F_i} satisfies Eq.6* **then**
    sort E_{F_i} into priority queue Q_F;
else
    flocking(F_i ,R_{Ni});
for *each F_i following the sequence in Q_F* **do**
    for *each available task T_k in task pool* **do**
        compute $\frac{1}{Pri_{ik}}$;
        insert $\frac{1}{Pri_{ik}}$ into priority queue Q_P;

    register F_i to T_j with the minimum element in Q_P: $ticks = Pri_{ij}$, $F_i \rightarrow Q_R$;
    robot F_i performs T_j;
    compute $Perf_{ij}$;
    update Pri_{bias};
for *each T_j in task pool not register by any robot* **do**
    $null \rightarrow Q_R$;
for *each F_i does not take any task* **do**
    flocking(F_i ,R_{Ni});
log information of time and cycles;

Algorithm 2. flocking(F_i,R_{Ni})

---
compute $\overline{V_i} = \frac{1}{n}\sum_{j=1}^{n} V_j$, $F_j \in R_{Ni}$;
compute $\overline{P_i} = \frac{1}{n}\sum_{j=1}^{n} P_j$, $F_j \in R_{Ni}$;
compute $\overline{\psi_i} = \frac{1}{n}\sum_{j=1}^{n} \psi_j$, $F_j \in R_{Ni}$;
drive robot F_i to $F_i(\overline{P_i}, \overline{V_i}, \overline{\psi_i})$;

---

4   Experiment

4.1   Experiment Setup

The standard platform for Robot Soccer Simulation Competition[1] is employed to conduct the simulation study. Following TFA framework, we design the strategy compatible to the simulation platform to test the performance and to compare the strategy with others. In this case, the work is specified to drive the ball to opponent field. Consider the real football game, we first design the roles(called tasks in TFA) generally used in Robot Soccer Competition following stability rule. These fundamental tasks are listed as follows,

- Marking(M): robot F_i runs to certain position in the field and waits for further tasks;
- Following(F): robot F_i follows an object as nearly as possible;
- Dribbling(D); robot F_i drives the ball along an trajectory;
- Interfering(I): robot F_i runs towards an object and drives it out of its desired moving direction.

In implement of the tasks, the control law for the robots is PID control for simplicity which can also give a satisfied performance. For the weighing parameters used in the algorithm, artificial weighting is not added and all the weighing parameters are set to the same. We only use these tasks to study the performance of the algorithm and design criterion, so the attack tasks such as shooting is not included.

4.2   Experiment Results

We load the TFA strategy on both sides and study how flocking energy changes and influences the task-selection criterion. Three continuous snaps(left part) taken and the corresponding flocking energy (right part) is shown in Fig. 1. The energy shown in the figure is normalized into $[0, 10]$ for a clear comparison and more contours represent more flocking energy. From the energy graph, we can see that the most active robots around the ball hold large flocking energy, which proves the rationality that the priority of the robots to choose the task is determined by the flocking energy. We also examined the performance feedback mechanism applied to the left-area robot members. The *bias* from performance

---

[1] Download at http://www.newneu.net/

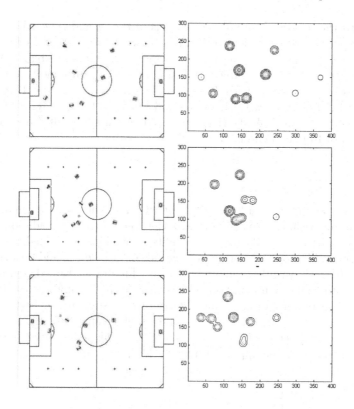

Fig. 1. Continuous snapshots and flocking energy

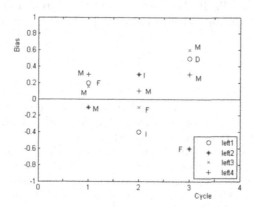

Fig. 2. Task switching and bias influence

feedback and the tasks the robots performs in each cycle are shown in Fig.2. From the data, we can conclude to what extent the performance feedback influences the task switching. Take the robot left1 as an example: in the first cycle, the robot tried to finish the task of "following" the ball, but robot right3 is faster and dibbled the ball out of its direction; thus in the second cycle, the selection of "following" task for robot left1 is "punished" with $Bias < 0$ and left1 turn to "Interfering" task since it was near right3 which is dibbling the ball; yet in the third cycle, the robot left1 switched to task "Dribbling" although the feedback of the last cycle is positive with $Bias > 0$. The framework of the TFA can be more powerful if featured with rich tasks and well-tuned parameters. The strategy designed based on TFA algorithm achieved very good results[2] in simulation competition of Robocup China Open, 2008.

5   Conclusion

In this paper, we proposed a task-based algorithm for the cooperation of mobile robots. This algorithm can make full use of the predominance of individuals than the formation-based algorithm. The stability of flocking is defined in sense of flocking energy. Moreover, we presented the rule of task design and task-selection criterion. Although we only simulated the algorithm in constrained environment, the results still hold in unconstrained scenario from the analysis of the team behaviors. The performance of this algorithm is desired to be better if each robot's task is implemented in separate thread which simulates the individuals in real life group. In conclusion, TFA is an effective algorithm for mobile robot cooperation which can be applied in real world applications.

Acknowledgments

This study was partially supported by Singapore National Research Foundation, Interactive Digital Media R&D Program, under research grant R-705-000-017-279 and National University of Singapore, under URC Funding: Mind Robotics (R263000490112).

References

1. Fua, C.H., Ge, S.S., Lim, K.W.: Fault tolerant task scheduling for multi-robot teams using self-organizing agents in formation. In: IEEE International Conference on Robotics and Automation, pp. 576–581 (2006)
2. Lee, G., Chong, N.Y.: Adaptive flocking of robot swarms: Algorithms and properties. IEICE Transactions on communication E91-B, 2848–2855 (2008)
3. Parker, L.E.: Alliance: An architecture for fault tolerant multi-rboot cooperation. IEEE Transactions On Robotics and Automation 14, 220–240 (1998)
4. Ren, W., Sorensen, N.: Distributed coordination architecture for multi-robot formation control. Robotics and Autonomous System 56, 324–333 (2005)
5. Balch, T., Arckin, R.C.: Behavior-based formation control for multi-robot teams. IEEE Transactions On Robotics and Automation (1999)
6. Ge, S.S., Fua, C.H.: Queue and artificial potential trenches for multi-robot formation. IEEE Transactions On Robotics 21(4), 646–656 (2005)

---

[2] http://ai.ustc.edu.cn/rco/rco08/scores.php, GNUS.

7. Lee, G., Chong, N.Y.: Decentralized formation control for small-scale robot teams with anonymity. Mechatronics 19, 85–105 (2008)
8. Paulikas, G., Rubliauskas, D.: Movement of flocked subpopulations in distributed genetic programming. Information Technology and Control 34 (2005)
9. Gervasi, V., Prencipe, G.: Coordination without communication: the case of the flocking problem. Discrete Applied Mathematics 144, 324–344 (2004)
10. Reynolds, C.W.: Flocking,herds,and schools: a distributed behavioral model. Computer Graphics (ACM SIGGRAPH 1987 Conference Proceedings) 21(6), 25–34 (1987)
11. Olfati-Saber, R.: Flocking for multi-agent dynamic systems: Algorithms and theroy. IEEE Transactions On Automatic Control 51, 401–420 (2006)
12. Crowther, B.: Rule-based guidance for flight vehicle flocking. Journal of Aerospace Engineering, 111–124 (2004)
13. Moreau, L.: Stability of multiagent systems with time-dependent communication links. IEEE Transactions on Automatic Control 50, 169–182 (2005)

Analysis of Spatially Limited Local Communication for Multi-Robot Foraging

Stephan Krannich and Erik Maehle

Institute of Computer Engineering
University of Luebeck
23538 Luebeck, Germany
{krannich,maehle}@iti.uni-luebeck.de

Abstract. This work presents a biologically inspired communication model for foraging swarms of cooperative mobile robots. In contrast to conventional, unrestricted local communication the exchange of messages is here spatially restricted to a nest-like area. The performance of the presented communication concept is evaluated using simulation and comparison to common forms of communication. An implementation on hardware robots allows to determine influences from the real world on the model. Results show that spatial limitation of communication to a single nest area can still speed up the performance of foraging swarms whereas further increasing the quantity of conventional local communication is less effective for the process of foraging.

1 Introduction

The task of retrieving objects from an unknown or unmapped area can be performed by robots that are capable of identifying the target objects and navigating autonomously. Swarm intelligence or biological processes are the basis of e.g. optimization of network traffic [1] or control of swarms of robots [2,3,4]. Robot swarms can be employed to perform tasks a single robot is not capable of or fulfill tasks in less time, although physically interference of swarm robots may lead to deadlocks [5]. Broken robots can be replaced by other swarm members.

The goal of this work is the analysis and evaluation of a biologically inspired communication model for foraging swarms of mobile, autonomous robots. In the process of foraging food items will be searched for. In case of success the information about paths to the items are to be distributed by the finder to other swarm members. Communication is only allowed in close range of a nest-like area. The concept is designed for large scale environments.

The basic idea is to combine principles found in foraging of honey bees and desert ants as the basis for the presented concept of swarm robots: Desert ants, e.g. the genus *Cataglyphis bicolor*, survive in hot regions where pheromone trails, laid by ants of e.g. genus *Lasius niger* for orientation and navigation to food sources, would instantly dry out. Instead, desert ants recognize prominent optical conditions (landmarks) in the environment while searching for food sources. The

J.-H. Kim et al. (Eds.): FIRA 2009, CCIS 44, pp. 322–331, 2009.

sequence of passed landmarks to a food source describes the path and enables the desert ants to relocate the food source. In experiments desert ants of genus *Melophorus bagoti* were capable of simultaneously keeping information of up to three ways to food sources and to orientate completely on the basis of optical landmarks [6]. In contrast to other studies ant pheromone based navigation [7] is not considered as the aim is to measure performance when communication relies only on message exchange restricted to one area.

Honey bees that successfully found a food source return to their nest and pass on information about the way to the target location to other bees in the nest. Returning bees encode information about the food source in a waggle dance. Surrounding bees in the nest can perceive the dance and gain information, directing them accurately to the destined food source [8]. Bees only transfer information on valid food source locations. Information on depleted sources is not spread. This principle is the basis for the local communication in this work: spatially limited local distribution of information about valid food source locations.

Forms of communication for mobile robots have been widely discussed. It has been shown in [9] that local communication is suitable for groups or swarms of cooperating robots. The transmission range of a robot is limited to mostly a circular area around the robot. An important factor is the determination of the range [10] [11]. In contrast to our model, communication may take place everywhere as long as robots are in communication range. Research in [12] showed that some multi-robot tasks, like foraging, can benefit from communication and be finished in less time while others, e.g. graze tasks, are hardly influenced. Increased complexity of messages turned out to have little or no influence on speeding up multi-robot tasks, compared to the simplest form of communication. In [13] light beacons are used for indirect communication in a multi-robot foraging scenario with real hardware. Time spent on communication is discussed as it seems to significantly affect the performance of real-world robot foraging processes. We presume that such a simple form of communication is not applicable in large scale real world environments with physical obstructions and mostly no line of sight and therefore rely on more complex messages, containing the description of a path to a food source by referring to landmarks.

2   Concepts

The overall concept of the foraging model is the same for the real hardware and simulation runs. Modifications are mentioned in the according paragraphs.

2.1   Hardware Robots

The used robots are E-Pucks[1]. This robot has a differential wheeled drive, infrared range sensors, a low resolution camera, a Bluetooth module, and a dsPIC controller with about 14 MIPS. It has further features, not utilized here.

---

[1] http://www.e-puck.org/

2.2 Simulation

A custom simulation model has been developed to evaluate the influence of the employed communication mode on the foraging performance of a swarm. Navigation and world representation are based on graphs entered into the simulator. Vertices represent landmark locations. Edges are unweighted and undirected. Simulated robots can move along all edges of the graph, food items can be located at every vertex. Three communication modes are are available: No communication, (common) local communication, and spatially limited communication. All actions in the simulation are divided into time slices. During one time slice each robot can move from one node to another, check the new position for food items and exchange messages with other robots.

2.3 Landmark-Based Navigation

Orientation and navigation are based on optical landmarks. A path to a food source is described by the sequence of landmarks a robot passed during the successful search. The robot saves the sequence of the detected landmarks. It can reuse this information later for navigating to the food source by simply locating the known landmarks and passing them in order. In contrast to simultaneous localization and mapping (SLAM) algorithms this approach does not lead to the generation of a map of the area where the robots operate.

The nest is the location where all robots start from and return to for all attempts of finding or checking food sources. It is considered to be a unique landmark.

To begin a forage, a landmark is randomly chosen from the set of visible ones and stored as the first way point. The robot will approach the chosen landmark. From there on, it will locate other landmarks or food sources and save the information about the next chosen way point. This process is performed recursively. If a food source is located the path is stored and marked as valid. The robot will then return to the nest. The maximum length of a path is limited by the number of landmarks the robot can save or by restrictions set for each single path. If no food source is found before reaching the maximum number of way points the search is cancelled and the robot will return to the nest, using the saved information to drive back to the nest. The validation of a known food source is performed correspondingly. The robot will drive according to a saved sequence and check if the last position contains food. In our test a robot can save up to three different paths, like the trained desert ants [6].

In the hardware experiments a landmark consists of two stacked different colored barrels. The depletion of a food source is emulated by manually removing it from the test area. For object detection with low computational power an adjusted object detection algorithm for mobile robots is used [14].

In the simulation visible landmarks are represented by adjacent vertices. An arbitrary number of food sources can be placed in the environment graph. Each food source has its own amount of prey that is reduced each time a robot visits it and consumes one item. If the amount becomes zero the source is depleted. If a robot checks the path to a depleted source the saved paths becomes invalid. Food sources can be set to reappear after a certain time since their depletion.

2.4 Communication

For comparing the influence of communication on the foraging performance of a swarm, we use the following three communication modes:

Spatially Limited Communication is the honey bee inspired model. Robots are only allowed to exchange messages while being in the nest area. Real hardware robots have to sojourn at the nest location for a certain amount of time as the Bluetooth modules require some time to establish communication links. In the simulation all robots located at the nest node are able to communicate.

Local Communication mode is utilized to compare performance improvements of a swarm where members are allowed to send messages at any place and a swarm with the spatially limited communication model. Message exchange may take place everywhere where robots are in close range of each other. This form of communication is solely implemented in the simulation.

No Communication is the control model. No messages are transmitted and the performance of the swarm only relies on the sum of all individual successes. This model is evaluated on both platforms.

It is common for the first two modes that messages contain only positive information on food source locations. Robots may only spread the information of their last validated path. If a robot receives a shorter path to a known food source the new path information is kept, otherwise the message is ignored. A model with unlimited communication range and the ability to reach all swarm members any time is not integrated here as it can be assumed not to be a realistic option for a real foraging scenario.

2.5 Program Concept

The complete program for a robot consists of the combination of the presented concepts. A state diagram of the program with the spatially limited communication model can be seen in Fig. 1. Each robot will start from the nest location. From there on, a robot has the choice between three different actions. A search for a new food source can be initiated or, if available, a robot can begin to check a known path to a food source. The third possibility is to wait at the nest for the return of another robot and receive foreign path information. If a certain amount of time has passed and the waiting robot does not receive a message it can begin a new search or check a known food source. This waiting step is skipped in the simulation, as it takes a simulated robot one time slice to enter or leave a vertex.

If the maximum number of optical landmarks per search is reached without finding a new food source or if a known path could not be validated this information will be stored (suc:=0) and the swarm member will drive back home to the nest and will not spread any information.

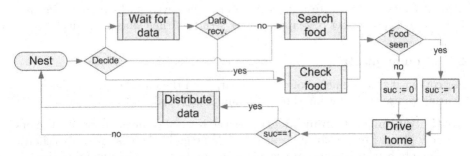

Fig. 1. Flowchart of the program for a robot with spatially limited communication

In case of successfully finding a new food source or validating the existence of a known path this information is stored (suc:=1) and the robot will drive back to the nest and try to spread the information to all robots waiting in the nest. The flowchart for foraging without communication just differs in that no waiting at the nest and no information distribution are incorporated. With the local communication model message exchange would be allowed in any state and waiting at the nest is skipped.

3   Evaluation Scenarios

Test runs on real hardware are captured on digital video and corresponding report messages from all robots are logged on a PC. A report message contains the unique ID of the robot and information on the latest events:

- No new way found or known food source not recovered
- New way to food source found or known food source successfully recovered
- ID of communication partner and transmitted content
- Error messages

All report messages about attempts of locating or checking food sources and propagation of path information contain the corresponding path information.

The test area size is about $5 \times 5m^2$. Landmarks and physical obstructions are placed in various layouts. Physical obstructions prevent transitions between locations and block sight onto landmarks, positioned behind the obstructions.

Test scenarios for the simulation differ in the size of the simulated environment which is represented by the quantity and arrangement of vertices (also called nodes) in the graph. Food sources vary in their distribution within the environment and in the amount of prey each source holds. For all scenarios, test runs are performed with each of the communication models. The simulator can log the following results, depending whether the task is to completely deplete all food sources or to forage for a predefined amount of time:

- Time it takes to deplete all food sources
- Number of sent messages during a depletion task

- Amount of prey collected during a given time frame
- Messages sent during given time frame
- Attempts of food searches
- Number of successful search attempts
- Attempts of food source checks
- Number of successful checks

4   Results

The simulation is run with various settings, each configuration performed at least 1000 times, to average out the influence of randomness. Simulated environments are sized from small environments with 20 vertices, up to large environments with 416 nodes. Simulated swarms consist of at least one and a maximum of 50 robots. The amount of prey per food source location varies from one item minimum and 150 maximum. All prepared setups are tested with the three presented communication modes. Two criteria are chosen: the task to completely clear a simulated environment from all food items and the task to collect prey for a given amount of time.

It turns out in results from the simulations that the presence of communication can have significant negative influence on the performance of a foraging swarm if the task is to clear an environment completely from all prey. Results from a test run with 15 robots in an environment of size 90 vertices and with 150 food items, equally distributed over 30 food locations, are shown in Fig. 2. The swarm with

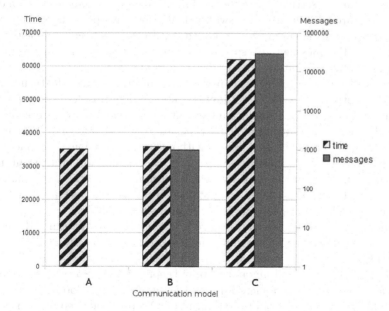

Fig. 2. Times and messages from simulation results for no communication (A), spatially restricted communication (B), and local communication (C)

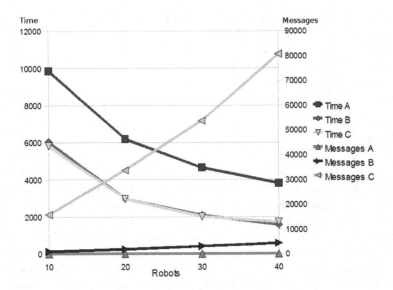

Fig. 3. Message and speed performances of various swarm sizes. No communication (A), spatially restricted communication (B), local communication (C).

restricted communication performs only slightly worse than a swarm without communication. The swarm with local communication needed nearly 1.8 times as long to fulfill the task and nearly 500 times as many messages were sent. Such results occur in any setting where the food locations contain only a small amount of prey compared to the size of a swarm (five prey items per source are placed in the the test run, shown in Fig. 2). Low quantity sources become depleted fast and false negative messages are spread more often in the swarm. This results in an increase of negative attempts to relocate a food source.

When food sources provide a greater amount of prey, swarms featuring communication outperform swarms without communication. Results from tests with different swarm sizes in an environment with rich food sources are shown in Fig. 3. The food sources contain 80 items per source. Both swarms with communication finish twice as fast as a swarm without communication. With increased swarm size the message increase in the local communication swarm outgrows the message increase in the spatially restricted model while the performance is nearly equal. Equivalent results are obtained from all other settings with rich food sources and the task to completely clear a simulated area.

Results from tasks where swarms have to collect for a predetermined amount of time in environments with recovering food sources yield that swarms with communication outperform a swarm without communication in most cases. Negative influences in cases of small amount of prey keeping food sources seem to affect the performance less than a clear all task with no recovering food sources.

Fig. 4 shows the amount of prey a swarms of 40 robots collected in a very large environment with 416 vertices after 3000 simulation steps. Both swarms with communication again outperform the no communication swarm. It is interesting that

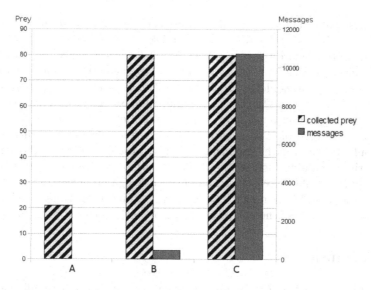

Fig. 4. Amount of retrieved prey and number of sent messages after 3000 simulated steps for. No communication (A), spatially restricted communication (B), and local communication (C).

the communication swarms perform almost similarly, with the prominent difference in the number of sent messages. Simulation runs with other setups for the same task yield similar results. In a test run with an environment consisting of 140 vertices, 15 robots foraged for 3000 simulation steps. Taking the collected prey with the no communication model as a reference, the swarm with the spatially limited communication harvested 30% more prey. The swarm with the unrestricted local communication retrieved 20% more prey than the one with spatially limited communication but 17 times as many messages were sent.

For appraising the functionality of the concept and the weaknesses in real world scenarios, report messages and videos from 50 test runs with up to 16 E-Pucks are examined. Small setups consist of a nest and three landmarks. Large test runs involve up to 24 landmarks. In general, the robots are able to locate and recover food sources. Successful robots also spread information about paths among other E-Pucks in the nest location with the spatially limited communication model, as expected. Robots that received messages are able to recover the target food sources.

It turns out that the major issue are hardware-related errors, resulting in performance loss in terms of time spent on actions without contributing results to the foraging performance. In about seven percent of all attempts of object detection with the camera the result is corrupted. The errors are either false positive or false negative detection of a food source or landmark. The false positive detection of a landmark has the major negative influence on the performance of the swarm. All robots that receive a message from a robot with a false positive detection try to check that food source later, spending operation time on a worthless path. The same phenomenon occurs when food sources become depleted. If n robots keep knowledge of a food source then all n robots will check a depleted source, before the information is invalid in the collective knowledge of the swarm. This effect can be seen as a disadvantage if the priority is to collect prey as fast as possible, for up to $n - 1$ additional useless attempts are made to relocate a food source. On the other hand, the absence of messages containing information about outdated food sources helps clearing all known food sources completely. In case of false negative detection the spreading of such information would kill valid paths from the swarms knowledge and cost additional time to find the remaining food items in that location by chance again. Also negative feedback would increase the number of propagated messages, which is contrary to the aim of this model. The existence of only positive feedback plays an important role in case the complete saving of food sources is of higher priority.

5   Conclusion

The test on low cost educational robots allowed to easily evaluate influences from the real world and communication model performance was analyzed in a simulation. The simulation results yield that spatially limited communication to a single nest-like location can speed up the process of foraging. Further increasing the number of places where communication may take place and, thus, increasing the number of messages like employing unrestricted local communication seems not to have significant influence on the foraging performance of a swarm. The restricted model outperformed the conventional communication approach in almost every test run, in terms of fewer sent messages and mostly a nearly equal amount of retrieved prey.

Besides the positive effects of spatially limited communication, in some cases the presence of communication decreases the foraging performance. This was mainly observed in environments where food sources kept a little amount of prey, resulting in the propagation of messages, containing outdated information. This leads to an increase of failing tries to rediscover a food source and slows down the foraging performance of a swarm. When an environment needs to be completely cleared from items, knowledge about the amount of prey in food locations would help to make the right decision whether communication should be incorporated. The same effect can arise from false positive detection of food sources, as shown in the real hardware test runs. In such cases the communication model has negative effects and suffers from the absence of messages containing information on depleted food sources.

As stated in [12] and [13], the complexity of messages is not the main factor to speed up a process of cooperative foraging robots. In addition, the results presented here allow the conclusion that the quantity of places where communication is allowed may be reduced to a single nest-like area to speed up the process of multi-robot foraging and reduce the amount of messages compared to common forms of communication.

Acknowledgment

This work was funded in part by the German Research Foundation (DFG) within priority programme 1183 under grant reference MA 1412/8-1.

References

1. Hsiao, Y.-T., Chuang, C.-L., Chien, C.-C.: Computer network load-balancing and routing by ant colony optimization. In: Proc. Networks, 12th IEEE International Conference on (ICON 2004), vol. 1, pp. 313–318 (2004)
2. McLurkin, J.: Distributed Algorithms for Multi-Robot Systems. Information Processing in Sensor Networks. In: 6th International Symposium on IPSN 2007, pp. 545–546 (2007)
3. Gerardo, B.: From Swarm Intelligence to Swarm Robotics Swarm Robotics, 1–9 (2004)
4. Dudek, G., Jenkin, M., Milios, E., Wilkes, D.: A taxonomy for swarm robots Intelligent Robots and Systems. In: Proceedings of the 1993 IEEE/RSJ International Conference on IROS 1993, vol. 1, pp. 441–447 (1993)
5. Lerman, K., Galstyan, A.: Mathematical Model of Foraging in a Group of Robots: Effect of Interference. Autonomous Robots 13(2), 127–141 (2002)
6. Sommer, S., von Beeren, C., Wehner, R.: Multiroute memories in desert ants. Proceedings of the National Academy of Sciences 105(1), 317–322 (2008)
7. Koenig, S., Szymanski, B., Liu, Y.: Efficient and Inefficient Ant Coverage Methods. Annals of Mathematics and Artificial Intelligence 31 (2001)
8. Dyer, F.C.: The biology of the dance language. Annu. Rev. Entomol. 47, 917–949 (2002)
9. Yoshida, E., Arai, T., Yamamoto, M., Ota, J., Kurabayashi, D.: Evaluating the efficiency of local and global communication in distributed mobile robotic systems. In: Proceedings of the 1996 IEEE/RSJ International Conference on Intelligent Robots and Systems 1996, IROS 1996, vol. 3, pp. 1661–1666 (1996)
10. Arai, T., Yoshida, E.: Design of Local Communication for Cooperation in Distributed Mobile Robot Systems. In: International Symposium on Autonomous Decentralized Systems (1997)
11. Ohkawa, K., Shibata, T., Tanie, K.: Method for generating of global cooperation based on local communication. In: IEEE/RSJ International Conference on Intelligent Robots and Systems. Proceedings, vol. 1, pp. 108–113 (1998)
12. Balch, T., Arkin, R.C.: Communication in reactive multiagent robotic systems. Autonomous Robots 1, 27–52 (1994)
13. Rybski, P.E., Larson, A., Veeraraghavan, H., Lapoint, M., Gini, M.: Communication strategies in Multi-Robot Search and Retrieval: Experiences with MinDART. In: Proc. 7th Int. Symp. Distributed Autonomous Robotic Systems, pp. 301–310 (2004)
14. Chang, P., Krumm, J.: Object Recognition with Color Cooccurrence Histogram. In: Proc. CVPR 1999 (1999)

AMiRESot – A New Robot Soccer League with Autonomous Miniature Robots

Ulf Witkowski[1], Joaquin Sitte[2], Stefan Herbrechtsmeier[1], and Ulrich Rückert[3]

[1] Electronics and Circuit Technology, South Westphalia University of Applied Sciences, Soest, Germany
witkowski@fh-swf.de
[2] Faculty of Science and Technology, Queensland University of Technology, Brisbane, Australia
j.sitte@qut.edu.au
[3] System and Circuit Technology, Heinz Nixdorf Institute, University of Paderborn, Paderborn, Germany
hbmeier@hni.upb.de

Abstract. AMiRESot is a new robot soccer league that is played with small autonomous miniature robots. Team sizes are defined with one, two, and three robots per team. Special to the AMiRESot league are the fully autonomous behavior of the robots and their small size. For the matches, the rules mainly follow the FIFA laws with some modifications being useful for robot soccer. The new AMiRESot soccer robot is small in size (maximum 110 mm diameter) but a powerful vehicle, equipped with a differential drive system. For sensing, the robots in their basic configuration are equipped with active infrared sensors and a color image sensor. For information processing a powerful mobile processor and reconfigurable hardware resources (FPGA) are available. Due to the robot's modular structure it can be easily extended by additional sensing and processing resources. This paper gives an overview of the AMiRESot rules and presents details of the new robot platform used for AMiRESot.

1 Introduction

The AMiRESot league has been defined for robot soccer matches with autonomous behavior using small robots. Team size can be 1, 3, or 5 robots. The robots have to percept their environment independently, e.g. no camera is used outside the play field for capturing the current scene from top view as done in some other robot soccer leagues. In addition, all processing has to be performed locally on the robot. For years KheperaSot was the main robot soccer league for miniature robots [1,2]. But due to the outdated processing hardware of the robot (Khepera type I and II) and low quality sensing KheperaSot is no more present at international championships. The AMiRESot aims to be the successor of the KheperaSot. Mayor enhancements are a new modular robot with up to date hardware for sensing and processing, and the increased team size of 3 vs 3 and 5 vs 5 in contrast to 1 vs 1 only. The AMiRESot tournament is intended

J.-H. Kim et al. (Eds.): FIRA 2009, CCIS 44, pp. 332–345, 2009.

for wheeled robots, because wheeled robots are easier and cheaper to build than humanoid walking robots. Wheeled robots are as useful as humanoid robots for developing all the necessary real time environment perception capabilities and cooperative behaviors required for a soccer game with a high level of realism.

The robots for the AMiRESot match are fully autonomous. There is no global vision system and communication with a field side computer is not necessary (and not allowed for robot control) during the game. The referee gives whistle signals for starting, halting and stopping the game, and for announcing penalties for rule infringements. The robots need to recognize these whistle signals. Although the intention is that the robots have on-board vision, there are no restrictions as to what sensors can be used as long as they do not interfere with similar sensors on the other players. The main characteristic of AMiRESot is the limitation on the size of the robot soccer players to a maximum diameter of 110 mm. This allows the field of play to be small enough to not require a large space. The play-field can be quickly and easily set up in the home, the school or small university laboratory or corridor. The ball is a Squash ball. The AMiRESot game can be played by teams of 1 or more robots each, as long as the contending teams have the same number of players. Reasonable team sizes would be 1, 3 or 5 players. The AMiRESot 2008 rules follow the FIFA Laws as close as it makes sense for the type of robot players and the field of play described herein. The name of the new AMiRESot league originates from the AMiRE symposium in 2007. The guidelines for the AMiRE Soccer Tournament (AMiRESot 2008) rules 2008 were conceived at the workshop held at the 4th AMiRE Symposium in Buenos Aires in October 2007 with the objective to push autonomous minirobot technology another quantum step further while keeping the tournament affordable. The paper is organized as follows. The next section gives an overview of the AMiRESot rules. Section 3 concentrates on the robot platform including processing devices and sensors for environment perception. Section 4 shortly introduces options of future robot soccer tournaments.

2   AMiRESot Rules

The AMiRESot 2008 rules follow the FIFA laws whenever it is possible. But some modifications are necessary to be useful in robot soccer. For more details of the robot soccer rules not completely described in the paper please refer to the document available in the web [3].

2.1   The Field of Play and the Ball

The playing field is 2000 mm long (touch lines) and 1400 mm wide (goal lines) with cut-off corners to avoid the ball getting cornered, cf. Fig. 1. The playing field is enclosed, except for the goal opening by a white frame 20 mm high and 20 mm wide. The surface of the playing field has a dark green felt cover. Markings on the field are white lines 10 mm wide. The playing field will have the following markings: The half-way line, parallel to the goal lines, that divides the playing

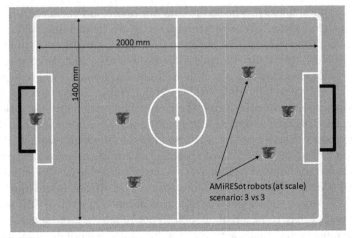

Fig. 1. Play field for AMiRESot. Team size is 1, 3, or 5 robots.

field into equal sized areas. The centre of the halfway line is marked by a filled circle of 15 mm radius. The goal area is a rectangular area in front of each goal that extends 110 mm to each side form the inside of the goal posts and 110 mm from the goal line into the play field. The centre circle has a radius of 200 mm. The goals have an inside width of 400 mm and a depth of 110 mm. The goals are enclosed from on the side and the rear by 40 mm high barriers painted flat black The goal barrier locks into the field barrier on each side so that the field barriers are flush with the goal opening.

The field can be indoors or outdoors. The playing field should be located in a well lit area. The lighting should be diffuse without casting any strong shadows such as provided by office style overhead fluorescent lighting. The ball is a white squash ball.

2.2   The Players

The players are fully autonomous robots. A player must fit into gauge cylinder of 110 mm internal diameter. There is no limitation on the height of the player. The game may be played by teams consisting of one, three or five players. In a game the number of players must be the same in each of the two teams. In games with more than one player per team, any player in the team may act as goalkeeper. The goalkeeper can only be changed during a stoppage in the match and with the permission of the referee. Players will be distinguished by their jersey. Jerseys are tubes of elastic fabric pulled over the robot with team. The jerseys are painted with a vertical stripe pattern of the team colours. The goalkeeper has a uniform coloured jersey.

2.3   The Player's Equipment

The players must not have any mechanism that protrudes from the convex hull of the player at any time during the game. Neither are they allowed to have any device that will interfere with the sensory system of other robots. Active infrared and ultrasound proximity sensors are allowed. The body of the player may not have a concave depression into the body's convex hull below 40 mm above ground level. A player may only exert a force on the ball through pushing with direct contact of the player's body. In other word no active kicking mechanism is allowed. Players must not be capable to inflict undue damage on collision with other players, at the discretion of the referee.

2.4   The Referee

Each match is controlled by a referee who has full authority to enforce the Laws of the game in connection with the match to which he has been appointed. The referee has the same function as in the FIFA rules:

- enforces the laws of the game
- ensures that the ball meets the requirements
- ensures that the players' equipment meets the requirements
- acts as timekeeper and keeps a record of the match
- stops, suspends or terminates the match, at his discretion, for any infringements of the laws
- stops, suspends or terminates the match because of outside interference of any kind
- allows play to continue when the team against which an offence has been committed
- restarts the match after it has been stopped

One or more assistant referees may be appointed for a match. The duties of the assistant referees, subject to the decision of the referee, can be reviewed in the complete rule document [3].

2.5   Duration of the Match

The match lasts two equal periods of 10 minutes. There will be a 5 minute interval between the two periods of play. The duration of the half-time interval may be altered only with the consent of the referee. Allowance is made in either period for all time lost through (1) substitution(s) or removal of inoperative players from the field of play, (2) assessment of faulty players and eventual repair in less than 30 seconds, and (3) any other cause.

The allowance for time lost is at the discretion of the referee. If a penalty kick has to be taken or retaken, the duration of either half is extended until the penalty kick is completed. There will be not extra time except for the allowance of lost time. If one team becomes unable to play a match due to malfunctioning of players the referee may (1) terminate the match and declare the other team to be the winner of the match, or (2) cancel the match and order a reply of the match at a later time.

2.6   Start and Restart of Play

A coin is tossed and the team which wins the toss decides which goal it will attack in the first half of the game. The other team takes the kick-off to start the match. The team that wins the toss takes the kick-off to start the second half of the match. In the second half of the match the teams change ends and attack the opposite goals. A kick-off is a way of starting or restarting play

- at the start of the match,
- after a goal has been scored,
- at the start of the second half of the match

Procedure for kick-off: All players are in their own half of the field, the opponents of the team taking the kick-off are outside the centre circle until the ball is in play. The ball is stationary on the centre mark. The referee gives the whistle signal. The ball has to be kicked forward into the side of the defending team. In this case kicking means literally kicking the ball not pushing it. The ball is in play when it has been kicked and moves forward. Unless the ball is in play all players stay on their own half of the field. The kicker does not touch the ball a second time until it has touched another player. A goal may be scored only when the ball was touched by another player. After a team scores a goal, the kick-off is taken by the other team. Infringements/Sanctions (for kick-off): If the kicker touches the ball a second time before it has touched another player: An indirect free kick is awarded to the opposing team to be taken from the place where the infringement occurred. For any other infringement of the kick-off procedure the kick-off is retaken. A dropped ball is a way of restarting the match after a temporary stoppage which becomes necessary, while the ball is in play, for any reason not mentioned elsewhere in the Laws of the Game. Procedure (for dropped ball): The referee puts the ball at the place where it was located when play was stopped. Play restarts when the referee gives the whistle signal.

2.7   The Ball In and Out of Play

The ball is out of play when: (1) it has crossed the barrier around the field of play, (2) play has been stopped by the referee, and (3) before kick-off. The ball is in play at all other times.

2.8   Method of Scoring

A goal is scored when the whole of the ball passes over the goal line, between the goalposts, provided that no infringement of the laws of the game has been committed previously by the team scoring the goal. The team scoring the greater number of goals during a match is the winner. If both teams score an equal number of goals, or if no goals are scored, the match is drawn.

2.9   Offside

It is not an offence in itself to be in an offside position. A player is in an offside position if it is nearer to his opponents' goal line than both the ball and the second last opponent (which is normally the goal keeper). A player is not in an off-side position if:

– it is in his own half of the field of play,
– or it is level with the second last opponent,
– or it is level with the last two opponents.

Offence: A player in an off side position is only penalized if, at the moment the ball touches or is played by one of his team, he is, in the opinion of the referee, involved in active play by:

– interfering with play,
– or interfering with an opponent,
– or gaining an advantage by being in that position.

No offence: There is no offside offence if a player receives the ball directly from: a goal kick, or a throw-in, or a corner kick. Infringements/Sanctions: For any offside offence, the referee awards an indirect free kick to the opposing team to be taken from the place where the infringement occurred.

2.10   Fouls and Misconduct

No direct free kicks are awarded. No penalty kicks are awarded. An indirect free kick is awarded to the opposing team if in the opinion of the referee or an assistant referee a player does any of the following:

– Deliberately runs towards and consequently collides with an opponent. There is no offence if the collision occurs due to the opponent accidentally crossing the path of the player.
– Pushes against another player for more than 5 seconds.

The indirect free kick is taken from where the offence occurred.

2.11   Free Kicks

Only indirect free kicks are awarded. The ball must be stationary when the kick is taken and the kicker does not touch the ball a second time until it has touched another player. The referee indicates an indirect free kick by two successive short whistle blows. When the free kick has been taken and the ball has touched another player. The referee signals the continuation of the game with a start of game whistle signal (one short blow). If the ball goes out of play the referee signals the halting of the game by a long whistle blow. If the ball enters the goal no goal is scored and the game is started by the dropped ball method at the position of the offence for which the indirect free kick was awarded. An indirect free kick awarded inside the goal area is taken from that part of the goal area line which runs parallel to the goal line, at the point nearest to where the infringement occurred.

3   Robot Platform

The new robot platform to be used in the AMiRESot league has been decided to be small in size, modular, extendable, and cost efficient. Following the rules, the maximum diameter of a robot is 110 mm, the height is not limited by the rules, but for practical use a too high robot reduces stability and limits the dynamics. The following sections give first insight into the prototype development of the robot and its features.

3.1   Robot Chassis

The AMiRESot robot has a round body and is covered with Plexiglas or other type of plastics. A prototype of the robot is depicted in Fig. 2. The robot has a diameter of 100 mm, a little bit less than the allowed maximum diameter of 110 mm. The lower part of the robot, i.e., drive system and cylindrical cover are considered to be the final version. But the depicted upper part is used for testing only. The final hardware will optimally fit into the robot body and the robot will be equipped with infrared sensors, an image sensor, and processing hardware integrating a powerful mobile processor and FPGA hardware resources. Holes for the infrared sensors have been already fabricated about 2 cm above floor (see Fig. 2).

Fig. 2. Prototype chassis of the AMiRESot robot (shown electronics is used for testing only, final version will be different)

3.2   Robot Drive

Before fixing the drive system three different solutions have been physically tested. These are a differential drive, a omni-directional system with three wheels, and a chain drive, see Fig. 3. The most robust drive is the chain drive, but it has limited dynamics and requires most electric power. The omni-directional drive offers the most flexible maneuvers, these are combined rotation and translation, but this drive is more difficult to control and path integration based on odometry has poor accuracy. Therefore we decided to realize a differential drive system with two wheels (Fig. 3 most left and Fig. 2). The achievable mobility is sufficient at high accuracy and high speed. Another advantage is a relatively simple control model. The integrated motors are DC type with gear box (1:22). The maximum speed is about 50 cm/s with wheels of 4 cm diameter. With our optimized controller the maximum speed is reached in less than 0.2 s at highest acceleration.

Fig. 3. Evaluated drive systems: differential drive, omni drive and chain drive

3.3   Robot Sensors

Basic sensors of the AMiRESot robots are active infrared sensors. I.e., short infrared pulses are emitted in a non overlapping scheme and the reflected signal is detected by the receiver. In total 12 sensors are equally arranged along the perimeter of the robot. The center of the sensing cone is about 2 cm above the ground (see holes in Fig. 2) Fig. 4 depicts the sensing principle when the robot is inside a corridor. This setup has been used for testing only; the two close walls are not used in AMiRESot. Fig. 4 (right part) shows the sensor characteristics. The highest sensitivity is in the range of 5 to 15 cm. The total range is up to 35 cm. Main objective for the infrared sensors are near range obstacle (opponent and wall) detection as well as the use of the sensors for realization of the ball pushing behavior.

The infrared sensors are attached at the inner surface of the chassis. To realize electric connections, a PCB has been developed and fabricated that consists of rigid and flexible sections as shown in Fig. 5. This technology ensures reliable connections at very small space.

For long range sensing in general, detection of opponents, identification of field lines, and ball recognition a tiny color image sensor is used, see Fig. 6. The maximum resolution is 1.3MPixel, but usually a smaller resolution is sufficient, e.g. 320 × 240 pixels, for object detection an recognition. This saves processing

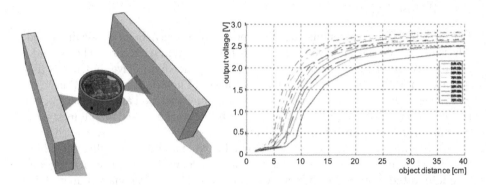

Fig. 4. Active infrared sensors for near range sensing (e.g. for obstacle and ball detection) and sensor characteristics: range up to 35 cm

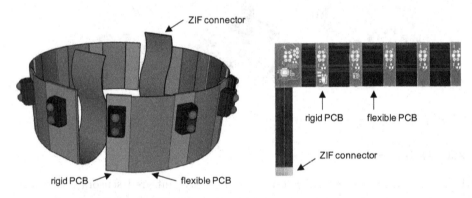

Fig. 5. Hardware realization of the infrared sensor ring with 12 sensors: two rigid/flexible PCBs with sensors and auxiliary devices

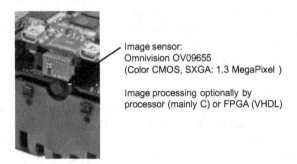

Image sensor:
Omnivision OV09655
(Color CMOS, SXGA: 1.3 MegaPixel)

Image processing optionally by
processor (mainly C) or FPGA (VHDL)

Fig. 6. Color image sensor for long distance sensing and detailed environment sensing: 1.3 MPixel color sensor with connection to the main processor and to the FPGA

resources of the robot, because processing has to be performed on the robot platform. The image sensor is placed at the front side of the robot approximately 5 cm above the floor. A frame rate of (up to) 30 fps (VGA, CIF) is supported. Resolution and frame rate can be easily selected by our camera driver available in the Linux operating system. For image processing, the robot's main processor or the (optionally) integrated FPGA can be used.

As an additional sensor an IMU (inertial measurement unit) is integrated into the robot's base module. The IMU can be used to increase the accuracy for short term path integration in addition to pure usage of incremental encoders of the motors. Another advantage is that passive movements can be detected if a robot is pushed by another robot. For advanced long range sensing a scanning laser range finder can be optionally used. We have prototypically integrated the Hokuyo URG-04LX.

3.4 Robot Information Processing

The hardware architecture has a modular structure. This reduces costs and eases the exchange of outdated components in order to support new developments of processing devices as well as sensors. As depicted in Figure 7 there are two main hardware levels. Level 0 is responsible for controlling the motors and capturing of sensor data. Several AVR micro-controllers are used in parallel to control motors and sensor devices and to preprocess sensor data. Main sensors of the bottom part are the active infrared sensors, incremental encoder and optionally a laser

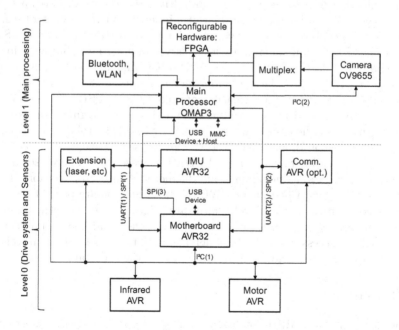

Fig. 7. Modular hardware architecture with level 0 and level 1 processing platforms

scanner. Usually the communication of robot is done via the OMAP processor (level 1), but it also possible to use a Bluetooth device that is connected at level 0 to the motherboard AVR.

At level 1 all higher processing tasks are executed. This comprises image processing of captured images (see block camera in Fig. 7), realization of the local behavior of a robot and overall strategy planning. Image processing can be done by using the OMAP processor or the FPGA. The advantage of using the processor for image processing is the availability of high level programming languages (e.g. C, C^{++}) including already available image processing libraries. In contrast, the FPGA is able to perform parallel image processing or every other type of complex processing that speeds up processing and may lead to increased overall performance.

3.5   Software: Operating System and Simulation

The software environment of the robot is a Linux operating system. It consists of a modified Linux kernel 2.6.24 [4], the standard GNU C Library and the device manager udev. The standard Unix tools were provided by the software application BusyBox [5]. BusyBox combines tiny versions of many common Unix utilities in a single small executable. The software building is done via OpenEmbedded [6]. This is a development environment which allows the creation of a fully usable Linux operating system. It generates cross-compile software packages and images for the embedded target. The existing software branch was extended with an overlay. This contains robot specific information, patches and additional software like the Player network server and drivers for the robot hardware. The Player project [7] is a language and platform independent robot control system. It consists of the Player client/server model and the platform simulators Stage and Gazebo. Player implements a Player server and a Player library which is used to build the Player client. Stage is a 2D robot simulator with a player interface. Together they allow platform independent simulations and code execution on real robots. The robot can be easily connected via an USB cable to any PC. This mode can be used for programming and simple debugging. For extended testing of the robot the cable hinders the maneuvers of the robot. In this case a wireless link based on Bluetooth or WLAN can be used. One option is to connect to the robot via a Bluetooth equipped PDA that eases for example set-up and calibration of the robot including some diagnosis. Another option for wireless access is TCP/IP via Bluetooth or WLAN, both options are part of the basic hardware configuration of the level 1 hardware (Fig. 7). In this case the robot can be easily integrated in existing networks with a PC using ad-hoc communication mode or via an access point. Various debugging options including transmission of captured images or remote (re)programming of the robot are supported.

4   Tournaments

Robot soccer in the AMiRESot league can be played at international tournaments like the annual FIRA Roboworld Cups [8]. But due to the fact that the

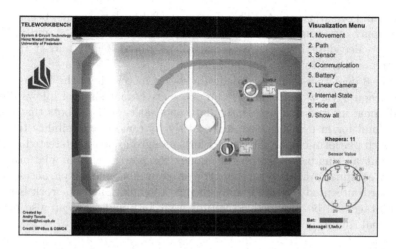

Fig. 8. Robot soccer (here KheperaSot setup) on the Teleworkbench platform. The robots can be programmed via the Internet. Matches can be started and monitored online via a web browser as well as recorded for detailed match analysis.

soccer rules are not limiting to a specific type of robot different types of robots can be used meeting the rules. There are only a few global limitations like the maximum diameter of the robot of 110 mm and the autonomous behavior of a robot during the match. These freedoms enable the use of relatively simple and also cheap robots as well as high-tech robots equipped with latest processing and sensing devices. Therefore, AMiRESot can be played on one side at schools with affordable robots. On the other side more powerful and also more expensive robots can be used at universities and research labs. It is planned to consider the new AMiRE robot platform as an open platform after the development has reached a stable version in mid of year 2009. I.e., construction details concerning chassis and electronic hardware and basic software will be available for free (announcement via the AMiRE symposium webpage: www.amiresymposia.org). In addition to the AMiRE soccer robots we will make available our Teleworbench platform for remote execution of AMiRESot matches and detailed match analysis [9,10]. The Teleworkbench system consists of a 2m × 2m field on which the robot experiments and soccer matches take place. In addition, the system comprises various cameras and servers for managing, recording, and analyzing the experiments. For AMiRESot the platform meets the field dimensions and shape as given in the rules. The system is accessible via the Internet, so that local or remotely located users are able to set up and run matches. Through the web based user interface, users can schedule matches and set programs to be downloaded to each robot individually. Via Bluetooth modules and WLAN robots can exchange messages to each other or to the Teleworkbench server wirelessly, that can be used for team play or for extended match analysis. These messages along with some other occurring events are logged and used later for analysis purpose. During matches, cameras are used to track the robots on the field to provide position and orientation data. This data is not fed back to robots, because they have to play autonomously, but it is useful for match recording and analysis. Also the video data from the cameras will be stored locally and at the same time streamed as live video via the Internet.

Figure 8 depicts a scene recorded of a soccer match (here the former KheperaSot league) via a top camera. Our interactive analysis tool allows display of additional information like sensor data and paths of the robots that can be very helpful for in depth match analysis. By its remote accessibility the Teleworkbench platform can be interesting for those group being interested in robot soccer but are unable to spend the money for powerful robot platforms like the fully equipped AMiRESot robot. For programming and simple access to the robot hardware used on the Teleworkbench a simulation framework based on Player/Stage will be provided. Programs used for simulation can also be used for the real robots.

5   Conclusion

We have presented first details of the new AMiRESot robot soccer league, mainly the rules of the match and construction details of the AMiRESot robot. Special to this league compared to other leagues is the small size (max. 110 mm diameter) of the robots and a fully autonomous behavior of the robots with team sizes ranging from 1, 3 to 5 robots. This paper aims to motivate robot soccer enthusiasts to force discussions concerning the AMiRESot league and to participate in the further development of this league comprising rule verification and feedback to the robot features. The idea is to arouse interest in robot soccer, not only at universities, but also at schools to enlarge the user group of robot soccer. To comply with the different requirements and budgets of the user several levels of extensions of the soccer robot will be available ranging from cost efficient solutions with publicly accessible construction manuals to powerful robots supporting latest sensing and processing techniques. Common for all robot platforms is the size and the drive system that already includes infrared sensing and motor control. First demonstrations of the new robot platform will be shown at the FIRA roboworld congress in August 2009. Due to the recent launch of the AMiRESot league there will be no international tournament in 2009, but it is intended to have test matches at the end of 2009 and a big championship at the next FIRA roboworld congress in 2010. Details on the progress of the AMiRESot league will be published at the AMiRE web page (http://www.amiresymposia.org/).

References

1. Chinapirom, T., Witkowski, U., Rückert, U.: Stereoscopic Camera for Autonomous Mini-Robots Applied in KheperaSot League. In: Proceeding of the FIRA Robot World Congress 2007, on CD San Francisco, USA (June 2007)
2. KheperaSot, http://www.fira.net/soccer/kheperasot/overview.html

3. AMiRE: AMiRESot rules, http://www.amiresymposia.org
4. The Linux Kernel Archives, http://www.kernel.org
5. BusyBox, http://www.busybox.net/
6. OpenEmbedded, http://www.openembedded.org
7. The Player Project: Free Software tools for robot and sensor applications, http://playerstage.sourceforge.net
8. FIRA Roboworld Cup, http://www.fira2009.org
9. Monier, E., Witkowski, U., Tanoto, A.: Soccer Teleworkbench for Development and Analysis of Robot Soccer. In: Proceeding of the FIRA Robot World Congress 2007, San Francisco, USA (June 2007) (on CD)
10. Tanoto, A., Witkowski, U., Rückert, U.: Teleworkbench: A Remotely-Accessible Robotic Laboratory for Education. In: Proceeding of the Spring 2007 AAAI Symposium on Robots in AI and CS Education-Robots and Robot Venues, Resources for AI Education (2007)

BeBot: A Modular Mobile Miniature Robot Platform Supporting Hardware Reconfiguration and Multi-standard Communication

Stefan Herbrechtsmeier[1], Ulf Witkowski[2], and Ulrich Rückert[1]

[1] System and Circuit Technology, Heinz Nixdorf Institute, University of Paderborn,
Paderborn, Germany
hbmeier@hni.upb.de
[2] Electronics and Circuit Technology, South Westphalia University of Applied
Sciences, Soest, Germany
witkowski@fh-swf.de

Abstract. Mobile robots become more and more important in current
research and education. Especially small 'on the table' experiments attract interest, because they need no additional or special laboratory
equipments. In this context platforms are desirable which are small, simple to access and relatively easy to program. An additional powerful
information processing unit is advantageous to simplify the implementation of algorithm and the porting of software from desktop computers to
the robot platform. In this paper we present a new versatile miniature
robot that can be ideally used for research and education. The small size
of the robot of about 9 cm edge length, its robust drive and its modular
structure make the robot a general device for single and multi-robot experiments executed 'on the table'. For programming and evaluation the
robot can be wirelessly connected via Bluetooth or WiFi. The operating
system of the robot is based on the standard Linux kernel and the GNU
C standard library. A player/stage model eases software development
and testing.

1 Introduction

Mobile robots are more and more in the focus of current research, with almost
every major university having one or more labs that focus on mobile robot research. Autonomous robot research is now gaining a broader base by spreading
from a few well resourced laboratories into many small university laboratories
and even to the hobbyist work bench. This will undoubtedly accelerate the advancement of the field. Alongside, the number of primary and secondary school
students participating in robot competitions has increased enormously, reflecting the high educational and entertainment value of low cost autonomous mobile
robots. Small robots are also becoming increasingly useful as a test bed for animal behavioural research and for small scale prototyping of larger engineering
systems. We at the Heinz Nixdorf Institute are using mobile miniature robots
in different applications in various fields like path planning, area explorations,

J.-H. Kim et al. (Eds.): FIRA 2009, CCIS 44, pp. 346–356, 2009.

map building and localization using SLAMs, also in multi-robot experiments like robot swarming, ad hoc networking, robot soccer, etc. Therefore we need a powerful but also small robot that should be compact in size, suitable for various environments even with slightly rough surfaces, and offers extendibility to other modules. All these factors lead to the design of the BeBot miniature robot.

In the following sections, the mini robot BeBot is introduced in detail. Section 2 explains the platform itself, i.e. chassis and electronic hardware architecture. Section 3 focuses on the software environment. Special features of the robot, these are its communication framework and dynamic reconfiguration option, are presented in section 4. Section 5 give some current applications were the BeBot is used. The last section finish the paper with a conclusion.

2   Platform

The miniature robot platform BeBot (figure 1) has been developed at the Heinz Nixdorf Institute, University of Paderborn. It has a size of approximately 9 x $9cm^2$ and a height of about 5cm.

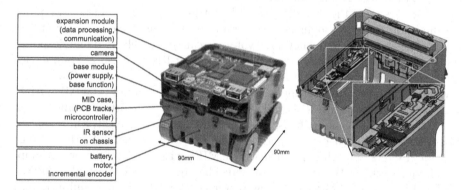

Fig. 1. BeBot mini robot (fully equipped)   **Fig. 2.** MID chassis of the BeBot robot

The chassis uses MID (molded interconnect device) technology and has traces directly on the surface [1] which offers new possibilities for the synergistic integration of mechanics and electronics. Figure 2 depicts the integration of electronic components on the plastic chassis. A microcontroller on the left and right side is used to control the infrared (IR) senders. Also the received IR signals are processed by the microcontroller and the digitized data is sent via I^2C link to the others processors used within the robot. The MID technology allows the assembly of electrical components directly on the device. This technique is used for mounting 12 infrared sensors and two microcontrollers, several transistors and resistors for preprocessing directly on the robot chassis. The actuation consists of a chain drive. Together with two 2W dc gear motors with built-in encoders the robot offers robust motion even on slightly rough ground. The complete system is supplied by a 3.7V / 3900mAh lithium-ion accumulator that allows a runtime of approximately 4 hours for a full equipped robot.

The robot uses a modular concept for information processing and has two slots for extension board. The lower board (base module in figure 1) implements basis functions like motor control and power supply. An ARM 7 based microcontroller allows low level behavior realization. The module also contains a three axis acceleration sensor, a yaw rate gyroscope and a sensor for battery monitoring. A possible application for the gyroscope and the 3D accelerometer are local navigation algorithms. Because of the high accuracy and low drift of these sensors they are able to provide lower errors than an odometry based on wheel encoders.

The upper slot (expansion module in figure 1) provides a more powerful information processing and wireless communication. It is equipped with a low power 520MHz processor, 64MB main and flash memory. An FPGA (field programmable gate array) enables the use of reconfiguration on hardware level. This allows the computation of complex algorithms through the use of dynamic co-processors. The integrated wireless communication standards ZigBee, Bluetooth and WLAN offer communication with various bandwidth and power consumption. The board provides a variety of additional interfaces, like USB, MMC / SD-card, audio, LCD and camera.

A new version of the expansion module optimized in terms of size and power consumption is in development. This supports different techniques for energy saving like dynamic frequency and voltage scaling as well as dynamic power down of non-used hardware components including RF processing. Main device of the new expansion module is Texas Instruments (TI) new OMAP 3 processor. This high-performance applications processor consists of a 600MHz ARM Cortex-A8 processor with NEON SIMD coprocessor. It supports dynamic branch prediction and has a comprehensive power and clock management, which enables high-performance, low-power operation via TI's SmartReflex adaptive voltage control. It offers more than 1200Dhrystone MIPS with maximal power consumption from less than 2W for the whole chip. The processor is connected to 512MB NAND Flash and 256MB mobile low power DDR SDRAM. It is equipped with the wireless communication standards Bluetooth and WiFi. Both have external antennas for better signal qualities and support power down to disable not needed communication devices. A coexistence solution ensures simultaneous operation of Bluetooth and WiFi. Both communication devices have a peak power consumption of less than 1W during continuous transmit over Bluetooth and WiFi. Additionally, the supported wired communication standards I^{2}C, SPI, UART and high speed USB allows variable expansion of the system. Via these interfaces other communication devices like ZigBee, Sub-1 GHz communication or UWB hardware or other components like sensors and actuators can be easily connected to the system enabling the robot meeting several demands. The main interface to a computer is a USB device interface.

From the information processing point a view the robot is an embedded system providing distributed processing. Besides the main processor (Marvel PXA270) three additional microcontrollers are available for distributed processing. For data exchange, the processor and microcontroller are connected by an I²C link. Closely coupled to the main processor an FPGA device (Xilinx XC3S1600E) has been integrated. It is connected to the memory bus enabling high bandwidth data exchange between FPGA, processor, and the memory devices.

The BeBot can be used in mainly two hardware configurations, not considering mechatronic extensions like a gripper or transporter at this point. In the minimal configuration, shown in figure 3, the robot is able to perform simple experiments. In this configuration the motors can be controlled - a speed controller based on integrated wheel encoders is already implemented. The microcontroller of the base board can get sensor data from the microcontrollers mounted on the chassis via the I²C link. Possible behaviors in this configuration are simple exploration strategies or behaviors known as Braitenberg behaviors [2]. Additionally, the robot can be remote controlled by using a Bluetooth wireless link. Control commands can be sent by a PDA as shown in figure 4. A feedback is given from the robot by sending the data from the infrared sensors to the PDA.

Fig. 3. Robot in minimal configuration equipped with base board

Fig. 4. Remote control of the robot via a PDA

Figure 1 depicts the BeBot robot fully equipped with hardware modules, both PCBs are inserted. In this configuration, the robot's hardware architecture corresponds to the architecture presented above. Powerful processing devices (processor and FPGA) are available for the implementation of complex algorithms. The advantage of using an FPGA device on this platform is discussed in the section 4.2.

3   Software Environment

The software environment of the robot is a Linux operating system. It consists of a modified Linux kernel 2.6.26, the GNU C standard library and the device manager udev. The standard Unix tools were provided by the software application

BusyBox. This combines tiny versions of many common Unix utilities in a single small executable. The software building is done via OpenRobotix [3]. This is an extension of the OpenEmbedded development environment which allows the creation of a fully usable Linux operating system. It generates cross-compile software packages and images for the embedded target. The existing software branch was extended to contain the robot specific information, patches and additional software like the Player network server and drivers for the robot hardware.

The Player project is a language and platform independent robot control system. It consists of the Player client/server model and the platform simulators Stage and Gazebo. Player implements a Player server and a Player library which is used to build the Player client. Stage is a 2D robot simulator with a player interface. Together they allow the platform and real / simulation independent robot programming.

The interface of the OS to the robot's hardware is divided into two parts. The first part is a kernel driver. This implements the low level I^2C communication and provides user space hardware control over virtual files in the Linux sys file system. Above, a Player driver uses these files and fulfills the second part. It implements the player driver class, makes number conversions and adds some additional information like sensor position and robot sizes. The whole system allows the controlling of the robot through the player interfaces. Additionally, a robot model for the Stage simulator allows the simulation of experiments with the same software interface.

The WiFi communication is directly supported through the Linux kernel and so supports all standard communication protocols. The Bluetooth communication is implemented by the BlueZ protocol stack and supports all standard Bluetooth protocols like RFCOMM and BNEP. Additionally, all Linux and platform independent or arm compatible protocol implantations can be ported to the robot platform. One example is the ad hoc wireless mesh routing daemon OLSRD. This implements the optimized link state routing protocol and allows mesh routing on any network device.

4   Special Features

The robot is supporting special features like wireless communication via several communication standards for synergetic combination. Additionally, partially dynamic reconfiguration is supported to maximize hardware utilization and to cope with restricted resources of the mobile robot. Both features are supported by the robot's operating system to ease program development.

4.1   Wireless Communication

Three types of wireless communication are supported by the mini robot BeBot. Bluetooth and ZigBee are directly integrated onto the robot's extension board. Communication via WiFi is realized by connecting a WLAN device to a USB connector. The implemented communication standards differ in network size, radio range, data rate and power consumption. Wireless LAN is suited to high data rate and high range communication at the cost of high power consumption. Bluetooth has a lower data rate and transmission range but in turn significantly lower power consumption. ZigBee is highly scalable with even lower power consumption but with a trade off for lower data rates. One major challenge in wireless ad hoc networks, particularly in mobile ad hoc networks, is the design of efficient routing algorithms. All integrated communication devices are supported by the Linux operating system running on the robot. It is possible to directly access the communication devices. But in order to ease the access to the communication devices, a network abstraction layer and a communication framework on top have been integrated.

4.2 Dynamic Reconfiguration

The robot supports dynamic reconfiguration of its hardware during runtime [4]. Reconfiguration capabilities are provides by the FPGA device. Several types of reconfigurations are supported. At startup of the robot, the FPGA is programmed by the processor with the contents of the Flash memory. This allows loading different hardware configurations for the FPGA at power-on of the robot to perform build-in routines e.g. self-test, demonstration mode etc. The FPGA is capable of dynamical reconfiguration, so that parts of the hardware design on the FPGA can be exchanged on demand by new modules, kept in Flash, SDRAM or received wirelessly by Bluetooth or WiFi via a network link.

Modern FPGAs are heterogeneous architectures constituted by programmable functional blocks and embedded application specific integrated hardware (e.g., embedded processors, SRAM memory, dedicated multipliers) interconnected by a reconfigurable network. The configuration and interconnection of the internal resources determine the functionality of the implemented design. This configuration is provided by a bitstream file, which is loaded at start-up. Some FPGAs can be partially reconfigured during run time. A partial bitstream, targeting a specific area of the FPGA is loaded while the rest of the FPGA can still operate without interruption. This process is known as dynamic reconfiguration, which can be used to enhance the resource-utilization of an FPGA by time-sharing logical resources among different designs (event-driven reconfiguration) or by time-multiplexing a design requiring a bigger amount of resources than available (virtual hardware) [5]. The event driven concept is explained in the following paragraph.

Given the resource limitations of FPGAs, it is not possible to realize a large number of algorithms using only static designs. Furthermore, it would be a waste of resources to implement these algorithms statically if they are not needed the whole time. Event-driven dynamic reconfiguration can be modeled as finite state machine where every state represents a different configuration of the hardware. Mutually exclusive configurations (e.g., designs that are not needed at the same time) time-share the same hardware slot on the dynamically reconfigurable area, where several slots can coexist (e.g., there are several non-mutually exclusive designs). Event-driven dynamic reconfiguration makes it possible to adapt, during run time, the behavior of the system without wasting resources.

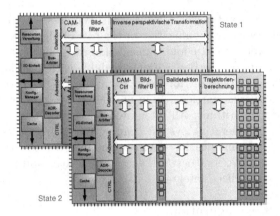

Fig. 5. Dynamic reconfiguration from state 1 to state 2

An example of the event-driven dynamic reconfiguration is shown in figure 5. Two simple states have been defined. The considered scenario is from robot soccer where the robot has the task to orientate itself in the soccer pitch, to detect the ball and to calculate trajectories to push the ball into the desired direction. The perception is done by 2D CMOS color camera. In state one, the FPGA is configured with a camera controller for image capturing, one image filter algorithm and a complex unit for calculating the inverse perspective transformation [6]. This calculation is done to get the robot's position based on the lines in the pitch. After the calculation of the robot's positions has been done, the FPGA is reconfigured, which means, that another filter is loaded and algorithms for ball detection and calculation of robot's trajectories are instantiated. This is state two in figure 5. If it's again necessary to calculate an updated position of the robot, the configuration is switched to state 1. By doing this reconfiguration, complex algorithms can be parallel executed in hardware with good resource usage.

5   Applications

The robot has been successfully used in education of students and research projects in the field of robotics. In the following paragraphs a student project is presented as well as the EU funded GUARDIANS research project. Furthermore, the robot BeBot is used as a platform in the SFB614 "Self-Optimizing Concepts and Structures in Mechanical Engineering".

5.1   Mechatronic Seminar

The aim of this student project was the development and implementation of a control strategy that makes it possible for a group of robots to drive collision free and as fast as possible through a gate from one side of a field to the other side, see figure 6. The challenge was to design an overall concept to solve the whole

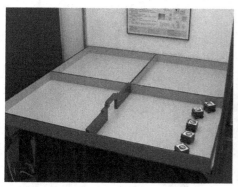

Fig. 6. Student project: robots have change the ends as fast as possible passing the narrow gate

Fig. 7. Student project: robots have to follow each other based on color recognition

problem and to split this in several tasks. The tasks were trajectory planning, model building of the robot, controller design, implementation and start-up of the whole system to verify the developed model and the control strategy.

The robot experiments were performed on our teleworkbench platform [7]. This platform allows the recording of multi robot experiments together with a live position tracking of the robots. The robots can ask for their current positions when acting on the teleworkbench. Based on this position data the robots have organized to drive from one side to the other. The group of students has successfully solved this multi robot problem.

5.2   Image Processing Project

Another educational project in Heinz Nixdorf Institute is to teach students how to use and program miniature robots. In this relationship a student project was to build a robot platoon follow each other based on image processing. One robot equipped with color markers drives randomly or based on Braitenberg behaviors throw a separate area. A second robot tries to follow the first one. Therefore it uses color recognition to find the color markers and by controlling the size and position of the color blob in the image the robot can control its distance to the leading robot. The challenge is to keep this behavior even on rough underground, see figure 7.

5.3   Research Project Guardians

Main disaster scenario covered by the GUARDIANS project (funded by Sixth Framework Programme of the European Union, no. 045269) is a large industrial warehouse on fire, where black smoke may fill large space of the warehouse that makes it very difficult for the firefighters to orientate in the building and

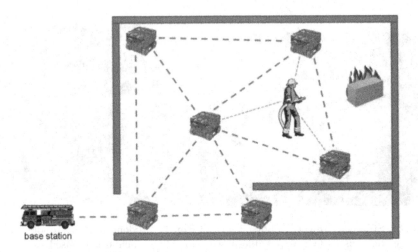

Fig. 8. Outline of a supporting communication network for fire fighters as proposed in the GUARDIANS project

thus limits the action space of the firefighters. During such mission, the robots navigate the site autonomously and serve as a guide for a human squad-leader in finding the target location or in avoiding dangerous locations or objects. They connect to a wireless ad hoc network and forward data to the squad-leader and the control station.

The ad-hoc network, which is actually a chain of robots equipped with wireless communication modules, is self-optimizing, adapts to connection failures by modifying its connections from local up to central connections [8, 9]. The autonomous swarm operates in communicative and non-communicative mode. In communicative mode, automatic service discovery is applied: the robots find peers to help them. The wireless network as depicted in figure 8 also enables the robots to support a human squad-leader operating within close range. In the

Fig. 9. Mission monitoring and remote control of selected robots via the base station in the mobile ad hoc network

case of loosing network signals, the robot swarm can still be functioning with non-communicative mode and continue serving the fire fighters.

Several robot platforms are used in this project, some for the down-scaled scenarios like Khepera III and the BeBot robot, others for the real scenarios provided by the Spanish partner Robotnik. As an example for the BeBot usage in the project, figure 9 depicts the BeBot used to build the mobile ad hoc network. The robots manage a TCP/IP based communication over different communication standards and enable the mission monitor and control in the base station.

6   Conclusion

We have introduced a new powerful and versatile mini robot optimized for small scale experiments supporting student project as well as research work. The miniature robot BeBot offers powerful information processing including dynamic reconfiguration via an integrated FPGA. Different wireless communication techniques enable comfortable wireless access to the robot. Via the Linux operating system the setup and maintenance of large communication networks is supported. A model of the robot using the Player/Stage framework merges simulations and real experiments. Due to its modular architecture the robot can be easily extended by new sensors and mechatronic modules to realize heterogeneous groups of robots. The robot has been successfully used in student projects as well as in research projects.

Acknowledgment

The development of the BeBot mini robot platform (MID chassis) is carried out in cooperation with the department of Computer Integrated Manufacturing (Prof. J. Gausemeier) of the Heinz Nixdorf Institute and the Corporate Technology Department of Siemens AG, Berlin. This work was supported by the Sixth Framework Program of the European Union as part of the GUARDIANS project (no. 045269, www.guardians-project.eu) and by the German Collaborative Research Center 614 - Self-Optimizing Concepts and Structures in Mechanical Engineering (SFB614, www.sfb614.de).

References

1. Kaiser, I., Kaulmann, T., Gausemeier, J., Witkowski, U.: Miniaturization of autonomous robot by the new technology molded interconnect devices (mid). In: Proceedings of the 4th International AMiRE Symposium, Buenos Aires (2007)
2. Braitenberg, V.: Vehicles: Experiments in Synthetic Psychology. MIT Press, Cambridge (1984)
3. OpenRobotix: Openembedded based open source linux distribution for mini robots, http://openrobotix.berlios.de/

4. Rana, V., Santambrogio, M., Sciuto, D., Kettelhoit, B., Köster, M., Porrmann, M., Rückert, U.: Partial dynamic reconfiguration in a multi-fpga clustered architecture based on linux. In: Proceedings of the 21th International Parallel and Distributed Processing Symposium (IPDPS 2007); Reconfigurable Architectures Workshop (RAW), Long Beach, California, USA (2007)
5. Paiz, C., Chinapirom, T., Witkowski, U., Porrmann, M.: Dynamically reconfigurable hardware for autonomous mini-robots. In: The 32nd Annual Conference of the IEEE Industrial Electronics Society (2006)
6. Witkowski, U., Chinapirom, T., Rückert, U.: Self-orientation of soccer robots on soccer pitch by identifying pitch lines. In: Proceedings of FIRA RoboWorld Congress, Dortmund, Germany, pp. 13–18 (2006)
7. Tanoto, A., Witkowski, U., Rückert, U.: Teleworkbench: A teleoperated platform for multi-robot experiments. In: Proceedings of the 3rd International Symposium on Autonomous Minirobots for Research and Edutainment (AMiRE 2005), Awara-Spa, Fukui, JAPAN (2005)
8. Witkowski, U., El Habbal, M.A.M., Herbrechtsmeier, S., Tanoto, A., Penders, J., Alboul, L., Gazi, V.: Ad-hoc network communication infrastructure for multi-robot systems in disaster scenarios. In: Proceedings of IARP/EURON Workshop on Robotics for Risky Interventions and Environmental Surveillance (RISE 2008), Benicassim, Spain (2008)
9. Witkowski, U., El Habbal, M.A.M., Herbrechtsmeier, S., Penders, J., Alboul, L., Motard, E., Gancet, J.: Mobile ad-hoc communication in highly dynamic environment optimized with respect to robustness, size and power efficiency. In: Proceedings of the International Workshop on Robotics for risky interventions and Environmental Surveillance (RISE 2009), Brussels, Belgium (2009)

System Design for Semi-automatic AndroSot

Yong Zhu, Zhimin Ren, Yin Xu, Linquan Yang, Zhongwen Luo, and Weixian Lv

Faculty of Information Engineering, China University of Geoscience, Wuhan
430074, China
luozw@cug.edu.cn

Abstract. A system design and implementation for semi-automatic AndroSot
was described. The system is divided into four subsystems, which are connected
with each other by some kind of network. Main difficult and key issue for the
effective implementation of each subsystem was described.

Keywords: AndroSot, Robot Soccer, Computer Vision.

1 Introduction

MiroSot has been proposed for more than 10 years. This has greatly stimulated
the research of fast computer vision, mechanical design, motion control and system
integration.

In recent years, with the development of Humanoid robot, Luo Zhongwen etc.[1]
and Hong Bingrong etc.[2] proposed a new robot competition system similar to Miro-
Sot with the replacement of car-like robot by human-like biped robot. And FIRA
announced a new category AndroSot for this kind of competition. RoboCup Japan
Open also proposed a similar competition called SSL-H [3].

In this paper, a system design for Semi-automatic AndroSot competition has been
described. Like the MiroSot, the competition system consist four sub-system. And
we will discuss the design of this part in the section 2 to section 5, which is Computer
Vision, Strategy, Communication and Humanoid robot. The overall system is
illustrated in Fig.1.

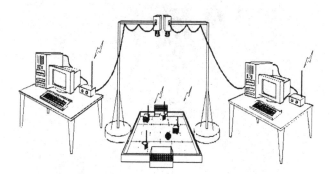

Fig. 1. Overall structure for semi-automatic AndroSot system

J.-H. Kim et al. (Eds.): FIRA 2009, CCIS 44, pp. 357–363, 2009.
© Springer-Verlag Berlin Heidelberg 2009

The system has the same architecture as described in [4], in which the software architecture is divided into four module. Each part is an independent software module, so that each module can be developed and implemented independently. This architecture is benefit for parallel development. The modules communication among modules are through a network[6][4].

1.1 Computer Vision

In this subsystem, a digital video is fixed on top of the soccer field and connected to computer through an IEEE Firewire cable. The computer gets the field image every 1/60 second. An image recognition algorithm is used to determine the position of the green baseball and robot, then the location and direction of the ball and robot was transmitted to the strategy system to make decision.

1.2 Strategy

The strategy system makes decision based on the field information. For example, we can find the robot which is nearest to the ball, and let that robot to move to the ball. To simplify the reasoning logic, we can divide the strategy system into several layers. In the top layer, a global strategy is used, in the lower layer, a basic motion method is chosen.

1.3 Communication

After the strategy system determine what to do for each of the robot. It will send command to the robot. The communication system consist two parts, one on the PC, the other on the robot. The PC side module sends command through 2.4 wireless digital signals to robot. And the robot side module receives the command and send to the robot through RS232 interface.

Fig. 2. Cap in a humanoid robot for semi-automatic AndroSot system

1.4 Humanoid Robot

The construct of humanoid system is time consuming. So currently, we base our system on the commercial available robot, such as KHR and Robonova. We just attach a cap on the robot head for localization of robot and identify the robot. Figure 2 shows the humanoid with a cap. Our work includes gait optimize and wireless communication module design.

2 Computer Vision Subsystem

The cap patch is same as MiroSOT. So we can use almost the same computer vision system for AndroSOT as MiroSot to identify the robot. The hardware and software architecture for both vision systems are same. The long history of MiroSOT development left a abundant resource of vision technology. That is really helpful for our vision system design.

Compared to the MiroSot vision system, the AndroSOT has some specific feature. One is that the Android has a relative slowly moving speed and not so flexible and also the accurate is not high. Another is that the AndroSOT's height is large.

The first feature has no affection to the design of vision system, for a fast vision can be used to a slow one with no changes. But for the second feature, this can make big difference.

The height of humanoid robot is over 30cm, and the ball's diameter is not more than 7cm. So the coordinate transform for the ball and the robot should be different. We based our AndroSOT vision system on an old MiroSOT vision system; the baseball diameter has little height difference with the MiroSOT's robot car. So we can directly use the old vision system for the ball recognition.

For the position of humanoid robot, we use a linear express to correct the change of height. The equation is as follow:

$$Pnew=Pold(1-k) \tag{1}$$

Where Pnew is the position vector after correct, Pold is the position vector before correct. k is a parameter which can get by simple measurement. For example, we place the humanoid at one boundary, and we know the Pnew=width/2, and we can get Pold, so substitute it to the formula; we can get the constant parameter k.

The height of humanoid robot also increases the possibility of the ball been hided. So that in some time cycle the ball is missing. One solution is to use a most recently ball position as current position. But in some case the humanoid robot may dribble the ball, so that the ball may be hided in a long time. In this case, we can assume that the ball is moving at the same speed and direction as that robot.

3 Strategy Subsystem

Compared to MiroSot, the AndroSot is really slow. This makes the design of strategy different from MiroSot. So two kinds of strategy system can be proposed. One is real-time decision strategy; another is target oriented plan strategy. The real-time strategy makes decision in each time cycle. The target oriented plan strategy makes decision only in some key cycles.

The real-time strategy scheme is same as the MiroSOT. Decision is made and a new command is sent at each time cycle. And a new primitive motion is performed by the robot. The primitive motion can be walk forward for one step, or do a primitive turn left motion.

The problem for real-time strategy scheme is that different primitive motion takes different time. We can not exactly predict the finish time of a motion, and send a command immediately after the finish of the old motion. Another difficulty is to smoothly and quickly connect two adjacent motion actions at emergent situation. Current treatment is to design each motion begins and ends with same home position, so that no connection is needed. But this greatly slow down the global speed of robot.

For the target oriented plan strategy, decision is made at some key cycles, and a complex motion or a series of primitive motion is planned to perform. For example if we want the robot to pursuit a ball at right forward, a series of motion such as: turn right, forward three step, and turn left may be performed. After that we will make a new decision.

In this scheme, the time between two adjacent commands is long. The robot movement of the scheme is smooth and so the robot can move quickly. The shortcoming is that if the situation changes, the planned motion can not be stopped and changed accordingly.

4 Communication Subsystem

The communication subsystems compose two parts. One is on the PC side; the other is on the humanoid robot.

The PC side hardware is a wireless box; it receives command from PC through a USB or RS232 serial port. Then the box will send the signal to the robot receiver through a 2.4GHz wireless module.

The humanoid robot side hardware receives the command through a 2.4GHz signal. And it then sent the command to the humanoid control board through a RS232 port.

The transceiver we choose to use is Nordic nRF2401[5]. A pair of nRF2401 is located one at robot side, the other at PC side.

4.1 PC Side Communication Module

For the PC side hardware, firstly we have chosen to use the original sending box for MiroSot, which connect to the PC through a RS232 interface. And then we have made a new sending box, which connect to the PC through USB. As we know, the USB is more popular and faster than RS232 interface. And its power is supplied by the PC, so omit a special power supply system.

Fig. 2. USB nRF241 signal transmitter on PC side

For the design of the USB wireless sending box, we choose to use a NXP LPC 2148 ARM7 CPU. This CPU have supplied USB interface and have a large memory.

4.2 Humanoid Robot Side Communication Module

For the humanoid robot side hardware design, the key issue is the size. So a small size SMT chip of ATMEGA8 is used for the wireless receiver. The receive module's size is 2cm*2cm, which is really small and can be easily mounted on one of the humanoid robot's shoulder.

To minimize the module size, the wireless module gets its power from the robot instead of designing a dependent power supply system.

5   Humanoid Robot

To fasten the design of the robot system, we choose to use the commercial Humanoid robot. There are three benefits for using commercial humanoid robot. The first benefit is comparative lower cost. The second benefit is that this robot is more mature and exist a vast resource, which can shorten development time. The third benefit is the robot has a good quality, because of industry standard control.

5.1   Main Feature of Humanoid Robot in the Market

Currently there are three kind of very popular humanoid robot at commercial market. They are Kondo's KHR-2HV, Robonova-I, Robotis' Bioloid. They are similar in function, but each one has some specific feature.

The Kondo's KHR series has a long history, it provided an easy to use icon based program interface. It provides two level of action, a primitive action is called motion, a complex action is called scenario, which is just an assembly of primitive motion. It also provide branch and loop. Except of 24 PWM servos IO, it also provides three digital IO and three analog IO.

Owing to its mechanical design, the Robonova-I robot has a very stable movement. It provides a program language called Robobasic, which may have more control ability. It also provides a script language, which is more easy to use.

The robotis' bioloid is a comparative newly product in the market. Its specific feature lies on its actuator, which use a RS485 interface instead of PWM. That makes the robot construction comparative easy.

5.2 Sensor for the Robot

To fit for the competing, some sensor must be added to the robot. The main sensors include an accelerator and a gyro. Fig 3. shows the wireless communication module and sensor that we have added to the robot.

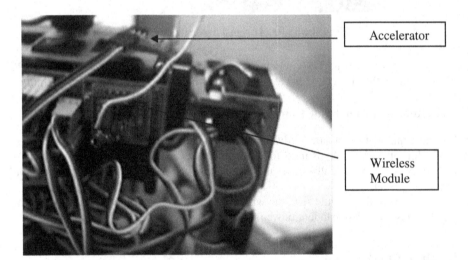

Accelerator

Wireless
Module

Fig. 3. A wireless module and accelerator on humanoid robot

An accelerator is used to check whether the robot has already fall down to the floor. So that when a fall down is detected, a get up motion should be awakened and performed. We have design a two axis accelerator, which can detect 4 kinds of fall down; they are backward, forward, left, right.

The gyro sensor is used to make dynamic stability. When a small unbalance of robot has been detected, an inverse movement of related servo is performed to balance the robot.

6 Conclusion and Future Development

A practical runable AndroSot system was designed and created. Some of the main idea has been discussed. Because of some specific feature discussed above, our team has won the Champion of fully-automatic Androsot 3:3 at Fira 2008. The main features are as follow:

The software architecture is divided into four independent components. This framework benefit for fast development and easy to debug, and makes parallel development possible.

There are much more things we can do to improve current system. For the vision system, we can make it more robust and pair-eye may added to get 3-D location information, which may be benefit for the detecting fall down in a more early time.

The robot should be more flexible. Currently the robots using in the competition are commercial available ones. The benefit for the commercial robot is cheap and good quality. But because this kind of robot is mainly for the entertainment market, the programming language provided is simple and easy to use, but not have the fully controllability for expert users. So a humanoid control board provide fully controllability should be developed.

References

1. Luo, Z.-w., Yin, X., Ying, Z., He, Z., Yong, Z., Jing, W.: A Research of Platform for Biped Robot Soccer. J. Harbin Institute of Technology 39(1), 22–25 (2007)
2. Hong, B.-r., Piao, S.-h., Liang, W., Zheng, G., Wen, Z.-m.: A scheme of humanoid robot soccer system. J. Harbin Institute of Technology 39(1), 1–3 (2007)
3. Robocup SSL-H information (2008),
   http://robocup-ssl-humanoid.org/index.html
4. Zhongwen, L., Yueqian, Z., Linquan, Y.: Research of Large Field Robot Soccer Parallel Vision System based on C/S model. J. Harbin Institute of Technology (New Series) 16(2), 151–155 (2008)
5. NORDIC, 4λ Printed Monopole Antenna For 2.45GHz, Norway (2003)
6. Wolf, J.C., Oliver, J.D., Robinson, P., Diot, C.: Multi-site development of a FIRA large league robot football system. In: The Proceedings of Third International Conference on Computational Intelligence, Robotics and Autonomous Systems, Singapore, CIRAS (2005)

Extended TA Algorithm for Adapting a Situation Ontology

Oliver Zweigle, Kai Häussermann, Uwe-Philipp Käppeler, and Paul Levi

Institute of Parallel and Distributed Systems,
Universität Stuttgart, 70569 Stuttgart, Germany
{zweigle,haeussermann,kaeppeler,levi}@informatik.uni-stuttgart.de
www.informatik.uni-stuttgart.de

Abstract. In this work we introduce an improved version of a learning
algorithm for the automatic adaption of a situation ontology (TAA) [1]
which extends the basic principle of the learning algorithm. The approach
bases on the assumption of uncertain data and includes elements from the
domain of Bayesian Networks and Machine Learning. It is embedded into
the cluster of excellence Nexus at the University of Stuttgart which has
the aim to build a distributed context aware system for sharing context
data.

1 Introduction

One of the central research topics of the cluster of excellence *Nexus* at the University of Stuttgart[2],[3] is distributed reasoning for the recognition of situations
using uncertain data. The basic concept of the developed Nexus platform is to
create in analogy to the WWW a World Wide Space, which provides the conceptual and technological framework for integrating and sharing context models.
The collection of context models is federated and leads to a large scale context
model, offering a global and consistent view on the context data. The federation allows for complex spatial queries, including continuous evaluation and
stream-based processing. Furthermore it is able to provide quality information
about context data gained from the federation. Based on the platform, different research areas make use of the provided technologies and information and
integrate them into their research efforts.

In a first approach different procedures for the recognition of situations were
examined and developed based on uncertain context data gained from the *Nexus*
federation respectively from so called *Nexus Context Servers*. A special focus was
on creating a general approach to use common methods for a large set of different
situations. During this work two concrete problems within this domain became
clear: on the one hand the possibility of uncertain context data, and on the
other hand the uncertainty of the actual inference of the situation recognition,
for example in the form of incorrect default values. The goal of this work is a
further enhancement of the situation recognition process presented in [1]. This is

J.-H. Kim et al. (Eds.): FIRA 2009, CCIS 44, pp. 364–371, 2009.

done by using an extended learning algorithm that sequentially refines so called *situation templates* (which are a subset of an ontology) and as a consequence the whole situation recognition process. The extended learning algorithm enriches the old approach with the possibility to change situation templates during the learning process precisely.

For the purpose of correcting potential errors in a situation template they have to be located first. As the system should run in a distributed way there is not only the possibility of errors in the context data but also during the transport of the data over different networks. One possibility of locating errors is to accomplish new measurements enough times until the fault location can be clearly determined. Corresponding work can be found in [4], [5], and [6]. Another interesting approach is the use of probability distributions described in [7] and [4]. In [8] an approach for recognizing and predicting context by learning from user behavior was described. Those results could be used as the basic idea of this work. Furthermore the approach presented in [9] can also be seen as a base for handling uncertainty data in a database system. In [10] and [11] other methods for ontology based learning are presented that are not directly used in this work.

1.1   Ontology Based Situation Recognition

With the presented extended algorithm a new technique for the automatic calculation and correction of a predefined situation including the precise adaption of situation templates based on a learning algorithm is introduced. A so called *situation template* will be created by a system designer and represents a subset of an ontology. This template will be used later to recognize certain situations automatically. The template is defined in a XML-file and is composed out of logical and temporal operators. Furthermore there is the possibility to apply sensor- or context data directly - for example from a lower hardware layer. Thus the template represents knowledge of the cohernces of context data which will be used to infer if a concrete situation is fulfilled or not.

In the next step the situation template is converted into a tree structure. Here the tree nodes represent the predefined operators and context data, and the edges in the tree represent the according cohernces between the operators. As different nodes can also be physically placed on different systems in a network an edge can also represent a communication link.

Aside from the possibility of a simple inference, we add for every node a certain probability value. This approach allows us to handle the tree according a kind of Bayesian network. Using the probability values of the context data and the operators, it will be possible to consider uncertainties within the inference. For that purpose we utilize the *Nexus Context Servers*, which provide the quality information of the context data automatically. As a consequence we will be able to make conclusions about the corresponding context quality and about the overall recognition quality of a situation recognition process. Using these techniques the primary goal of this work is finally an enhancement of the situation recognition process. This is done in a process that sequentially refines a situation template and as a consequence the whole situation ontology.

2   Definition of Situation Recognition Process

A situation template is an abstract description of different constraints and their coherences for automatically recognizing a certain situation. As a consequence a situation template indicates a form of explicit knowledge. In the Nexus scenario the initial description of a template is done by an expert. Based on his implicit knowledge the expert models the coherences and operators as well as the constraints of the different context data. In a first version the implemented operators can be logical operators as well as temporal operators [12]. But there is also the possibility for further flexible extensions. The concrete specification of a template is done within a XML-structure which will be automatically transformed into a tree structure (described in chapter 2.1).

2.1   Situation-Aggregation-Tree (SAT)

The representation of the described situation recognition process can be modelled mathematically, using a directed graph. Because of the resulting tree structure the graph is called *Situation Aggregation Tree (SAT)*. The graph is defined as $SAT = (O, E)$. An important fact is that the tree is not branched top-down from a root element. The tree aggregates its branches bottom-up. As a consequence all paths are joined in the top node. The set of all used operators and context nodes is defined as $O = \{o_1, ..., o_n\}$. $E = \{e_1, ..., e_n\}$ is defined as the set of directed edges or links between the operators. Furthermore $e_i = (o_j, o_k)$ is a directed edge from node o_k to node o_k. In other words, o_j is the parent node of o_k, thus we will say formally $pa(o_k) = o_j$.

The set of nodes O can be differentiated in a more detailed way:

Top Node t: The node t, which is not a parent node of another node $o \in O$.
Formally: $\nexists o : pa(o) = t$, for $o, t \in O$.
Leaf Nodes C: The set of nodes $C = \{c_1, c_2, ..., c_n\} \in O$ without any parent node. Formally: $pa(c) = \emptyset$, for $c \in O$ is valid.
Hidden Nodes H: The set of nodes $H = \{h_1, h_2, ..., h_m\} \in O$, which are neither Top nor Leaf nodes. Formally: $h \in O \backslash \{t \cup C\}$.

Transferred to the Nexus environment the predefined conditions of context data correspond to the leaf nodes C. The logical or temporal operators are in accordance with the nodes t and H in the SAT. The edges represent the corresponding links of the certain conditions and operators.

Using the information about the uncertainty of context data from the *Nexus Context Servers* the quality of every single context data can be extracted initially. This information can further be used for the calculation of the situation recognition process. Because of the SAT structure and the stochastic independence of the context data, the quality of the data respectively the uncertainty is propagated in the tree structure bottom-up to the top node.

For the implementation we have to add to the current structure $SAT = (O, E)$ a set of probability distributions P_M. So we get the new extended structure, where:

$$SAT = (O, E, P_M) \tag{1}$$
$$P_M = \{CPT_{o_1}, ..., CPT_{o_n}\} \tag{2}$$

CPT_{o_i} equates to the Conditional Probability Table (CPT)[13] of a node o_i and makes assertions about the conditional probability of the correctness of a SAT Node o_i in correspondence to its parent node $pa(o_i)$. Using this extension a Bayesian Belief approach is implemented. The aim of this approach is to make a conclusion about the probability of correctness of the recognized situation.

3 Basic Principle of TAA

TAA is the abbreviation for Template Adaption Algorithm presented in [1]. The basic principles will be shown here as an introduction. After the situation has been specified by an expert in a situation template using a XML structure, it is automatically transferred into the SAT structure we already described above. In addition the CPTs of each leaf nodes C are filled up with the quality information of the corresponding context data using a query to a Nexus context server.

In the following step a *Joint Combine Operation* is executed, which eliminates all Hidden Nodes H in the SAT. In order the edges E of the leaf nodes C are linked to the top node t. Due to this modification of the new graph SAT^*, we have a new definition of the new top node t as $pa(t) = \{C\}$. See figure 1b) for an example.

Subsequently the Conditional Probability Table of the top node CPT_t in the structure SAT^* is defined using the previously calculated CPT_c for $c \in C$ based on the quality information of the context data. Based on the assumption that there are no errors in the connections between the operators (which could be also a network link), respectively in the operators itself and the fact that the context data is stochastic independent the following relation is given:

$$P(t|pa(t)) = P(t|l_1, ..., l_n) = \prod_{i=1}^{n} P(t|l_i) = \prod_{i=1}^{n} P(l_i) \tag{3}$$

Where $n = |\{C\}|$ is considered.

In the next step we create a new table, called *Global Control Table (GCT)*. The GCT consists of entries which represent the binary switching states $\{0, 1\}$ of the conditions $c \in C$ as well as the probability value p of each switching combination. The probability value p of the according switching combination (under the assumption that no errors occurred) can be achieved using a simple mapping algorithm based on the formerly calculated CPT_t.

For the unknown switching combinations or if there are no quality values available (e.g. the context server did not provide them), random values or the trivial value of $p = 0.5$ can be used instead.

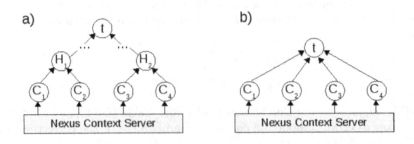

Fig. 1. a) SAT b) SAT*

With the help of a feedback module the probability value p_i of row i of the GCT depending on the switching state, can be updated according to a simple Delta-Δ-Rule in every recognition step (episode). It appears that for faulty template values and small enough Δ the probability p_i converts with a growing number of episodes towards 0. Formally: $\lim_{\#episodes \to \infty} p_i = 0$.

An important fact is that the feedback module is exchangeable. Its concrete implementation is independent from the rest of the system framework. The feedback module could be implemented using supervised learning algorithms [14] with a teacher giving correct values. Alternatively other approaches like Neuronal Networks [14] or Reinforcement Learning[15] could be used instead. In our approach supervised learning was used. The fundamental concept is that the user can voluntarily give the system binary answers (yes/no) if the automatic recognition of a situation was successful or not.

For every p_i which is under a predefined threshold ϵ, the corresponding row i in the GCT is selected. Afterwards for each of the selected rows a new vector with the switching status will be created. Every created vector represents a potentially faulty template value. Afterwards statistics, counting the number of all switching states in the vectors is used. For that we define a function $\psi(value)_{C_i}$ accordingly. Furthermore a helping function $set(X, value)$ is defined. This function returns the number of elements $x \in X$ which has the given switching state $value$. We can say formally:

$$set(X, value) = \{x | (valueOf(x) = value) AND (x \in X)\}$$
$$\psi(value)_{C_i} = |set(C_i, value)| \tag{4}$$

In order to be able to correct errors we first locate the potential error using the previously calculated statistics. It is clear that the vector with the highest value of same switching states in the statistics ψ represents the template value of the condition C_j that has to be corrected. Afterwards an adaption of the (faulty) template value with a predefined d (random or fixed value) can be executed. The result is the automatic correction of wrong or imprecise template values over n learning episodes. In case that $m > 1$ vectors $(C_j..C_k)$ have the same maximum amount of switching states compared to another vector two cases have to be distinguished.

- Case 1: The complete set of m selected template values is faulty.
- Case 2: The error is caused by a faulty operator (which aggregates the m vectors) or if we assume that the operators (nodes of the SAT^*) are distributed on several different systems the error can also be caused by a faulty network-connection between the nodes.

To detect and distinguish the two error cases a resolving strategy was developed: At first a random template value (which symbolizes C_i) within the m selected vectors is chosen and updated with the previously described Delta-Δ-Rule. If the statistics of the adjusted vector i improves, the process is continued gradually with the remaining $m - 1$ vectors. If the statistics remains unchanged the error is located in one of the operators which joins the m underlying conditions.

3.1 Improvement

The basic idea of the improvement of the above described algorithm is not to change the template values with a fixed value or a random factor but change them selectively. For that extension a new table T_{C_i} has to be created for every condition. In the case of n context data a set of n tables T_{C_i} for i=1..n have to be created. The newly added tables include as columns the calculated global binary return value of the situation recognition process called *result*, the binary value of the feedback of the user called *feedback* and the context data queried from the context server called *value*. See also table 1 for an simple example.

Table 1. Example of an extension table T_{C_i} with fictitious context data

row i	result	feedback	value
1	1	0	16
2	0	0	15
3	0	1	17
4	1	1	18

Furthermore it has to be distinguished between the amount of so called "Error Cases" (EC) and "Non Error Cases" (NEC). These are defined as:

$$EC = \{row_i \in T_{C_i} | (result_i \ XOR \ feedback_i) == true\}$$
$$NEC = \{row_i \in T_{C_i} | (result_i \ XOR \ feedback_i) == false\}$$

In order to adapt the values in the template a case distinction is necessary. This fact is presented with the example of the logical operator *greater-than* ">".

Example: Greater-than-Operator ">"

- Case 1: maximum value is unequal to the current template value $\in EC$.
  $\longrightarrow$ The template value is updated using the value with the smallest difference to the value V = max(NEC).
- Case 2: maximum value is equal to the current template value $\in NEC$.
  $\longrightarrow$ The template value is updated using the value with the smallest difference to the value V = min(NEC).

The complete framework for the algorithm is shown in figure 2.

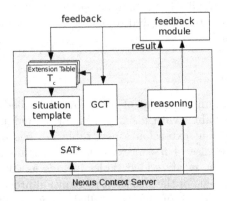

Fig. 2. System-Architecture

4 Conclusion

In this article we presented an extended approach for the TA algorithm [1], a learning algorithm that is automatically adapting situation templates which are a subset of a situation ontology. The approach was extended Furthermore the method accounts to the improvement of the quality of a situation recognition system. In an implemented testbed the results of the situation recognition process sufficiently improved when enough feedback from users was available. For the purpose of the improvement of situation templates a graph based algorithm with ideas of Bayesian probabilities properties was used. Based on this basic situation recognition method we developed a learning algorithm called TAA that automatically adapts values of situation templates using supervised learning. As a result the quality of the overall situation recognition is enhanced what leads to an improvement of the resulting quality value of a recognized situation. That benefits to applications which are using the dynamic situation recognition component especially if they are in need for information with a high quality.

This work is partially funded by the German Research Foundation within the Collaborative Research Center 627.

References

1. Zweigle, O., Häussermann, K., Käppeler, U.-P., Levi, P.: Learning algorithm for automatic adaption of a situation ontology using uncertain data (2009), http://www.ipvs.uni-stuttgart.de/abteilungen/bv/forschung/projekte/SFB627-Folder/Basic_TA_Algorithm_BV_Stuttgart.pdf
2. Lange, R., Cipriani, N., Geiger, L., Großmann, M., Weinschrott, H., Brodt, A., Wieland, M., Rizou, S., Rothermel, K.: Making the world wide space happen: New challenges for the nexus context platform. In: Proceedings of the 7th Annual IEEE International Conference on Pervasive Computing and Communications (PerCom 2009), Galveston, TX, USA (2009)
3. Rothermel, K., Bauer, M., Becker, C.: Sfb 627 nexus umgebungsmodelle für mobile kontextbezogene systeme. In: Molitor, P., Küspert, K., Rothermel, K. (eds.) It - Information Technology, vol. 45(5) (2003)
4. de Kleer, J., Williams, B.C.: Diagnosing multiple faults, 372–388 (1987)
5. Hou, A.: A theory of measurement in diagnosis from first principles. Artif. Intell. 65(2), 281–328 (1994)
6. Reiter, R.: A theory of diagnosis from first principles. Artif. Intell. 32(1), 57–95 (1987)
7. de Kleer, J.: Using crude probability estimates to guide diagnosis, 118–123 (1992)
8. Mayrhofer, R., Radi, H., Ferscha, A.: A Recognizing and predicting context by learning from user behavior, 25–35 (2003)
9. Cheng, R., Prabhakar, S.: Managing uncertainty in sensor database. SIGMOD Rec. 32(4), 41–46 (2003)
10. Maedche, A., Staab, S.: Ontology learning for the semantic web. IEEE Intelligent Systems 16(2), 72–79 (2001)
11. Apted, T., Kay, J.: Automatic construction of learning ontologies. In: Proceedings of International Conference on Computers in Education, December 2002, vol. 2, pp. 1563–1564 (2002)
12. Allen, J.F.: Maintaining knowledge about temporal intervals. ACM Commun. 26(11), 832–843 (1983)
13. Pearl, J.: Probabilistic reasoning in intelligent systems: networks of plausible inference. Morgan Kaufmann Publishers Inc., San Francisco (1988)
14. Riedmiller, M.: Int. journal of computer standards and interfaces special issue on neural networks (5), 1994 advanced supervised learning in multi-layer perceptrons-from backpropagation to adaptive learning algorithms (1994)
15. D'Esposito, M.: Journal of cognitive neuroscience 11(1), 126–134 (1999)

An Integer-Coded Chaotic Particle Swarm Optimization for Traveling Salesman Problem

Chen Yue[1], Zhang Yan-duo[1], Lu Jing[1], and Tian Hui[2]

[1] Hubei Key Lab of Intelligent Robot, School of Computer Science and Engineering,
Wuhan Institute of Technology, Wuhan 430073, China
[2] Wuhan National Laboratory for Optoelectronics, School of Computer Science and
Technology, Huazhong University of Science and Technology, Wuhan 430073, China

Abstract. Traveling Salesman Problem (TSP) is one of NP-hard combinatorial optimization problems, which will experience "combination explosion" when the problem goes beyond a certain size. Therefore, it has been a hot topic to search an effective solving method. The general mathematical model of TSP is discussed, and its permutation and combination based model is presented. Based on these, Integer-coded Chaotic Particle Swarm Optimization for solving TSP is proposed. Where, particle is encoded with integer; chaotic sequence is used to guide global search; and particle varies its positions via "flying". With a typical 20-citys TSP as instance, the simulation experiment of comparing ICPSO with GA is carried out. Experimental results demonstrate that ICPSO is simple but effective, and better than GA at performance.

Keywords: Particle Swarm Optimization, Chaotic, Traveling Salesman Problem, Genetic Algorithm.

1 Introduction

TSP is an ancient difficult problem, which can be traced back to the problem of knight's tour studied by Euler in 1759. Initially introduced by RAND Corporation in 1948, TSP now has gradually became one of the most intensively studied problems in optimization with increasing popularity of the application and research of combinatorial optimization problems. The significance of studying TSP lies in the fact that many problems can be abstracted as TSP by slight modifying, such as traffic management, network routing, large-scale production process and etc. Meanwhile, the application of TSP extends to many other industries, such as transportation, logistics service industry and so on. However, TSP experiences a "combinatorial explosion" by requiring enormous computational resources when the problem goes beyond a certain size. Therefore, finding a practical and efficient algorithm for TSP is particularly important.

Particle Swarm Optimization (abbreviated as PSO) is a random search algorithm based on individual evolution as well as the collaboration and competition of the population. With some prominent characteristics of operation process, such as simplicity, feasibility and high efficiency of computation, PSO is widely acknowledged as an effective algorithm which can parallel with Genetic Algorithm (abbreviated as GA)

J.-H. Kim et al. (Eds.): FIRA 2009, CCIS 44, pp. 372–379, 2009.

and has been widely adopted in many fields. In this paper, Integer-coded Chaotic PSO (abbreviated as ICPSO), on the basis of the original velocity-position model, is put forward to solve TSP by encoding particles with integer, guiding global search with chaotic sequences and describing flying of particles as well as updating the position of particles. Meanwhile, simulation experiments are conducted in TSP with a size of 20 cities [3], comparing ICPSO with GA. Experimental results demonstrate that ICPSO is simple and effective and better than GA at performance for solving TSP.

2   The Mathematical Model of TSP

2.1   General Mathematical Model

TSP can be described as follows: a traveling salesman has to travel through n cities for sale promotion, the distance between any two cities (denoted by city i and city j) is d_{ij} ($i, j = 1, 2... n$). The task is to find a shortest possible tour that visits each city exactly once and ends up in the same city in which the salesman started.

From the perspective of graph theory, TSP is simplified as searching for a Hamiltonian cycle with the least weight in an undirected complete graph in which weight is assigned to each of its edge. Let $G = (V, E)$ be a weighted graph, the set of vertices and the set of edges in Graph G are denoted by $V = \{1, 2... n\}$ and $E = \{e_{ij} \mid i, j \in V, i \neq j\}$, respectively. d_{ij} ($i, j \in V$) is the distance between two vertices i,j (the length of e_{ij}), where $d_{ij} > 0$ and $d_{ij} < \infty$; $\forall\ i, j \in V$ and $i \neq j$, $d_{ij} = d_{ji}$. Then the classical mathematical model of TSP is:

$$\begin{cases} \min F = \sum_{i \neq j} d_{ij} x_{ij} \\ s.t. \begin{cases} (a) \sum_{i=1}^{n} x_{ij} = 1, i \in V \\ (b) \sum_{j=1}^{n} x_{ij} = 1, j \in V \\ (c) x_{ij} = 0 \end{cases} \end{cases} \tag{1}$$

Where x_{ij} is the decision variable, i.e., if the path from city i to city j constructs the salesman's tour, then $x_{ij} = 1$; otherwise, we set $x_{ij} = 0$. The matrix consisting of x_{ij} is the solution matrix; the constraint conditions (a) and (b) represents the salesman leaves city i once and the salesman arrives at city i once, respectively. Therefore, conditions (a) and (b) ensures the salesman visits every city exactly once. The set of edges $E^* \subset E$ which satisfies the above conditions is the optimal path that we required.

2.2 TSP Model Based on Permutation and Combination

The former model is proposed from the perspective of linear programming, which is not directly applicable in using intelligent algorithm for problem solving. To fit the needs for solving the task presented in this paper, we introduce an equivalent definition of TSP: a traveling salesman has to travel through n cities for sale promotion (all the cities are denoted by 1, 2, ... n), every pair of distinct cities is connected by a path and the distance between any two cities (denoted by city i and city j) is d_{ij} ($i, j = 1, 2... n$). What we require is an array of all the cities, so that:

$$\sum_{i=1}^{n-1} d_{S(i)S(i+1)} + d_{S(n)S(1)} \tag{2}$$

gets the minimum value.

3 ICPSO Algorithm for Solving TSP

3.1 Particle Encoding

According to the TSP model based on permutation and combination, we can direct encode the population as an integer set $X=(x_1, x_2... x_n)$ of solution space I, where $\forall$ i, $j = 1, 2... n$ and $i \neq j$, $x_i \neq x_j$. This coding method denotes that the salesman travels all the cities following the sequence of $x_1 \rightarrow x_2 \rightarrow ... \rightarrow x_n \rightarrow x_1$ and ends the tour at where he started.

3.2 Velocity-Position Model

In ICPSO Algorithm, particles are coded with integers, chaotic sequence is used to direct global search, and particle varies its positions via "flying", consequently, positions are updated. Update equations can be described as follows:

$$v_{ij}(t+1) = w(t) \times v_{ij}(t) + c_1 \times r_1 \times \phi(p_{ij}, x_{ij}(t)) + c_2 \times r_2 \times \phi(g_j, x_{ij}(t)). \tag{3}$$

$$\phi(p_{ij}, x_{ij}(t)) = \varphi(p_{ij}, x_{ij}(t)) \times \frac{\left| f(p_{ij}) - f(x_{ij}(t)) \right|}{f_{max} - f_{min}} \times (b-a). \tag{4}$$

$$\phi(g_j, x_{ij}(t)) = \varphi(g_j, x_{ij}(t)) \times \frac{\left| f(g_j) - f(x_{ij}(t)) \right|}{f_{max} - f_{min}} \times (b-a). \tag{5}$$

$$\varphi(x, y) = \begin{cases} 0 & x = y \\ 1 & x \neq y \end{cases}. \tag{6}$$

$$x_{ij}(t+1) = \begin{cases} mutation, If \ R_{ij} \le Sigmoid(v_{ij}(t)) \\ x_{ij}(t), \quad If \ R_{ij} > Sigmoid(v_{ij}(t)) \end{cases}. \tag{7}$$

$$Sigmoid(v_i(t)) = \frac{1}{1+e^{-v_i(t)}}. \tag{8}$$

In equation (3), $w(t)$ is the inertia weight which has great influence on the convergence of the algorithm and has tendency to extend search space by maintaining the inertia of the particles. The traditional linear decreasing control model, however, can hardly reflect the variation of search space. Therefore, chaotic sequence is adopted in the adaptive generating of inertia weight, that is, the equation below is used in the iterative generation of inertia weight:

$$w(t+1) = u \times w(t) \times (1 - w(t)). \tag{9}$$

Where $u = 4.0$, and $w(0)$ are random numbers in interval $(0, 1)$.

Function $f(x)$ is fitness value for every position, where f_{max} and f_{min} are the maximum and minimum value of all the available fitness values, respectively. The specific definition of f_{max} and f_{min} will be given in following section. Equation (4) (or equation (5)) represents $\phi(p_{ij}, x_{ij}(t))$ (or $\phi(g_j, x_{ij}(t))$) $=0$ if current position equals to individual extremum value (or global extremum value), otherwise, their values are located in the interval $[a, b]$ to ensure the velocities do no exceed the maximum value. In equation (7), the calculation of function Sigmoid uses the formula (8) [5], which is generally $v_{max} \in [-6.0, 6.0]$. Correspondingly, we set $a = 0$, $b = 2.0$.

INPUT: positions of particle i : $Xi=(x_{i1}, x_{i2}..., x_n)$, current individual extremum value $P_i =(p_{i1}, p_{i2}..., p_n)$, current global extremum value $G =(g_1, g_2..., g_3)$

OUTPUT: updated positions of particle i

1: **For** positions x_{ij} ($j = 1,2,...n$) in every dimension of particle i
2: {
3:    calculate velocities $v_{ij}(t+1)$ in all dimensions according to formula (2-14);
4:    generate a random number R;
5:    If($R < Sigmoid(v_{ij}(t))$)
6:    {
7:       choose one dimension of the particle randomly, exchange their positions;
8:       update the positions Xi of particle i;
9:    }
10: }
11: **Return** Xi;

Fig. 1. Updating algorithm for particles' positions

Equation (7) represents that particles remain unchanged under the probability of 1-$Sigmoid(v_{ij}(t))$ and mutation occurs under the probability of $Sigmoid(v_{ij}(t))$. "Mutation" means to change the current position of particles according to certain rules. As to TSP, all elements in encoding of particles are mutually exclusive (two identical digits do not exist). Thus, we define mutation operation as: if the criterion of mutation is satisfied, we randomly choose another particle which differs from this one among the population and do the exchange operation. Fig. 1 shows the updating algorithm for particles' postions.

3.3   The Definition of Fitness Function

There are two types of fitness functions in ICPSO algorithm: one is the typical fitness function of particles (denoted by $F(X)$); the other one is fitness function used to direct the particles' variety of coordinates in all dimensions, which is specifically owned by ICPSO (denoted by $f(x_i)$, $i = 1, 2, \ldots, n$).

According to the general rule of determining fitness function in evolutionary algorithm, the fitness function of particle $X=(x_1, x_2\ldots, x_n)$ is defined as below:

$$F(X) = \sum_{i=1}^{n-1} d_{x_i x_{i+1}} + d_{x_n x_1} . \tag{10}$$

The fitness function used to direct the particles' variety of coordinates in all dimensions is defined as below:

$$f(x_i) = \begin{cases} \dfrac{d_{x_{i-1} x_i} + d_{x_i x_{i+1}}}{2}, 1 < i < n \\[3mm] \dfrac{d_{x_1 x_n} + d_{x_1 x_2}}{2}, i = 1 \\[3mm] \dfrac{d_{x_1 x_n} + d_{x_{n-1} x_n}}{2}, i = n \end{cases} . \tag{11}$$

It means the fitness value of a particle in every dimension is equal to the average distance of two neighboring cities. Besides, f_{max} and f_{min} in formula (4) and (5) are the maximum and minimum values among all the $f(x_i)$ evaluating so far, respectively.

3.4   The Solving Process of TSP Based on ICPSO

Through the above definition and analysis, a solving process of TSP based on ICPSO is obtained as follows:

STEP1: Initialize the population by generating an array of particles with random positions.

STEP2: Evaluate the fitness function for each particle according to equation (10) and equation (11).

STEP3: To any particle i, compare the evaluated fitness value with that of individual extremum value P_i (the personal best position of particle i), set the current position as the individual extremum if former one is better;

STEP4: To any particle i, compare its fitness value with that of global extremum value G (the best position discovered by any of the particles so far), set current position as the global extremum if former one is better;

STEP5: Stop if one of the stopping criteria is satisfied; otherwise, go on to the next step.

STEP6: Update individual velocities on all dimensions according to equation (3) and position according to the algorithm shown in Fig. 1, respectively. Loop to Step 2.

4 Simulation Experiment and Analysis

In this section, simulation experiments are conducted in TSP with an instance of 20 cities [3] (the coordinate information related to this problem is listed in Tab. 1.). Meanwhile, comparison of performance between ICPSO and GA algorithm in solving TSP is carried out. The encoding method proposed in this paper is adopted by this algorithm, and the substance of individual evolution is the adoption of ordered crossover (OX) operation based on path display: OX operation maintains the arrangement and combines ordered structural units with different arrangement. When two parents crossover, a child is generated by choosing a segment of parent 1 and maintaining the according order of the city numbers in parent 2.

For example, randomly choose two crossover points "|" in parents presented below:

$$p_1: (1\ 2\ 3\ |\ 4\ 5\ 6\ 7\ |\ 8\ 9)$$

$$p_2: (4\ 5\ 2\ |\ 1\ 8\ 7\ 6\ |\ 9\ 3)$$

Firstly, keep the segments between two crossover points remaining unchanged:

$$o_1: (X\ X\ X\ |\ 4\ 5\ 6\ 7\ |\ X\ X)$$

$$o_2: (X\ X\ X\ |\ 1\ 8\ 7\ 6\ |\ X\ X)$$

Secondly, recording down the parent 2's sequence of city number which starts at the second crossover point, while reaching the end of the list, return to the head of the list to continue recording city number until the second crossover point ends, hence, parent 2's sequence of city number which starts at the second crossover point is 9—3—4—5—2—1—8—7—6. Remove city number 4,5,6,7, which exist in the sequence of parent 1, from the city numbers sequence of parent 2, the sequence of 9—3—2—1—8 is obtained. Duplicate the sequence to parent 1 starting from the second crossover point to determine the unknown code X in child 1, therefore, child 1 generated as follows:

$$o_1: (2\ 1\ 8\ |\ 4\ 5\ 6\ 7\ |\ 9\ 3)$$

Similarly, child 2 is generated as:

$$o_2: (3\ 4\ 5\ |\ 1\ 8\ 7\ 6\ |\ 9\ 2)$$

Table 1. Coordinates of 20 cities

CITY	x	y	CITY	x	y
1	5.294	1.558	11	4.399	1.194
2	4.286	3.622	12	4.660	2.949
3	4.719	2.774	13	1.232	6.440
4	4.185	2.230	14	5.036	0.244
5	0.915	3.821	15	2.710	3.140
6	4.771	6.041	16	1.072	3.454
7	1.524	2.871	17	5.855	6.203
8	3.447	2.111	18	0.194	1.862
9	3.718	3.665	19	1.762	2.693
10	2.649	2.556	20	2.682	6.097

In the experiment, the parameters of GA are set as follows. Total individuals are 300, crossover probability is set to 0.45, and the largest evolutionary generation is 300; the parameters of ICPSO are set as well: total individuals are 300, the largest evolutionary generation is set to 300. Each experiment the simulation runs 1000 times, each time optimal solution (1—3—12—2—9—17—6—20—13—5—16—18—7—19—15—10—8—4—11—14) can be obtained both by GA and ICPSO, the length of the shortest tour is 24.38. The panorama of the shortest tour is shown in Fig.2. Besides, Tab. 2 is the performance comparison results of two algorithms.

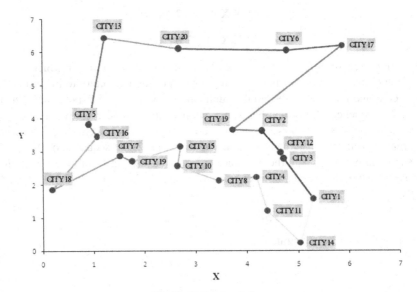

Fig. 2. Optimal solution

Table 2. Comparison results of two algorithms

Method TestingComponent	ICPSO	GA
	Parameters: total individuals:300;the largest evolutionary generation:300	
SIZE OF SOLUTION SPACE	$20! = 2432902008176640000$	
EXECUTIVE COUNTS	1000	
CONVERGENCE COUNTS	1000	1000
AVERAGE CONVERGENT GENERATIONS	136.32	213.77
MAXIMUM CONVERGENT GENERATIONS	269	300
MINIMUM CONVERGENT GENERATIONS	1	1
SIZE OF AVERAGE SOLUTION SPACE	$136.32 \times 300 = 40896$	$213.77 \times 300 = 64131$
$\dfrac{\text{SIZE OF AVERAGE SOLUTION SPACE}}{\text{SIZE OF SOLUTION SPACE}}$	$\dfrac{40896}{20!} = 1.68 \times 10^{-14}$	$\dfrac{64131}{20!} = 2.64 \times 10^{-14}$

Tab. 2 shows GA [3] and ICPSO proposed in this paper are both effective in solving TSP. However, in the instance of 20-citys TSP, the average search space in GA is as 1.57 times as big as that in ICPSO. In other words, ICPSO searches out optimal solution quickly in a relatively smaller search space, with performance better than GA [3].

5   Conclusion

In this paper, ICPSO, on the basis of original velocity-position model, is put forward to solve TSP is by encoding particles with integer, guiding global search with chaotic sequences and describing flying of particles as well as updating the position of particles. Meanwhile, simulation experiments are conducted in TSP with a size of 20 cities [3], comparing ICPSO with GA. Experimental results demonstrate that GA and ICPSO are both effective in solving TSP, however, ICPSO is better than GA at performance for searching out the optimal solution in a relatively smaller search space quickly.

References

1. Yan, C., Wang, Z.-j.: Study on Combinatorial Optimization Problem represented by TSP: Recent Research Work and Perspective. Computer Simulation 24(6), 171–174, 247 (2007)
2. Eberhart, R., Kennedy, J.: A New Optimizer Using Particle Swarm Theory. In: Proc. 6th Int. Symposium on Micro Machine and Human Science, pp. 39–43 (1995)
3. Wang, X.-p., Cao, L.-m.: Genetic Algorithm- Theory, Application and Implementation. Press of Xi'an Jiaotong University (2002)
4. Huang, R.-s.: Chaos and Applications, pp. 128–140. Press of Wuhan University, Wuhan (2002)
5. Kennedy, J., Eberhart, R.C.: A Discrete Binary Version of the Particle Swarm Algorithm. In: Proceedings of the 1997 Conference on Systems, Man and Cybernetics, Piscataway, NJ, pp. 4104–4109. IEEE Service Center, Los Alamitos (1997)

USAR Robot Communication Using ZigBee Technology

Charles Tsui[1], Dale Carnegie[2], and Qing Wei Pan[3]

[1,3] School of Electrical Engineering, Manukau Institute of Technology,
Private Bag 94006, Manukau 2240, New Zealand
`charles.tsui@manukau.ac.nz, qing.wei@manukau.ac.nz`
[2] School of Engineering and Computer Sciences, Victoria University of Wellington,
PO Box 600, Wellington 6140, New Zealand
`dale.carnegie@vuw.ac.nz`

Abstract. This paper reports the successful development of an automatic routing wireless network for USAR (urban search and rescue) robots in an artificial rubble environment. The wireless network was formed using ZigBee modules and each module was attached to a micro-controller in order to model a wireless USAR robot. Proof of concept experiments were carried out by deploying the networked robots into artificial rubble. The rubble was simulated by connecting holes and trenches that were dug in 50 cm deep soil. The simulated robots were placed in the bottom of the holes. The holes and trenches were then covered up by various building materials and soil to simulate a real rubble environment. Experiments demonstrated that a monitoring computer placed 10 meters outside the rubble can establish proper communication with all robots inside the artificial rubble environment.

Keywords: USAR Robots, ZigBee, Wireless Routing Network, Artificial Rubble.

1 Introduction

Prof. Carnegie of Victoria University of Wellington proposed the concept of "Robot family to help at disasters" in 2007 [1]. The "Robot Family" is a system of USAR robots which consists of a hierarchy of "Grandmother, Mother and Daughter robots" used to assist the search for survivors trapped in the rubble of collapsed buildings following earthquake or possible terrorist activity.

A previous study [2] on testing various ZigBee [3] (IEEE 802.15.4) modules selected the XBee-PRO [4] to form the communication network. Each module was attached to a micro-controller in order to model a wireless USAR robot. This paper reports the experiments using four simulated robots to form a wireless mesh network in an artificial rubble environment which was built by connecting holes and trenches that were dug in 50 cm deep soil.

The holes and trenches were then covered up by various common building materials and soil to simulate a real rubble environment. Results demonstrated that a monitoring computer 10 meters outside the rubble can establish proper communication with the USAR robots deep inside the artificial rubble environment.

J.-H. Kim et al. (Eds.): FIRA 2009, CCIS 44, pp. 380–390, 2009.
© Springer-Verlag Berlin Heidelberg 2009

Results of RF (radio frequency) attenuation in real rubble environment from other researchers were used as reference to verify experimental results. Comparison has been done with a project that used WiFi repeaters in another artificial rubble scenario.

2   Attenuation of RF Signal in Rubble

High frequency communication, such as the ZigBee technology in the range of MHz to GHz, is subject to significant attenuations in a rubble environment. The National Institute of Standards and Technology (NIST) have carried out experiments on RF signals before, during, and after the implosion of three large building structures [5]. Their measurements showed a 20 to 80 dB of attenuation for RF signals in the frequency range 50 MHz to 1.8 GHz after the collapse, depending on the building type and location of the transmitter. This high attenuation is a major impediment to using these RF signals for direct point-to-point communication between devices inside the rubble and rescue persons outside. A routing wireless network can overcome this problem.

The link margins for a ZigBee network and a WiFi network can be estimated using the 80 dB attenuation measured by NIST as follows:

2.1   Link Margin for ZigBee Device

A ZigBee device, XBee-PRO [4], with a typical output power of +18 dBm and receiver sensitivity of -100 dBm will have a link margin of 38 dB.

2.2   Link Margin for WiFi Access Point

A WiFi access point, D-Link DWL-2100AP [6], with a typical output power of +15 dBm and receiver sensitivity of -89 dBm will have a link margin of 24 dB.

The link margin seems good for both ZigBee and WiFi cases. However, the experiment done by NIST was up to 1.8 GHz but ZigBee and WiFi is at a higher frequency of 2.4 GHz that will experience greater attenuation. This high level of attenuation coupled with environmental noise indicates that a direct link for continuous data transfer may not be viable. The solution is to build an ad-hoc digital network using multiple nodes. Communication can be established by routing packets of data from one node to another until they arrive at the receiving end.

3   WIFI versus ZIGBEE

WiFi is one of the technologies that can provide routing of data across a mesh wireless network. The Network-Centric Applied Research (NCAR) Team of Ryerson University in Canada has demonstrated communication range extension using WiFi repeaters in an artificial rubble environment [7]. The NCAR team used the D-Link DWL-2100AP WiFi access point for data routing. Compared with the XBee-PRO data router (Figure 1), the access point is double the size, cost three times more and requires ten times more power.

The manufacturer of XBee-PRO provides firmware [8] which allows configuration of the module as a coordinator, router or end-node. A mesh wireless network can be formed by one coordinator and multiple routers and multiple end-nodes.

The module can be programmed to work alone as a router and powered directly by two AA-size batteries. This makes a very low cost and small size routing node, such that it can be placed or dropped into any opening of the rubble. Each module provides a RS232 TTL interface for easy integration with any micro-controller, such as the ATMEGA16L [9] that was used in the experiments described in this paper.

4 Experiment Equipments

A data transmitter was made by attaching an XBee-PRO module to the serial port of a micro-controller (Figure 1). The micro-controller was programmed to continuously

Fig. 1. Data Transmitter & Data Router

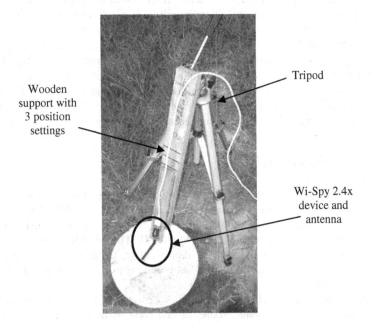

Fig. 2. Tripod and Wi-Spy 2.4x Device

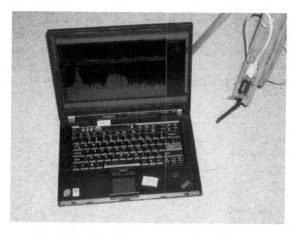

Fig. 3. Laptop and Spectrum Analyser Software

send out blocks of 500-byte data with 100 ms breaks between blocks. Two data routers (Figure 1) were made by XBee-PRO modules powered by four AA-size batteries with a 3 V regulator.

An XBee-PRO module was configured as a coordinator and attached to the RS232 port of the monitoring laptop computer. A tripod (Figure 2) with a wooden support was built to hold a Wi-Spy 2.4x device [10] and antenna to facilitate measurement of received signal strength. The Wi-Spy 2.4x device was connected to a laptop computer (Figure 3) with spectrum analyser software that will be used for measuring the received RF signal power at various locations of the artificial rubble.

5 Soil Environment

The test setup was arranged in the backyard of a residential in the eastern suburb of Auckland, New Zealand. The geology of the residential area is "Alternating SANDSTONE and MUDSTONE of the Waitemata Group" [11]. It is a bottom layer of about 10 m deep of clay with a layer of silt on top. The backyard was further filled up with a layer of organic soil to make an even surface, on which tough lawn was grown.

Soil can act as a lossy wave guide [12] when its moisture level is above 25%. That is the reason for choosing ground soil as the background for the experiments. The bottom layer of clay forms a good layer for retaining water in the top soil layer. The soil around the artificial rubble will be kept moisturised by plenty of water. Measurements will be taken during the experiments to verify that the RF signal will propagate through the under soil trench instead of going above ground.

6 Attenuation of Materials

In order to establish an accurate understanding of the effects on the 2.4 GHz RF signals introduced by various materials in the soil environment, a series of experiments were carried out. A 50 cm diameter hole 50 cm deep was dug in the backyard, at a location such that there was no underground piping or cables, or any other structure within 2 meters, except soil with lawn on top (Figure 4). The data transmitter was placed in the bottom of the hole and then the hole was filled up with various building materials; any cracks and openings were filled by the soil that was dug out from the hole.

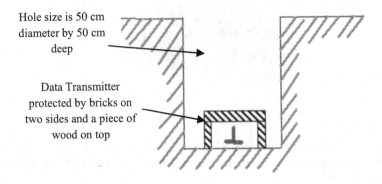

Hole size is 50 cm diameter by 50 cm deep

Data Transmitter protected by bricks on two sides and a piece of wood on top

Fig. 4. Measurement of Material Attenuation

To get reliable results and to even out the effect of multi-path signals, the measuring device was fixed onto a wooden support which allows three position settings. Each position setting is 3 cm apart (about one quarter wavelength). The wooden support was then fixed onto a tripod, which will be positioned at the centre and 20 cm on top of the hole (Figure 2).

For each set up three readings were taken by shifting the measuring device to the three position settings on the wooden support. The average of the three readings was recorded as the final measured result for that setting. Attenuation was found by comparing measured results before and after the hole was filled by various building materials. Table 1 lists the attenuation measured for the various materials. These results show that moisturised soil can produce enough attenuation to stop the RF signals going above ground.

Table 1. Attenuation of Materials

Description of materials	Attenuation (dB)
30 cm deep moisturised soil	59
2 layers of concrete slab, 7 cm thick	15
8 layers of concrete slab, 28 cm thick	36
4 layers of bricks	23
8 layers of pave-stone	24

7 Construction of Artificial Rubble

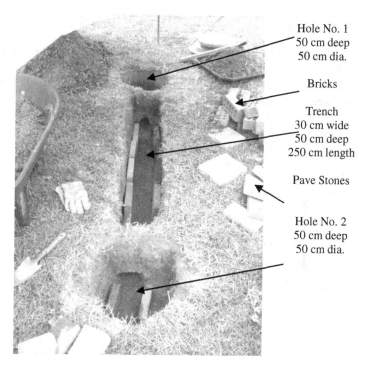

Hole No. 1
50 cm deep
50 cm dia.

Bricks

Trench
30 cm wide
50 cm deep
250 cm length

Pave Stones

Hole No. 2
50 cm deep
50 cm dia.

Fig. 5. The artificial rubble was built by digging a trench which linked two holes in the soil

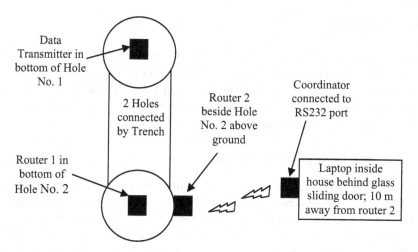

Data Transmitter in bottom of Hole No. 1

2 Holes connected by Trench

Router 2 beside Hole No. 2 above ground

Coordinator connected to RS232 port

Router 1 in bottom of Hole No. 2

Laptop inside house behind glass sliding door; 10 m away from router 2

Fig. 6. Routing Experiment

8 Data Routing Experiments

8.1 Experiment Setup

To verify data routing by the mesh wireless network, four XBee-PRO modules and a laptop computer were used. The experiment was setup as Figure 6.

One module worked as a data transmitter and was placed in the bottom of Hole No. 1. Two data routers were used for data routing. Router 1 was placed in the bottom of Hole No. 2. Router 2 was placed beside Hole No. 2 above ground. The forth module was programmed as a coordinator and attached to the laptop. Various materials and soil were used to fill up the holes and trench.

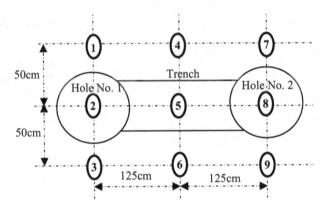

Fig. 7. Nine spots above ground for measuring signal strength in surrounding area

8.2 Monitoring Program

As described in section 4 above, the data transmitter is programmed to continuously send blocks of 500-byte data with 100 ms breaks between blocks. An in-house designed monitoring program was used on the laptop to record and analyse data received by the coordinator. The monitoring program displays statistics of the received data and raw data packets for analysis.

8.3 Experiment Description and Results

Step 3 showed the attenuation introduced by the rubble was 90.3 to 101.3 dB which is higher than the 80 dB measured by NIST for the collapsed buildings. With a 10 meter path loss about 60dB in free space, the coordinator will receive signal at -132.3 to -143.3 dBm which is well below the receiver sensitivity of -100 dBm. This explained why a direct link cannot be established between the transmitter and the coordinator as shown by Step 2.

8.4   Routing Reconnection Tests

Further tests were carried out by switching Router 2 off for 5 minutes and then back to on again. The monitoring program showed no data at the instant of switching off. Then after about 1 minute bytes of data reappeared but missing bytes were reported. This is back to step 5 in Table 2.

Table 2. Experiment Steps. The following sequence of experiment steps was carried out to verify routing of data in the mesh wireless network.

	Experiment Description
1	• Data Transmitter placed in bottom of Hole 1 • Hole 1 filled up with moisturised soil • Trench and hole 2 not filled • Data received by coordinator without error
2	• A tunnel of 19 cm height by 13 cm width was built in the trench by pave stones • The trench was covered with moisturised soil • Received signal strength measured at bottom of hole 2 was -50 dBm; received spectrum shown on Figure 8 • Both routers were switched off • Laptop cannot receive any data
3	• Signal strength received above ground at nine spots as illustrated in Figure 7 were measured with values between -72.3 dBm to -83.3 dBm; graphed results shown on Figure 9
4	• Router 1 placed in bottom of hole 2 and switched on • Data received by laptop without error
5	• 8 pieces of concrete slab placed on top of router 1 • Bytes missing in data received by coordinator
6	• Router 2 placed beside hole 2 above ground and switch on • Data received by coordinator without error

About 1 minute after Router 2 was switched back on all 500 bytes of data were displayed. This is back to step 6 in Table 2.

The tests demonstrated that the routers can re-establish the network automatically. Several runs of this reconnection test were carried out. Results showed that the reconnection time varied between 40 seconds to 90 seconds. This variation depends on the power on and reconnection mechanism built-in on the ZigBee firmware provided by the manufacturer.

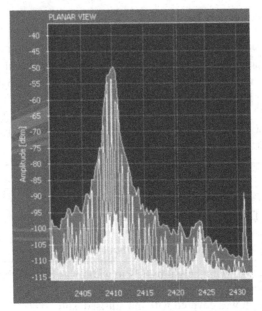

Fig. 8. Signal Spectrum for experiment step 2

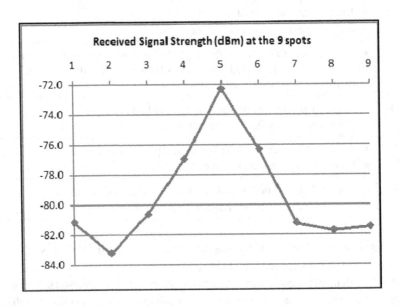

Fig. 9. Received Signal Strength at the nine spots for experiment step 3

9   Conclusions

A set of simulated USAR robots were constructed and placed in an artificial rubble environment. Automatic routing wireless network for the robots in the rubble environment was successfully developed.

It was demonstrated that the artificial rubble in the soil environment is suitable for testing and verifying the wireless network for USAR missions.

Attenuations of soil and building materials on 2.4 GHz ZigBee RF signals were measured. The experiments indicated that ZigBee technology implemented by the XBee-PRO modules can form a useful mesh wireless network for USAR robots. The modules can automatically reconnect after network interruption.

The next phase is to build bigger and more complex artificial rubble environments and conduct tests using groups of mobile robots on the wireless mesh network using ZigBee technology.

Acknowledgements

This project is supported by the School of Electrical Engineering and Trades, Manukau Institute of Technology and the School of Engineering and Computer Sciences, Victoria University of Wellington.

References

1. Carnegie, D.A.: A Three-Tier Hierarchical Robotic System for Urban Search and Rescue Applications. In: IEEE International Workshop on Safety, Security and Rescue Robotics, Rome, Italy (2007)
2. Tsui, C., Jennings, L., Carnegie. D.A.: Is ZigBee a suitable communication link for the 'Robot Family' at disasters? In: Proceedings of ENZCON (2007)
3. ZigBee® Alliance, http://www.zigbee.org
4. Digi International Inc., XBee® & XBee-PRO® DigiMesh™ 2.4 RF Modules, http://www.digi.com/products/wireless/zigbee-mesh/xbee-digimesh-2-4specs.jsp
5. Holloway, C.L., Koepke, G., Camell, D., Remley, K.A.: Radio Propagation Measurements Before, During, and After the Collapse of Three Large Building Structures. In: Proceedings of the General Assembly of the International Union of Radio Science (Union Radio Scientifique Internationale-URSI) (2008)
6. D-Link Corporation, DWL-2100AP, High Speed 2.4GHz (802.11g) Wireless 108Mbps1 Access Point, http://www.dlink.com/products/resource.asp?pid=292&rid=912&sec=0
7. Ferworn, A., Tran, N., Tran, J., Zarnett, G., Sharifi, F.: WiFi repeater deployment for improved communication in confined-space urban disaster search. In: IEEE International Conference on System of Systems Engineering, pp. 1–5, 16-18 (2007)
8. Digi International Inc., XBee/XBee-PRO Series 1 ZigBee Module Firmware Revision History, http://www.digi.com/support/kbase/kbaseresultdetl.jsp?id=2182

9. Atmel Corporation, ATmega16 and ATmega16L product document, http://www.atmel.com/dyn/resources/prod_documents/2466S.pdf

10. MetaGeek, L.L.C.: Wi-Spy 2.4x Specifications, http://www.metageek.net/products/wi-spy-comparison

11. GHD Ltd., Part 1 of 2001 (March) Pakuranga Creek Catchments Comprehensive Catchment Discharge Consent Application - Comprehensive Catchment Study and Management Plan Options, http://www.manukau.govt.nz/tec/catchment/pakuranga_pages/pdf/pakuranga_creek_ccdc1_low.pdf

12. Holloway, C.L., Hill, D.A., Dalke, R.A., Hufford, G.A.: Radio wave propagation characteristics in lossy circular waveguidessuch as tunnels, mine shafts, and boreholes. IEEE Transactions on Antennas and Propagation 48(9), 1354–1366 (2000)

Author Index